SEMICONDUCTORS AND SEMIMETALS

VOLUME 13

Cadmium Telluride

Semiconductors and Semimetals

A Treatise

Edited by R. K. WILLARDSON　　　*ALBERT C. BEER*

　　　　　ELECTRONIC MATERIALS DIVISION　　BATTELLE MEMORIAL INSTITUTE
　　　　　COMINCO AMERICAN INCORPORATED　　COLUMBUS LABORATORIES
　　　　　SPOKANE, WASHINGTON　　　　　　　COLUMBUS, OHIO

Volume　1　**Physics of III–V Compounds**

Volume　2　**Physics of III–V Compounds**

Volume　3　**Optical Properties of III–V Compounds**

Volume　4　**Physics of III–V Compounds**

Volume　5　**Infrared Detectors**

Volume　6　**Injection Phenomena**

Volume　7　**Applications and Devices (in two parts)**

Volume　8　**Transport and Optical Phenomena**

Volume　9　**Modulation Techniques**

Volume 10　**Transport Phenomena**

Volume 11　**Solar Cells**

Volume 12　**Infrared Detectors (II)**

Volume 13　**Cadmium Telluride**

SEMICONDUCTORS AND SEMIMETALS

VOLUME 13   Cadmium Telluride

Kenneth Zanio

HUGHES RESEARCH LABORATORIES
MALIBU, CALIFORNIA

1978

ACADEMIC PRESS   New York   San Francisco   London
A Subsidiary of Harcourt Brace Jovanovich, Publishers

COPYRIGHT © 1978, BY ACADEMIC PRESS, INC.
ALL RIGHTS RESERVED.
NO PART OF THIS PUBLICATION MAY BE REPRODUCED OR
TRANSMITTED IN ANY FORM OR BY ANY MEANS, ELECTRONIC
OR MECHANICAL, INCLUDING PHOTOCOPY, RECORDING, OR ANY
INFORMATION STORAGE AND RETRIEVAL SYSTEM, WITHOUT
PERMISSION IN WRITING FROM THE PUBLISHER.

ACADEMIC PRESS, INC.
111 Fifth Avenue, New York, New York 10003

United Kingdom Edition published by
ACADEMIC PRESS, INC. (LONDON) LTD.
24/28 Oval Road, London NW1

Library of Congress Cataloging in Publication Data

Willardson, Robert K
    Semiconductors and semimetals.

    Vol. 13-   by K. Zanio.
    Includes bibliographical references.
    CONTENTS: v.   Physics of III-V compounds.
—v. 3   Optical properties of III-V compounds.—v. 5.
Infrared detectors.—v. 6.   Injection phenomena.
[etc.]
    1. Semiconductors—Collected works. 2. Semi-
metals—Collected works. I. Beer, Albert C.,
joint ed. II. Title.
QC610.9.W54   537.622   65-26048
ISBN 0–12–752113–5 (v. 13)

PRINTED IN THE UNITED STATES OF AMERICA

Contents

FOREWORD vii
PREFACE ix
CONTENTS OF PREVIOUS VOLUMES xiii

Chapter 1 Materials Preparation

I. Phase Relationships 1
II. Crystal Growth 11
III. Purification 38
IV. Summary 51

Chapter 2 Physics

I. The Zinc Blende Structure 53
II. Bonding 58
III. Lattice Dynamics 69
IV. Band Structure 77
V. Excitons 90
VI. Transport Properties 103
VII. Summary 113

Chapter 3 Defects

I. Defects at High Temperature 116
II. Defects at Low Temperature 129
III. Summary 161

Chapter 4 Applications

I. Gamma-Ray and X-Ray Spectrometers 164
II. Modulators 186
III. Optical Elements 191
IV. Solar Cells 197
V. CdS/CdTe Heterojunction in Liquid Crystal Imaging 204

v

VI.	Miscellaneous Applications	.	.	.	.	.	.	.	.	.	.	205
VII.	Summary	.	.	.	.	.	.	.	.	.	.	209

References 211

INDEX 231

Foreword

This book is the second instance where an entire volume of "Semiconductors and Semimetals" has been devoted to one subject, written by a single author. The first example, Volume 11, concerned with solar cells and authored by Harold Hovel, proved to be very successful in meeting the needs of workers in the field. The subject of the present work, namely cadmium telluride, also represents an area where detailed coverage is desirable, due in part to the overall increased activity in II–VI semiconducting compounds. The editors consider it fortunate that Dr. Zanio, who has made numerous contributions in this field, has been able to provide this valuable addition to this Treatise.

R. K. WILLARDSON
ALBERT C. BEER

Preface

In the last eight years, single crystals of cadmium telluride have become useful as a nuclear radiation detector, an electrooptical modulator, and as an optical material in the infrared. The availability of higher quality CdTe and an impending energy crisis suggest examining this material further for terrestrial solar cells. Alloys of $Cd_xHg_{1-x}Te$ have become important as infrared detectors. A review of these and other recent CdTe developments is appropriate at this time so that existing technology can be exploited more effectively.

Since CdTe will not impact on the semiconductor market as silicon has, a review to the extent of a full text would not appear to be appropriate. However, the interest in the above important applications, along with other not so promising applications using CdTe (electroluminescence and microwave devices, high power laser windows, etc.), has stimulated considerable basic research on single crystals of this material. Consequently, a considerable wealth of information on CdTe single crystal studies has been accumulated and has contributed to the knowledge of semiconductor physics and defects in semiconductors. This is in contrast to earlier work on the other wide band gap II–VI compounds where phosphorescence applications stimulated research on thin films.

In the past eight years, more extensive defect studies have been undertaken on CdTe. Older theories assumed that the defect structure is frozen in upon cooling crystals from elevated anneal temperatures. To the contrary, due to rapid diffusion and subsequent association and precipitation, defects are not frozen in. Consequently and unfortunately, the defect structure must be evaluated from a more complicated vantage point. Further significant technological progress is unlikely unless research proceeds from the latter viewpoint. The insight that the recent defect studies on CdTe have provided makes an extensive review on CdTe appropriate.

Also in the past seven years, the dielectric bonding theory of Phillips and Van Vechten has been developed and applied to the physical properties of CdTe and other semiconductors. During the same period the empirical pseudopotential model of the band structure has been developed and applied to the optical properties of CdTe and other semiconductors. Both of these models are self-consistent in explaining the bands and bonds and in evaluating the degree of ionicity of CdTe and other semiconductors. The physics of the group IV elemental semiconductors and the group III–V compound semiconductors had been studied in greater depth and at an earlier date than the wide band gap II–VI compounds. The extended work on CdTe provides information as to just how the more ionic II–VI compounds fit into the general pattern of bands and bonds in semiconductors. Consequently, throughout this review, CdTe will, wherever possible, be examined within the framework of its degree of ionicity. This approach is intended to further the general understanding of semiconductors.

Early work on CdTe is summarized in "The Physics and Chemistry of II–VI Compounds" (1967) edited by M. Aven and J. S. Prener, "II–VI Semiconducting Compounds" (1967) edited by D. G. Thomas, "Cadmium Telluride" (1968) edited by V. S. Vavilov and B. M. Vul in Russian, and "II–VI Compounds" (1969) by B. Ray. The Proceedings of the First (1971) and Second (1976) International Conference on CdTe at Strasbourg emphasize work on CdTe as a nuclear radiation detector. The present text is intended to cover the general topics of materials preparation, physics, defects, and applications of CdTe. The first chapter on materials preparation, in addition to covering Bridgman and vapor phase growth, includes the more recent subject of the growth of crystals from Te-rich solutions. Solution growth is an approach toward improving the purity of the material while still permitting the growth of crystals of reasonable size. The second chapter tabulates most of the physical, optical, and transport properties of CdTe. Wherever possible, these properties are discussed relative to those of the more covalent group IV elemental and III–V compound semiconductors. Although the empirical pseudopotential model and the dielectric theory of bonding provide much physical insight into the physics of CdTe, first principle calculations are not always available. The approaches used to analyze the phonon dispersion relationship are a case in point. The resulting force constants are difficult to interpret. However, some insight into the bonding forces is available when CdTe is analyzed within the framework of the Sn–InSb–CdTe series. The third chapter separates defect studies into two parts. The first examines the equilibrium concentration of atomic defects and their interplay with foreign atoms at elevated temperatures. Keeping in mind the rapid diffusion and precipitation that occur during cooling to room temperature, the second part examines the evidence for the various defects and complexes that are metastable at room temperature and below. The room temperature structure is complicated and by no means does this chapter pretend to establish a final model. The last chapter reviews the applications of CdTe. Although the text is intended primarily as a reference book useful in developing those applications, it is hoped that the physics and defect studies will be beneficial to understanding other compounds.

The author's familiarity with the subject is a consequence of support provided by the Energy Research and Development Agency and earlier by the United States Atomic Energy Commission to Hughes Research Laboratories for development of CdTe as a room temperature gamma-ray and x-ray spectrometer. The author is therefore thankful to Messrs. J. Hitch, R. Butenhoff, and H. Wasson of these agencies for their continued support. Extended discussions and criticism by Professor J. W. Mayer, Dr. H. Winston, Dr. J. Lotspeich, Mr. L. DeVaux, Dr. C. Barnes, Professor T. McGill, Dr. L. Fraas, Dr. A. Gentile, and Professors G. Ottaviani and B. Wagner, Jr. were important to the development of the text. The author also appreciates the assistance of Joanne Ordean and Carol Zanio in preparing the manuscript.

Semiconductors and Semimetals

Volume 1 Physics of III–V Compounds

C. Hilsum, Some Key Features of III–V Compounds
Franco Bassani, Methods of Band Calculations Applicable to III–V Compounds
E. O. Kane, The $k \cdot p$ Method
V. L. Bonch-Bruevich, Effect of Heavy Doping on the Semiconductor Band Structure
Donald Long, Energy Band Structures of Mixed Crystals of III–V Compounds
Laura M. Roth and Petros N. Argyres, Magnetic Quantum Effects
S. M. Puri and T. H. Geballe, Thermomagnetic Effects in the Quantum Region
W. M. Becker, Band Characteristics near Principal Minima from Magnetoresistance
E. H. Putley, Freeze-Out Effects. Hot Electron Effects, and Submillimeter Photoconductivity in InSb
H. Weiss, Magnetoresistance
Betsy Ancker-Johnson, Plasmas in Semiconductors and Semimetals

Volume 2 Physics of III–V Compounds

M. G. Holland, Thermal Conductivity
S. I. Novkova, Thermal Expansion
U. Piesbergen, Heat Capacity and Debye Temperatures
G. Giesecke, Lattice Constants
J. R. Drabble, Elastic Properties
A. U. Mac Rae and G. W. Gobell, Low Energy Electron Diffraction Studies
Robert Lee Mieher, Nuclear Magnetic Resonance
Bernard Goldstein, Electron Paramagnetic Resonance
T. S. Moss, Photoconduction in III–V Compounds
E. Antončik and J. Tauc, Quantum Efficiency of the Internal Photoelectric Effect in InSb
G. W. Gobeli and F. G. Allen, Photoelectric Threshold and Work Function
P. S. Persham, Nonlinear Optics in III–V Compounds
M. Gershenzon, Radiative Recombination in the III–V Compounds
Frank Stern, Stimulated Emission in Semiconductors

Volume 3 Optical Properties of III–V Compounds

Marvin Hass, Lattice Reflection
William G. Spitzer, Multiphonan Lattice Absorption
D. L. Stierwalt and R. F. Potter, Emittance Studies
H. R. Philipp and H. Ehrenreich, Ultraviolet Optical Properties
Manuel Cardona, Optical Absorption above the Fundamental Edge

Earnest J. Johnson, Absorption near the Fundamental Edge
John O. Dimmock, Introduction to the Theory of Exciton States in Semiconductors
B. Lax and J. G. Mavroides, Interband Magnetooptical Effects
H. Y. Fan, Effects of Free Carriers on the Optical Properties
Edward D. Palik and George B. Wright, Free-Carrier Magnetooptical Effects
Richard H. Bube, Photoelectronic Analysis
B. O. Seraphin and H. E. Bennett, Optical Constants

Volume 4   Physics of III–V Compounds

N. A. Goryunova, A. S. Borschevskii, and D. N. Tretiakov, Hardness
N. N. Sirota, Heats of Formation and Temperatures and Heats of Fusion of Compounds $A^{III}B^{V}$
Don L. Kendall, Diffusion
A. G. Chynoweth, Charge Multiplication Phenomena
Robert W. Keyes, The Effects of Hydrostatic Pressure on the Properties of III–V Semiconductors
L. W. Aukerman, Radiation Effects
N. A. Goryunova, F. P. Kesamanly, and D. N. Nasledov, Phenomena in Solid Solutions
R. T. Bate, Electrical Properties of Nonuniform Crystals

Volume 5   Infrared Detectors

Henry Levinstein, Characterization of Infrared Detectors
Paul W. Kruse, Indium Antimonide Photoconductive and Photoelectromagnetic Detectors
M. B. Prince, Narrowband Self-Filtering Detectors
Ivars Melngailis and T. C. Harman, Single-Crystal Lead-Tin Chalcogenides
Donald Long and Joseph L. Schmit, Mercury–Cadmium Telluride and Closely Related Alloys
E. H. Putley, The Pyroelectric Detector
Norman B. Stevens, Radiation Thermopiles
R. J. Keyes and T. M. Quist, Low Level Coherent and Incoherent Detection in the Infrared
M. C. Teich, Coherent Detection in the Infrared
F. R. Arams, E. W. Sard, B. J. Peyton, and F. P. Pace, Infrared Heterodyne Detection with Gigahertz IF Response
H. S. Sommers, Jr., Microwave-Biased Photoconductive Detector
Robert Sehr and Rainer Zuleeg, Imaging and Display

Volume 6   Injection Phenomena

Murray A. Lampert and Ronald B. Schilling, Current Injection in Solids: The Regional Approximation Method
Richard Williams, Injection by Internal Photoemission
Allen M. Barnett, Current Filament Formation
R. Baron and J. W. Mayer, Double Injection in Semiconductors
W. Ruppel, The Photoconductor–Metal Contact

Volume 7   Application and Devices: Part A

John A. Copeland and Stephen Knight, Applications Utilizing Bulk Negative Resistance
F. A. Padovani, The Voltage-Current Characteristic of Metal-Semiconductor Contacts

P. L. Hower, W. W. Hooper, B. R. Cairns, R. D. Fairman, and D. A. Tremere, The GaAs Field-Effect Transistor
Marvin H. White, MOS Transistors
G. R. Antell, Gallium Arsenide Transistors
T. L. Tansley, Heterojunction Properties

Volume 7   Application and Devices: Part B

T. Misawa, IMPATT Diodes
H. C. Okean, Tunnel Diodes
Robert B. Campbell and Hung-Chi Chang, Silicon Carbide Junction Devices
R. E. Enstrom, H. Kressel, and L. Krassner, High-Temperature Power Rectifiers of $GaAs_{1-x}P_x$

Volume 8   Transport and Optical Phenomena

Richard J. Stirn, Band Structure and Galvanomagnetic Effects in III–V Compounds with Indirect Band Gaps
Roland W. Ure, Jr., Thermoelectric Effects in III–V Compounds
Herbert Piller, Faraday Rotation
H. Barry Bebb and E. W. Williams, Photoluminescence I: Theory
E. W. Williams and H. Barry Bebb, Photoluminescence II: Gallium Arsenide

Volume 9   Modulation Techniques

B. O. Seraphin, Electroreflectance
R. L. Aggarwal, Modulated Interband Magnetooptics
Daniel F. Blossey and Paul Handler, Electroabsorption
Bruno Batz, Thermal and Wavelength Modulation Spectroscopy
Ivar Balslev, Piezooptical Effects
D. E. Aspnes and N. Bottka, Electric-Field Effects on the Dielectric Function of Semiconductors and Insulators

Volume 10   Transport Phenomena

R. L. Rode, Low-Field Electron Transport
J. D. Wiley, Mobility of Holes in III–V Compounds
C. M. Wolfe and G. E. Stillman, Apparent Mobility Enhancement in Inhomogeneous Crystals
Robert L. Peterson, The Magnetophonon Effect

Volume 11   Solar Cells

Harold J. Hovel, Introduction; Carrier Collection, Spectral Response, and Photocurrent; Solar Cell Electrical Characteristics; Efficiency; Thickness; Other Solar Cell Devices; Radiation Effects; Temperature and Intensity; Solar Cell Technology

Volume 12   Infrared Detectors (II)

W. L. Eiseman, J. D. Merriam, and R. F. Potter, Operational Characteristics of Infrared Photodetectors
Peter R. Bratt, Impurity Germanium and Silicon Infrared Detectors

E. H. Putley, InSb Submillimeter Photoconductive Detectors
G. E. Stillman, C. M. Wolfe, and J. O. Dimmock, Far-Infrared Photoconductivity in High Purity GaAs
G. E. Stillman and C. M. Wolfe, Avalanche Photodiodes
P. L. Richards, The Josephson Junction as a Detector of Microwave and Far-Infrared Radiation
E. H. Putley, The Pyroelectric Detector—An Update

CHAPTER 1

Materials Preparation

I. Phase Relationships

Crystal growth and subsequent processing of cadmium telluride requires precise knowledge of the existence regions of the solid and liquid phases with respect to pressure, temperature, and composition. The relationships between these variables are expressed in temperature versus composition (T–x), component pressure versus temperature (p–T), and total pressure versus temperature (P–T) diagrams. The effect of total pressure on the phase boundaries is small for pressures up to many atmospheres; consequently the conventional T–x diagram is useful over a wide range of processing conditions. In fact, the T–x diagram shows that the growth of single crystals is possible by combining the components in any proportion. However, the largest crystals are prepared by solidifying the components in their near atomic equivalents. Although the deviation from stoichiometry by electrically active native defects is small (i.e., less than 10^{-5} mole fraction) compared to the extent of the phase diagram, it is extremely important to control these deviations when considering the semiconducting properties. The defect structure influences strongly the electrical, optical, and transport properties of CdTe. This structure is related to the component pressure used during crystal growth and subsequent processing.

A. Temperature versus Composition Diagram

The shape of the Cd–Te phase diagram is quite simple; it has a maximum melting point of $1092° \pm 1°C$ at approximately 50 at% Te and eutectic temperatures of $324° \pm 2°$ and $449° \pm 2°C$ on the Cd-rich and Te-rich sides of the phrase diagram (Kulwicki, 1963). The Te-rich eutectic was found to be at about 99 at% Te, whereas the Cd-rich side was found to be nearly pure Cd. Experimental data for the phase diagrams are plotted in Fig. 1.1. Liquidus data have been obtained by observing the onset of initial freezing (de Nobel, 1959; Kobayashi, 1911), thermal analysis (Kulwicki, 1963; Lorenz, 1962a; Steininger et al., 1970), and by using an optical technique to measure the partial pressure of Te vapor in equilibrium with solid–liquid mixtures

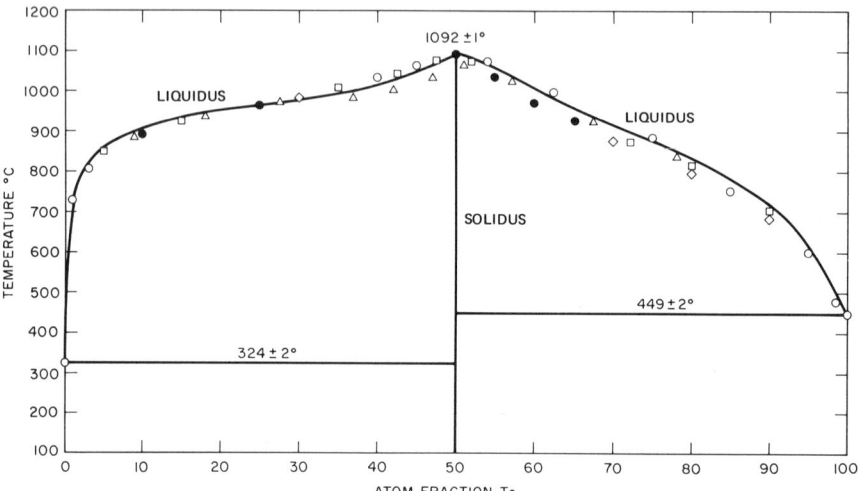

FIG. 1.1. Temperature versus composition (T–x) diagram for the Cd–Te system. △, de Nobel (1959); □, Lorenz (1962a); ○, Kulwicki (1963); ◇, Steininger et al. (1970); ▽, Brebrick (1971); ●, overlapping data points; —, Jordan (1970).

(Brebrick, 1971). The most extensive data are the differential thermal analysis of Kulwicki. The data between various investigators are in good agreement; however the high temperature data of de Nobel on the Cd-rich side is slightly lower and that of Kulwicki on the Te-rich side is slightly higher than that reported by others.

The phase diagrams of GaAs and other III–V systems show broad parabolic liquidus curves with a large radius of curvature near their melting points (Stringfellow and Greene, 1969). However, for CdTe and other II–VI compounds, the liquidus is almost pointed at the stoichiometric composition. The abnormal increase in the liquidus temperature results from a much stronger interaction between unlike atoms of the II–VI compounds than of the III–V compounds. This is not surprising since the ionic contribution to the cohesive energy in the II–VI compounds is greater than in the III–V compounds. Because of this difference, a greater increase in the free energy of the system is realized by ordering or solidification in the vicinity of the stoichiometric composition than at an extreme deviation from stoichiometry. Regardless of the degree of association (i.e., concentration of unlike pairs) in the melt, the interchange energy would be expected to be at most only a slowly varying function of composition. The interchange energy is defined as

$$\omega = (2\omega_{\text{Cd–Te}} - \omega_{\text{Cd–Cd}} - \omega_{\text{Te–Te}})/2 \qquad (1.1)$$

where $\omega_{\text{Cd–Te}}$, $\omega_{\text{Cd–Cd}}$, and $\omega_{\text{Te–Te}}$ are, respectively, the energy released when

I. PHASE RELATIONSHIPS 3

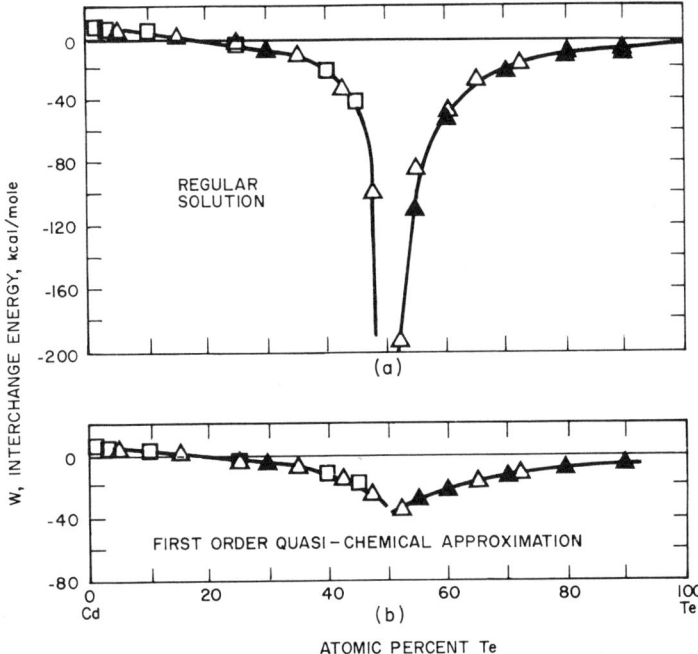

FIG. 1.2. Interchange energy parameters for (a) regular and (b) quasi-chemical approximation in the Cd–Te system: ▲, (Steininger et al., 1970); □, Kulwicki 1963; △, Lorenz (1962a).

forming Cd–Te, Cd–Cd, and Te–Te bonds. Calculations of the interchange parameter versus composition using the regular solution theory (random pairing) and the quasi-chemical approach (the distribution of pairs weighted exponentially with interchange energy), have been made for the Cd–Te system and are plotted in Fig. 1.2 (Steininger et al., 1970). The regular solution model shows a steep minimum (in excess of 200 kcal/mole) near the stoichiometric composition. For the quasi-chemical model, the slopes near the stoichiometric composition are more gentle with a minimum value of less than 35 kcal/mole for the interchange parameter.

In order to account for nonrandom pairing, a thermodynamic model for the liquidus curve for a binary system AB with a congruent melting point was developed (Jordan and Zupp, 1969; Jordan, 1970). Defined as a regular association solution (RAS) theory, it assumes a regular ternary solution with species A, B, and AB in equilibrium. The equation for the liquidus curve is given by

$$\frac{\frac{RT}{2}\ln\left(\frac{1 + \sqrt{1 - 4x_1 x_2 (1-\beta)^2}}{(1+\beta)16x_1^2 x_2^2}\right)^2 - \Delta H_f + T \Delta S_f}{(x_2 - 0.5)^2} = \alpha, \quad (1.2)$$

where x_1 and x_2 are the mole fractions of the components, ΔH_f and ΔS_f are, respectively, the heat and entropy of fusion, and α and β are semiempirical constants. The symbol β represents the degree of association and α is related to the interchange energy. For complete association $\beta = 0$ and the liquid is made up of either AB and B or AB and A species depending upon which component is in excess. For $\beta = 1$, the liquid is completely disassociated and the AB species does not exist. Values of 10.9 kcal/mole and 0.055 were used for α and β on the Cd-rich side; values of 3.1 kcal/mole and 0.055 were respectively used on the Te-rich side. With these values for α and β the liquidus equation [Eq. (1.2)] is plotted in Fig. 1.1 and is in good agreement with the experimental results.

Because the regular and quasi-chemical models fail to differentiate between the energies of metal–metal and chalcogen–chalcogen bonds, they predict the liquids to be symmetrical about the stoichiometric composition. However, the II–VI and CdTe phase diagrams are unsymmetrical and the regular associated solution theory can be made to fit the CdTe experimental results by separating the phase diagram into metal-rich and chalcogen subsystems.

The relative flatness of the Cd-rich liquidus in the region $x_{Te} \approx 0.2$ indicates the tendency of the liquid to separate into two compositions. This trend is more distinct for CdSe and ZnTe. In fact, a miscibility gap exists on the Cd-rich side of the cadmium–selenium system (Reisman et al., 1962). The regular associated theory also predicts a miscibility gap on the Zn-rich side of the Zn–Te system.

A number of thermodynamic properties of CdTe have been determined. A best value of 12.0 ± 1.9 kcal/gm-mole is recommended by Kulwicki as the heat of fusion of CdTe at the congruent melting point. The standard free energy of formation of the compound from pure solid Cd and Te at the melting point was calculated to be 24.4 kcal/mole. Values of the free energy of formation at the melting point from liquid Cd and Te are also in good agreement. The value of -12.6 kcal/mole from regular association theory compares favorably with the value of -15.6 kcal/mole from thermochemical data. The latter number uses -5.3 kcal/mole for the energy of formation of liquid CdTe from the vapor phase and -10.3 kcal/mole for the transformation of Cd and Te gaseous molecules to the liquid phase. Positive and negative values for the heat of mixing at infinite dilution were found for Te in Cd and Cd in Te (Kulwicki, 1963). These results are consistent with the positive and negative deviations from Raoult's law for excess Cd and excess Te.

Determining the thermodynamic properties from more basic principles is not straightforward. Predictions are successful where central bonding forces predominate such as in metals. However, success is limited when directional (covalent) forces are present such as CdTe. For example, the thermochemical molecular theory of Pauling predicts 4.4 kcal/mole for the heat of formation of CdTe (Phillips and Van Vechten, 1970), which is far from the experimental value of 23 kcal/mole (Kulwicki, 1963; Wagman et al., 1968). Phillips (1973b) predicts a value of about 23 kcal/mole based on the scaling of the bond energies in the dielectric theory of $A^N B^{8-N}$ crystals, described in a later section. Even though a scaling factor is required, this theory is remarkable in that the same scaling factor predicts the heats of formation within 10% for other tetrahedral structures. Dielectric theory also predicts the melting point of CdTe to be 1358°C far from the actual value of 1092°C (Van Vechten, 1972). The approach to the dielectric theory assumes the liquid phase to be metallic. This assumption is known to be correct for Si and Ge. However, it is not valid for all III–V and II–VI compounds.

Since liquid CdTe has a positive slope for the conductivity with respect to temperature in the region of the solid phase, Rud' and Sanin (1971b) conclude that a semiconductor–semiconductor melting transition occurs. A more sophisticated approach to dielectric theory is required.

The homogeneity region of CdTe is represented by the vertical line in Fig. 1.1, and its extent may be as high as 1 at% on both the Cd-rich and Te-rich sides of the stoichiometric composition (Woodbury and Hall, 1967; Medvedev et al., 1972). Smith (1970) shows the concentration of electrically active centers to be about 10^{-3} at% on the Cd-rich side and less than 10^{-3} at% on the Te-rich side of the stoichiometric composition at about 900°C. The carrier concentrations were directly determined by Smith through Hall measurements at elevated temperatures. By extrapolating the electron concentration versus p_{Cd} to the values of p_{Cd} over Cd-saturated conditions, the Cd solidus was located. Placement of the solidus in terms of deviation from stoichiometry in Fig. 1.3 requires the assumption of a doubly ionized defect. Similar extrapolations from conductivity studies also give comparable results (Whelan and Shaw, 1968; Matveev et al., 1969b; Zanio, 1970a). Different results were found under Te-saturated conditions. For samples held under near Te-saturated conditions, Smith found differences in the hole concentration. Such differences between samples indicate that residual impurities dominate the conductivity. If a one-to-one relationship between carrier concentration and acceptor native defect is assumed, the carrier concentration probably represents an upper limit as to the deviation from stoichiometry on the Te-rich side of the phase diagram. The data in Fig. 1.3 represent the sample containing a lower carrier concentration. The data of de Nobel (1959) are also plotted and must be considered as even more difficult to interpret because carrier concentrations were determined from

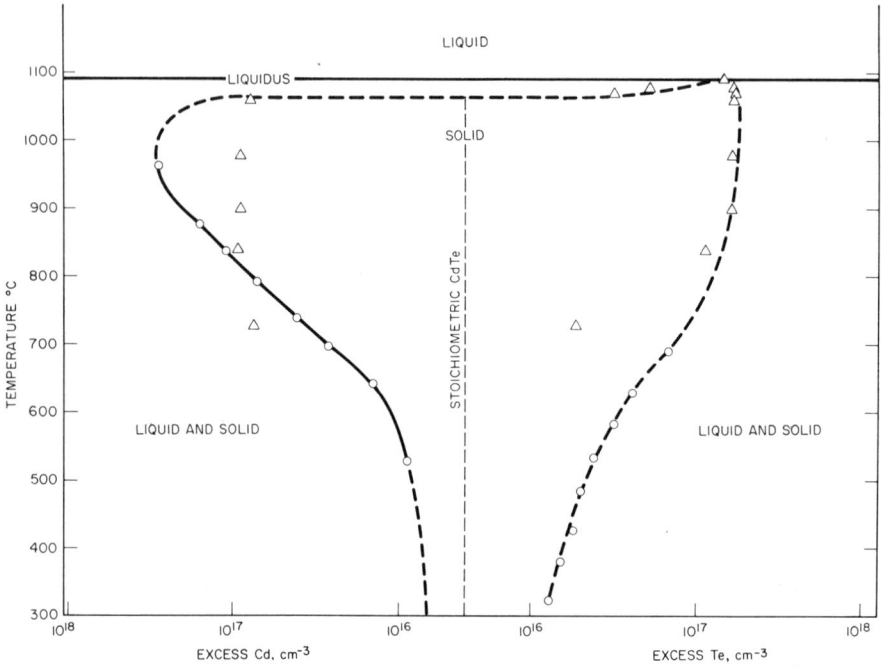

Fig. 1.3. Temperature versus composition diagram on a logarithmic scale, showing the extent of the *electrically active* homogeneity region as interpreted through defect models. Most reliable data indicated by solid line. ○, Smith (1970); △, de Nobel (1959).

samples quenched from elevated temperatures and measured at room temperature. Simple point defect models, to be discussed in Chapter 3, must be invoked to interpret the electrical measurements. Consequently, the border in terms of deviation from stoichiometry is not precisely known.

B. COMPONENT PRESSURE VERSUS TEMPERATURE DIAGRAM

Although the extent of the homogeneity region is not known precisely in terms of composition, the border is a strong function of component pressures and is well defined in terms of these pressures and temperature. At the three-phase boundary and within the existence region the formation of solid from the vapor is represented by the following reaction equation:

$$Cd(g) + \tfrac{1}{2}Te_2(g) \leftrightarrows CdTe(s) \tag{1.3}$$

where $Cd(g)$ and $Te_2(g)$ represent respectively monatomic Cd and diatomic gaseous molecules. The corresponding mass action equation is given by

$$K_{CdTe}(T) = p_{Cd} p_{Te_2}^{1/2}, \tag{1.4}$$

where p_{Cd} and p_{Te_2} are the partial pressures of Cd and Te, and $K_{CdTe}(T)$ is the equilibrium constant which is only a function of temperature. In a general study of vapors over II–VI compounds by effusion measurements, CdTe and other molecules of the II–VI compounds in the vapor were not found (Drowart and Goldfinger, 1958; Goldfinger and Jeunehomme, 1963). Since the concentration of gaseous CdTe species is negligible, the sublimation of CdTe can be effectively stopped by applying an overpressure of one of the components over the solid. This technique is used in crystal growth and subsequent heat treatments.

Several methods have been used to define the component vapor pressures at the solid–liquid interfaces. The most recent and highly successful is the optical method of Brebrick (1971). In this method the optical densities of coexisting Cd and Te vapors in equilibrium with liquid and solid CdTe are determined. These values of the optical densities are then compared against calibration runs where the partial pressure of the pure components are known. Earlier measurements of the three-phase boundary were also made using the two-zone technique (de Nobel, 1959; Lorenz, 1962a). In this technique a CdTe charge and reservoir of Cd are positioned at opposite ends of a two-zone furnace. The lower temperature of the reservoir furnace determines the Cd vapor pressure over the charge. The charge is heated above the melting point and subsequently cooled. From the arrest point on the temperature–time curve Lorenz determined the freezing point. The same procedure was followed by de Nobel, but visual methods were used to determine the temperature of crystallization. The corresponding vapor pressure of Te can be determined by calculation through Eq. (1.4).

These results have been conveniently summarized by Strauss (1971) in Fig. 1.4, in a plot of component vapor pressure versus $1/kT$. The three-phase boundary or loop is shown for vapor pressures of Cd (upper) and Te (lower) coexisting in equilibrium with solid and liquid CdTe. Solid and dashed lines in each loop correspond to Cd-saturated and Te-saturated conditions. The lines p_{Cd}^0 and $p_{Te_2}^0$ tangent to the upper portion of the loops correspond to the vapor pressures of pure Cd and pure Te. At lower temperatures on the Cd-rich (Te-rich) side of the loop the vapor pressure of Cd(Te) in equilibrium with solid and liquid is identical to the vapor pressure of pure Cd(Te). An examination of the upper portion of the p_{Cd} loop shows that with increasing temperature and correspondingly a larger fraction of Te, the vapor pressure of Cd increases reaching a maximum value of 7 atm at about 1030°C and decreases rather abruptly to 0.65 atm at the maximum melting point. The same pattern is observed for the Te loop; however, the corresponding partial pressures of Te are appreciably less. The maximum vapor pressure of Te is approximately 0.185 atm at 930°C. A value of 5.5×10^{-3} atm is found at the maximum melting point. Optical measurements (Brebrick, 1971) and

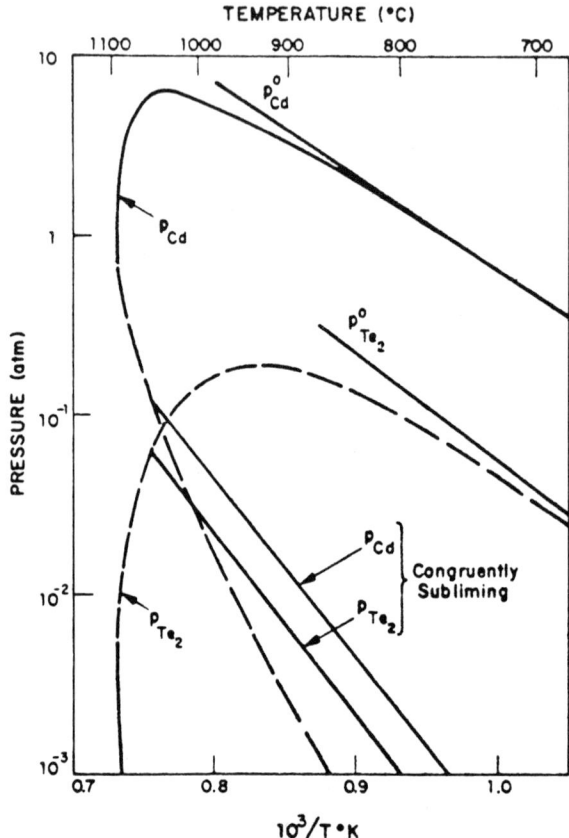

FIG. 1.4. Three-phase boundary or loop for vapor pressures of Cd (upper) and Te (lower) coexisting in equilibrium with solid CdTe and liquid. Regions inside and outside of the loops correspond to the vapor phase in equilibrium with, respectively, solid CdTe and liquid. The vapor pressure of pure Cd and Te are denoted by p^0_{Cd} and $p^0_{Te_2}$ (Strauss, 1971).

thermal measurements (Lorenz, 1962a) were used to construct the Cd loop. The Te loop is plotted for data obtained only from the optical technique. Strauss (1971) considers the values obtained by thermal methods to be much too high. Portions of the three-phase loop are not accessible by measurements; i.e., $p_{Cd}(p_{Te_2})$ cannot be directly measured for Te-saturated (Cd-saturated) conditions. Instead, the lower pressure component is calculated using Eq. (1.4). By making optical measurements over congruently subliming CdTe, the relationship between component pressures as a function of temperature in Eq. (1.3) was determined (Brebrick, 1971; Brebrick and Strauss, 1964). Over the temperature range 780°–939°C, the Cd pressure was confirmed to be twice the Te pressure and the Te pressure was expressed by:

$$\log p_{Te_2}(\text{atm}) = [(-1.00 \times 10^4)/T] + 6.346. \quad (1.5)$$

The log of the pressure product is given by:

$$\log(p_{Cd}p_{Te_2}^{1/2})(\text{atm}^{3/2}) = -15.0(10^3/T) + 9.820. \quad (1.6)$$

Vapor pressures for congruently subliming CdTe are also indicated within the three-phase loops. The total pressure in the vapor phase at a fixed temperature is a minimum at the congruent composition, and is about 1.8×10^{-1} atm at the melting point; much less than the maximum value of 7 atm with the Cd overpressure.

Other methods used to investigate the solid–vapor equilibrium of CdTe are in good agreement with the optical technique. By combining emf cell measurements and data on the thermodynamic properties of the elements, Brebrick and Strauss found agreement to within 9%. Results for the Lorenz (1962b) technique are about 40% higher, but have the same temperature dependence. In this technique a known volume of the vapor phase in equilibrium with solid CdTe is sealed off. The vapor is condensed and analyzed colorimetrically for Te. By calculation p_{Te_2} is determined. It is possible that the ratio of the weight of the CdTe to the volume of the vapor was too high for the congruently subliming state to be reached. Drowart and Goldfinger (1958) obtained values of p_{Te_2} from mass spectrographic analyses of the sublimation products between 507°C and 642°C. Considering that the composition of the vapor and solid inside the effusion cell differ from each other and from the equilibrium congruently subliming composition, values of p_{Te_2} calculated from effusion data were about 30 to 40% higher than the optical technique. The optical technique is the most reliable method of investigating the solid–vapor equilibrium because it permits the partial pressures to be measured simultaneously under equilibrium conditions.

Measurement of both pressures permit precise determination of the Gibbs free energy of formation:

$$\Delta G°[2Cd_{1-x}Te_x(s)] = RT \ln p_{Cd}p_{Te_2}^{1/2} + 2(\tfrac{1}{2} - x)RT \ln[p_{Cd}p_{Te_2}^{1/2}] \quad (1.7)$$

where x represents the deviation from the stoichiometric composition. Considering the vapor pressures at the extreme of the homogeneity region and the deviation from stoichiometry to be 10^{-5} or less, the last term can be neglected. Using $p_{Cd} = 2p_{Te_2}$ for the congruently subliming melt, Brebrick and Strauss obtained for the free energy between 780°C and 939°C:

$$\Delta G[CdTe(s)] = -68.64 + 44.94(10^{-3})T \quad \text{kcal/mole}. \quad (1.8)$$

A calculation of the free energy by the other methods results in good agreement.

C. Total Pressure versus Temperature Diagram

At higher total pressures, shifts in the phase boundaries occur, and the solid undergoes transitions to different crystallographic forms. The melting point of CdTe as determined by differential thermal analysis over the pressure interval 0 to 50 kbar is plotted in the pressure versus temperature (P–T) diagram of Fig. 1.5 (Jayaraman et al., 1963). The pressure in this case represents the total pressure exerted on the system. The melting point decreases with a slope of approximately $-5°C/kbar$ from 1092°C under atmospheric pressures to a triple point near 19.2 kbar and 996°C where the zinc blende–rocksalt transition occurs. The slope of the fusion curve of the rocksalt structure is about $+10°C/kbar$. At room temperature this phase change occurs at approximately 33 kbar. At 90 kbar a further transition to the tetragonal white tin structure occurs. These transitions were indicated by the volume discontinuity method (Jayaraman et al., 1963), and resistivity measurements (Samara and Drickamer, 1962), and verified by x-ray analysis (Mariano and Warekois, 1963; Smith and Martin, 1963; Owen et al., 1963). The zinc blende to rocksalt transition appears to be quite sluggish (Jayaraman et al., 1963), and has been attributed to transitions which are diffusion controlled. Considering the ease with which Cd atoms could shift from tetrahedral to octahedral sites in an anion sublattice, this is unlikely (Rooymans, 1963). It is more likely that two phases coexist due to the difficulty of maintaining a completely hydrostatic system.

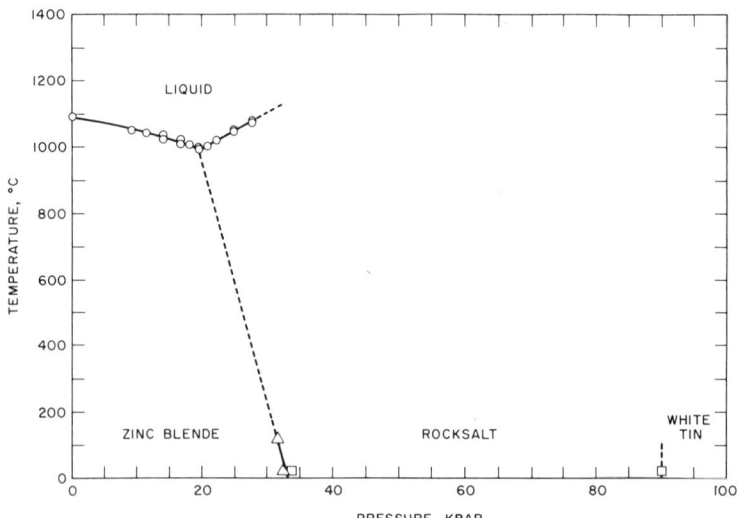

Fig. 1.5. Total pressure versus temperature (P–T) diagram for CdTe. ○, DTA: Jayaraman et al. (1963); △, Volume discontinuity: Jayaraman et al. (1963); □, x ray and resistivity: Samara and Drickamer (1962), Mariano and Warekois (1963), Owen et al. (1963).

II. Crystal Growth

Bulk single crystals of CdTe have the zinc blende structure. Three general approaches have been used to grow CdTe crystals. Growth from congruent or near-congruent melts has been the most popular technique because large single crystals can be easily obtained within a few days. However, due to the high congruent melting point (1092°C), and possible contamination from quartz crucibles at high temperatures, growth from Te-rich solutions at lower temperatures has recently been pursued more actively. Crystal growth from the vapor is an alternative method to avoid high temperatures. Thin films as well as bulk single crystals have been obtained through vapor growth. Deposition of thin films by vacuum evaporation is discussed separately since the conditions of growth are far from equilibrium. Thin films exist in the stable cubic form and the metastable hexagonal form. Growth from foreign hosts at significantly reduced temperatures has also been undertaken and will be briefly discussed at the end of this section.

A. Congruent and Near-Congruent Melt Growth

1. *Bridgman Techniques*

Growth by the Bridgman technique has been pursued more actively not only because large single crystals are easily obtained and growth rates are much faster, but also because the crystal grower has to some extent the freedom of controlling the type and the conductivity by controlling the deviation from stoichiometry. In the Bridgman method (Mizuma et al., 1961), atomic equivalents of Cd and Te are loaded into a fused silica ampoule which is lowered through a vertical furnace with a gradient that produces directional freezing from the bottom up. Ideally, the growing portion of the boule at the solid–liquid interface is smooth and its shape is characteristic of temperature isotherms. The upper portion of the growth chamber is usually maintained at a hotter temperature than that of the vapor–melt interface to prevent sublimation and vapor growth at the top of the ampoule. Usually the ampoule is pointed to seed a single crystal. Variations of the Bridgman method exist. Yamada (1960) used a capillary for the purpose of seeding a single crystal. Prior to crystal growth, the elements were reacted at approximately 800°C to synthesize CdTe. The ampoule was then held at 1150°C to homogenize the melt and lowered through the furnace at a rate of approximately 2 cm/hr. With this procedure crystals were about 1 cm in diameter and 5 cm in length. Numerous workers report the growth of CdTe by the simple Bridgman technique in their studies (Segall et al., 1963; Inoue et al., 1962; Brodin et al., 1968, 1970; Smith, 1970; Kyle, 1971; Ichimiya et al., 1960). A variation of this method is to alternatively maintain the ampoule at a fixed position and cool the furnace. In this way Sagar and Rubenstein (1966) routinely obtained Hall samples 1 cm in length.

Growth from congruent melts using liquid encapsulation techniques has not resulted in large single crystals. However, using B_2O_3 as a liquid encapsulant under argon pressures of approximately 60 psi, Bridgman crystals were grown having dislocation densities as low as 10^4 cm^{-2} (Carlsson and Ahlquist, 1972). Boron oxide has also been used as an encapsulant in the Czochralski technique (Meiling and Leombruno, 1968). It was possible to grow centimeter-size single crystals. However difficulties were encountered for dimensions above 5 mm. In another case, only polycrystalline rods 5 to 8 mm in diameter and several centimeters long were obtained (Vandekerkhof, 1971). The difficulty in pulling CdTe is attributed to the inability of the B_2O_3 to wet the CdTe and the slight solubility of CdTe in B_2O_3. Klausutis et al. (1975) grew mixed crystals of CdTe$_{1-x}$Se$_x$ with values of $x = 0.05, 0.10$, and 0.15 by the Czochralski technique using B_2O_3 as an encapsulant and argon pressures of 20 atm. Clouding of the encapsulant layer by dissolved CdTe and continuous loss of Cd vapor from the melt was not found. Vapor losses were found at pressures below 13 atm and it was suggested that earlier difficulties by Vandekerkhof were due to an insufficient overpressure. The liquid encapsulation technique should be applicable to the growth of CdTe. However, it is unlikely that the crystal quality will be comparable to that of the Bridgman techniques. The mixed crystals of Klausutis et al. were at most 1 mm in diameter and 1 cm long, and were inferior to Bridgman type crystals which were grown in an open ampoule with a 20 atm Ar overpressure.

In many studies crystals are intentionally grown off stoichiometry to control the properties of the crystal during growth without the necessity of subsequent annealing treatments. Adding an excess of Cd(Te) to the starting material is more likely to result in an excess donor (acceptor) concentration. Crowder and Hammer (1966) and Morehead and Mandel (1964) added an excess of Te to the charge to study the properties of foreign acceptors. In some cases this method is complicated by the presence of compensating impurities. Yamada (1960) failed to make crystals n-type by adding an excess of Cd to the starting material.

It is important to be fully aware of the pressure–temperature–composition relationships discussed in the last section. As an excess of Te or Cd has a strong effect on the melting point and the vapor pressure (Fig. 1.4), slight excess of Cd results in a substantial increase in the total vapor pressure even though the growth temperature is reduced. A slight Te excess has been used by Rowe et al. (1974) to prevent the buildup of an excessive Cd overpressure that might rupture the ampoule. Lorenz and Blum (1966) intentionally added an excess of Cd to the charge for the purpose of studying the segregation of impurities at different temperatures (and vapor pressures of Cd. Large excesses are detrimental to crystallinity because of either constitutional supercooling or bubble formation. With large excesses of Cd the crystals of Yamada became porous due to bubble formation. If the melt is non-stoichiometric, either excess Cd or Te accumulates at the solid–liquid interface, the excess resulting in a lowering of the melting point, dendritic growth, and subsequent entrapment of the melt. Constitutional supercooling can be avoided by growing from stoichiometric or near-stoichiometric melts at reduced growth rates. Borsenberger and Stevenson (1968) took care to prepare crystals from stoichiometric melts to obtain crystals for diffusion studies by presubliming material in a dynamic vacuum under the condition $2p_{Cd} = p_{Te_2}$.

In general, the component vapor pressures over the crystal are not well defined during growth due to severe temperature gradients and changes in the composition of the melt as the crystal grows. To establish a better defined system and attain solidification under a constant vapor pressure of Cd, the temperature at the top of the ampoule is held constant (Triboulet and Rodot, 1968; Triboulet and Marfaing, 1973). In this way the deviation from stoichiometry of the crystal is better controlled. A further and more precise extension of this technique is to maintain the vapor pressure by a separate reservoir of Cd in an additional furnace located above the growth furnace (Kyle, 1971), as shown in Fig. 1.6. A further modification to increase crystal size is to position the nose of the ampoule on a metal rod which acts as a heat sink and results in a more favorable solid–liquid interface shape. With a convex curvature at the solid–liquid interface, crystal growth is more likely to proceed from the substrate rather than from the wall of the crucible. This method has resulted in untwinned single crystals up to 100 cm^3 in

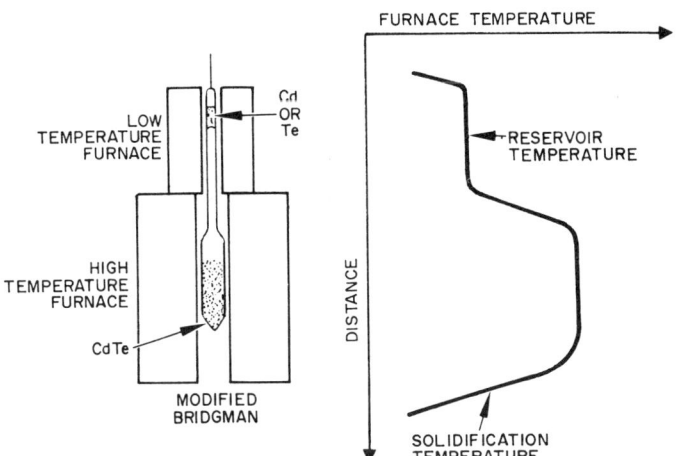

FIG. 1.6. Schematic illustration of modified Bridgman growth system and associated temperature profile (Kyle, 1971). The solidification and resevoir (p_{Cd}) temperatures are approximately 1090°C and 900°C, respectively.

14   1. MATERIALS PREPARATION

FIG. 1.7. Wafers of CdTe cut from a 5 cm diameter input grown by the modified Bridgman technique. (Reprinted with permission from Gentile et al., 1973a,b. © 1973, Pergamon Press Ltd.)

volume. Figure 1.7 shows several wafers cut from a 5 cm diameter ingot. These methods do not permit growth at or close to p_{min} since excessive sublimation to the cooler reservoir would occur.

2. Vertical Zone Refining

The second broad area of crystal growth from congruent or near-congruent melts is the sealed-ingot vertical zone refining method. Material is zone refined, as in the method of Pfann (1966), except that the ampoule is held vertically and lowered through the zone. It has the advantage over the Bridgman technique in that the charge can be purified before crystal growth. Halsted et al. (1961) first obtained single crystals by lowering charges through rf-heated graphite sleeves at rates of 3 to 6 mm/hr. The technique was quite successful and consequently merits further discussion (Lorenz and Halsted, 1963). Elemental Cd and Te corresponding to 110 gm of CdTe are placed in $\frac{1}{16}$ in. i.d. quartz tubes previously coated with pyrolytic graphite. After evacuating and sealing the ampoule the charge is reacted and heated to about 1150°C. The ampoule is then slowly lowered through the furnace as in the Bridgman technique to condense the CdTe into a solid rod approximately 7 in. in length. The ampoule is cut open and a quartz plug is placed within the ampoule on the ingot before resealing. The ampoule is sealed at the plug preventing the vaporization of CdTe during sealing, thus eliminating major voids and mass transport during zone refining and crystal growth. Again the ampoule is lowered through the narrow hot zone of an induction furnace, the details of which are shown in Fig. 1.8. Multiple passes for the purpose of purification were made at a speed of approximately 2 cm/hr. The effectiveness of this purification technique is discussed in the next section. For the final pass the length of the susceptor ring is increased and the rate of zone travel is reduced to 5 mm/hr or less. Single crystals of several centimeters in length are found in every ingot. Many small improvements

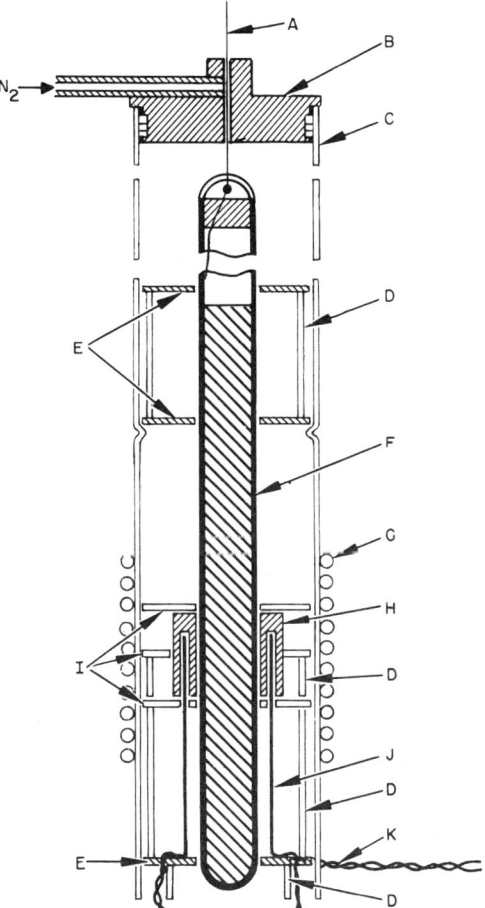

FIG. 1.8. Vertical zone refining and crystal growth apparatus. Key components are: quartz sealed-ingot of CdTe (F), rf induction coil (G), and graphite susceptor (H) (Woodbury and Lewandowski, 1971).

have accumulated over the years using the sealed-ingot zone refining method. Woodbury and Lewandowski (1971) grew crystals with volumes up to 10 cm^3 by imposing vertical oscillations (i.e., 4 mm at 100 cpm) on the ingot during the last zone pass. Their studies indicate that the heat flow pattern at the liquid–solid interface or shape of the interface and not the melting–recrystallization phenomenon are important to obtain large single crystals.

A correlation has been observed between the susceptor temperature and the type and resistivity of the final ingot. Ingots were more *p*-type (*n*-type) with a higher (lower) susceptor temperature. However, contamination from the ampoule by *p*-type impurities is possible at higher processing temperatures. Triboulet (1971) and Triboulet and Marfaing (1973) grew single crystals about 2 to 3 cm^3 in volume using this technique of crystal growth. They did not find a correlation between susceptor temperature and conductivity type. Instead the type was found to be consistent with the initial state of the material imposed by the vapor pressure of the Cd reservoir during the growth step used to synthesize the CdTe. Rud' and Sanin (1969a) find radial distributions in the conductivity and in some cases a central *n*-type region and a peripheral *p*-type region. Marple (1963), Segall *et al.* (1963), Lorenz and Woodbury (1963), Ludwig and Lorenz (1963), Slack and Galginaitis (1964), Lorenz *et al.* (1964), Cusano and Lorenz (1964), Halsted and Aven (1965), Slack *et al.* (1966), Ludwig (1967), Oliver *et al.* (1967), Hall and Woodbury (1968), Woodbury and Aven (1968), Rud' and Sanin (1969a,b), Slack *et al.* (1969), and Cornet *et al.* (1970b), among others, either have grown crystals by this technique or have evaluated such materials in the course of their studies.

3. *Horizontal Techniques*

The third broad area of crystal growth from congruent or near-congruent melts are the horizontal techniques. The horizontal methods were used in the early CdTe crystal growth programs and were replaced by the vertical sealed-ingot and Bridgman techniques. The experimental configuration is similar to that used in the postannealing of CdTe. Analogies exist between the vertical and horizontal methods. The horizontal counterpart of the modified Bridgman technique corresponds to withdrawing the charge through a temperature gradient and controlling the overpressure with a reservoir located in a second furnace. A Cd reservoir of 1 atm is sufficient to prevent dissociation of the melt and transport to the cooler regions of the ampoule (Kröger and de Nobel, 1955). Starting with a seed and withdrawing the charge at a rate of 1 cm/hr results in single crystalline rods containing only twins. Ichimaya *et al.* (1960) also used the seeding technique. Rather than withdraw the charge, directional cooling has been attained by reducing the temperature of the melt (Matveev *et al.*, 1969a). Three zones (Lorenz, 1962c;

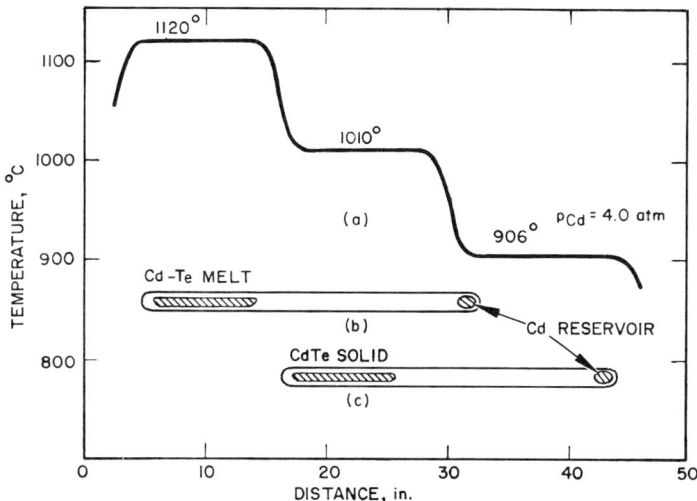

FIG. 1.9. Crystal growth from near stoichiometric melts by the horizontal method. (a) Furnace profile. (b) Position of tube before growth. (c) Position of tube after growth (Lorenz, 1962c).

Medvedev et al., 1968a) have also been used to grow single crystals (see Fig. 1.9). The low temperature furnace is used to control the vapor pressure of the Cd. The ampoule is initially inserted into the furnace at a position corresponding to (b) in Fig. 1.9. After several hours at this position the melt is passed through the gradient between the two hotter zones at a rate of 3 mm/hr. After reaching position (c) the ampoule is removed from the furnace and water quenched. Single crystals up to 5 cm in length could be obtained; more typically the crystals were 1 to 2 cm on a side. Twinning was evident in nearly all crystals. Lorenz (1962c) obtained crystals of poor quality by holding the system at a Cd vapor pressure of 5.9 atm and heating the melt from 970°C to 1025°C. As the charge passes through the three-phase loop in Fig. 1.4 solidification occurs. Using a growth rate of approximately 1 mm/hr resulted in single crystals up to 1 cm in length which occasionally contained porous inclusions. Since the composition of the melt is far from stoichiometric (61 at% Cd and 39 at% Te) the growth rate is limited by the removal of the Cd from the solid–liquid interface. Heating the melt to form single crystals is not of technical importance, but it has demonstrated the usefulness of the three-phase loops.

By implementing the methods of Pfann, CdTe has been zone refined in a horizontal manner prior to crystal growth (de Nobel 1959; Matveev et al., 1969a; Prokof'ev, 1970; Prokof'ev and Rud', 1970). Similar to the previous horizontal technique CdTe is loaded in a graphite or silica boat and held well below the melting point. With an auxiliary furnace a narrow (1.5 to

2.0 cm) molten zone is passed through the charge. A reservoir of Cd maintained at a lower temperature in a separate furnace prevents transport of the molten zone to cooler regions of the ampoule. A Cd vapor pressure of 1.5 atm is high enough to prevent sublimation and yet low enough to prevent excessive deviations from the stoichiometric composition. By operating at very low zone speeds (0.5 to 0.8 cm/hr) bubbles formed by the excess of Cd can rise to the surface of the zone before becoming entrapped and forming porous material. Prokof'ev and Rud' (1970) grew untwinned single crystals up to 15 cm in length by this method using a Cd vapor pressure from 0.8 to 1.2 atm and a zone rate of 0.5 cm/hr. Lawson et al. (1959, 1960) also grew single crystals by the movement of a zone. Instead of using a separate reservoir the charge was maintained at a uniformly high temperature. It is likely that the excess of one of the components prevented decomposition of the molten zone.

B. Growth from Tellurium-Rich Solutions

In order to alleviate the contamination problems associated with high temperature growth, crystals have been grown from Te-rich solutions at significantly lower temperatures ($<900°C$). Sixty wt% and 95 wt% Te-rich solutions correspond to liquidus temperatures of, respectively, 900°C and 550°C. Possible advantages of the lower growth temperatures are

    a. a reduction in impurity pickup from the crucible material,
    b. the opportunity to use alternative crucible materials which soften at lower temperatures,
    c. a reduction in the concentration of native defects in the as-grown crystals,
    d. a lower solubility for impurities in the growing solid phase,
    e. more favorable values for the segregation coefficients of residual impurities.

Technically, the near-congruent melt techniques in the preceding section are also solution growth techniques, since the crystals are grown from small excesses of Te or Cd. The definition of solution growth, therefore, is arbitrary. It is considered here as the addition of an excess of Te or Cd to intentionally lower the freezing point to consequently obtain an advantage not possible by congruent or near-congruent melt growth.

The disadvantages of solution growth should not be dismissed. It is important to point out that solution growth crystals are about an order of magnitude less in size, have a higher inclusion content, and grow at a slower rate than crystals grown from the congruent melt.

In principal crystals can be grown from both Te-rich and Cd-rich solutions. Medvedev et al. (1972) grew 5 mm × 7 mm × 0.3 mm single crystal platelets

II. CRYSTAL GROWTH

FIG. 1.10. Cluster of CdTe single crystals (a) growth from solution at approximately 550°C in a static system. Separate crystals (b) having well-defined {100} planes and {111} planes (Zanio, 1974).

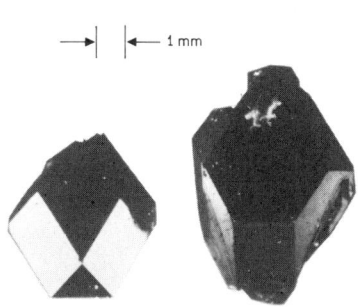

from Cd solutions at 800°C–850°C. However, because of the restricted solubility of Te in Cd-rich liquids ($\approx 1\%$ at 800°C) the growth of crystals on the Cd side of the T–x diagram is impractical. Growth from Te-rich nearly static isothermal systems also results in single crystals, only a few millimeters on a side (Zanio, 1974; Medvedev et al., 1972). Crystals grow in clusters on the walls of the ampoule with well-defined {110}, {100}, and {111} facets. Figure 1.10 shows such a cluster (a) and two crystals (b) separated from that cluster, having well-defined {100} and {111} planes. Infrared examination showed the crystals to contain large inclusions. The top of Fig. 1.11 shows an infrared photograph of random inclusions contained within a crystal. Below are enlarged infrared (left) and visible (right) photographs of an inclusion at the surface of the CdTe. These inclusions are a consequence of constitutional supercooling. As CdTe grows, the solution at the solid–liquid interface is depleted of Cd and becomes more Te-rich. Because the liquidus temperature at the interface is less than that of the rest of the melt, dendritic growth proceeds into the undepleted regions. Subsequent lateral growth at some distance away from the original solid–liquid interface results in entrapment of Te-rich liquid.

Temperature gradients are necessary to grow inclusion-free CdTe crystals from solution. Large relatively precipitate-free crystals grow from solution by the traveling heater method (THM) (Bell et al., 1970a; Bell and Wald,

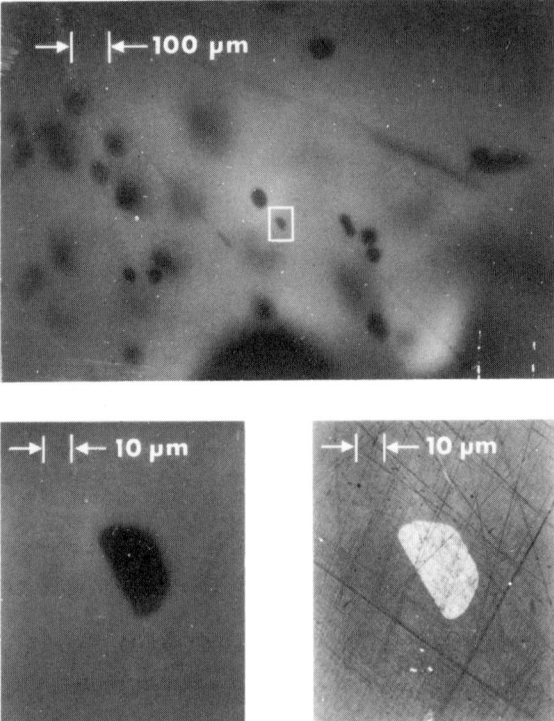

FIG. 1.11. Infrared photograph (a) of random inclusions. Enlarged infrared (right) and visible (left) photographs of an inclusion (b) at a lapped and polished surface (Zanio, 1974).

1972; Wald and Bell, 1975; Bell, 1974; Triboulet et al., 1973, 1974; Tranchart and Bach, 1976; Martin et al., 1975; Taguchi et al., 1977) and the cold finger technique (Zanio, 1974; Zanio et al., 1974). Broder and Wolff (1963) initially suggested the growth of GaP by THM (Figs. 1.12 and 1.13). This technique represents a variation of the temperature gradient zone melting method (Pfann, 1955).

In THM, as in the sealed-ingot zone refining method of Halsted et al. described earlier, a molten zone passes through an ingot by the slow movement of a heater relative to the charge. Dissolution and crystallization of feed material occurs respectively at the hotter and cooler solid–liquid interfaces. However, the THM process is different from simple zone refining in that the zone acts as a solvent through which dissolving Cd must pass through before recrystallization at the solid–liquid interface can take place. Cadmium telluride ingots up to 15 cm long and 1.5 cm in diameter containing crystals several cubic centimeters have been grown. Growth rates range from 0.3 to 2 cm/day, with the best crystallinity attained at the slowest speeds. Bell

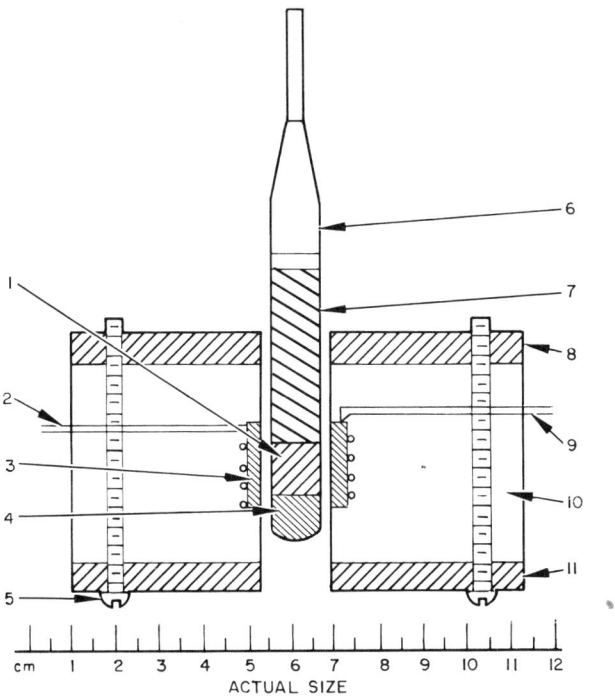

FIG. 1.12. Schematic of traveling heater growth furnace (Wald and Bell, 1975).

1 SOLVENT ZONE
2 HEATER LEADS
3 NICKEL LINER
4 REGROWN MATERIAL
5 HOLDING SCREW
6 QUARTZ AMPOULE
7 FEED MATERIAL
8 ALUMINUM PLATE
9 THERMOCOUPLE LEADS
10 DYNAFLEX INSULATION
11 ALUMINUM SUPPORT PLATE

et al. (1970a) give a brief account of the preparation of the CdTe feed material and crystal growth procedure. Initially 250 gm of Cd and Te in near-stoichiometric amounts are reacted in a carbon-coated fused quartz ampoule. The cast material and approximately 10 gm of pure Te are placed in a second ampoule which is lowered through a 650°C to 750°C zone heater. Such an apparatus is shown in Fig. 1.12. A bottom or after-heater is sometimes used to subject the crystals to gradual annealing. Vertical zone melting has been used as an intermediate step to purify the material previous to crystal growth by THM (Triboulet *et al.*, 1973, 1974). Bridgman grown material has also been used for feed material (Taguchi *et al.*, 1977).

The slopes of the liquidus and solidus, viscosity of the solution, heat transport coefficients of the ampoule and solid, and geometry of the liquid

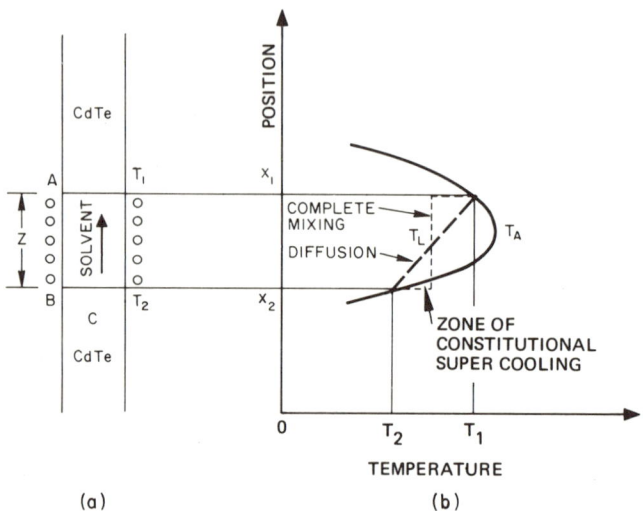

FIG. 1.13. (a) Schematic of the traveling heater method (THM) process for the solvent traveling upward relative to the ingot. (b) For diffusion limited transport the liquidus temperature (T_L) is always below the actual temperature (T_A). However, for complete mixing, T_L is above T_A resulting in constitutional supercooling and possible dendritic crystal growth (based on Wald and Bell, 1975).

zone determine the rate of crystal growth and quality of resulting crystals. Tranchart and Bach (1976) introduced a gas bearing system which permits rotation and translation motions during growth by THM without vibration. Superior crystallinity was obtained presumably due to the absence of spurious nucleation from vibration. An analysis of the temperature distribution under static conditions shows that the ideal interface shape is planar (Bell, 1974). However during crystal growth convection currents are present. If the interface is convex upward convection currents provide a sweeping action at the center of the ampoule. Although the perimeter is starved of Cd and entrapment of Te is likely this condition is more desirable than a convex downward interface where entrapment is likely at the center. The large number of variables prevents a detailed analysis of the growth kinetics. However the process is idealized in Fig. 1.13. Here the liquid zone moves upward relative to the growing crystal C. The thickness of the zone Z is nearly independent of the rate of travel of the zone. The liquid has a composition of C_1 at A and a composition of C_2 at B. Transport through the average concentration gradient $(C_1 - C_2)/Z$ tends to increase the concentration of Cd at B and reduce it at A. This tendency is prevented by crystal growth at B and solution at A. Although practical analysis of the actual transport process is complicated, two limiting cases have been discussed;

diffusion and complete mixing (Wald and Bell, 1974, 1975). From conductivity measurements Rud' and Sanin (1971b) calculate the diffusion coefficient of Cd atoms in liquid CdTe to be $(1-8) \times 10^{-5}$ cm^2/sec at 1100°C. Unfortunately, the diffusion coefficient of Cd in liquid Te has not been measured at lower temperatures. Estimates of the growth rate using a diffusion model have been made. It was calculated, however, that the materials transport process is 50 times larger than predicted using the self-diffusion coefficient of copper in liquid Te. The validity of this analysis depends upon whether or not instabilities develop. In the diffusion case the actual temperature T_A is likely to be everywhere above the liquidus temperature T_L and supercooling is unlikely. This assumes that the diffusion coefficient is nearly independent of temperature and the composition gradient from A to B is nearly constant. It is likely that mixing by natural convection occurs. If complete mixing could be achieved the liquidus temperature would be constant and a supercooled zone would exist at the growing solid–liquid interface near X_2. During actual crystal growth the liquidus line probably occupies a position intermediate to that of the diffusion and complete mixing cases and stable crystal growth occurs. However, at excessive growth rates the liquidus temperature near the growing solid–liquid interface drops below the actual temperature and unstable conditions might develop. Additional studies have also shown that rotation of the crucible can improve the quality of the crystal. To grow crystals free from solvent inclusions, it is necessary to assure that the growth rate does not exceed the materials transport rate whether they be diffusion or convection limited.

The second approach of obtaining large relatively precipitate-free crystals from Te-rich solutions is by imposing a sharp temperature gradient on the growth system. This circulates the solution and restricts Cd depletion at the solid–liquid interface (Zanio, 1974). The system, shown in Fig. 1.14 consists of a 2-in. bore vertical furnace, a flat bottomed quartz ampoule, and a stainless steel cold finger. The cold finger consists of two concentric tubes through which coolant flows. Rather than lowering the ampoule through the furnace, the ampoule remains stationary and the temperature of the furnace is reduced at approximately 4°C/hr. Nucleation occurs at the cold finger and growth proceeds with a convex solid–liquid interface due to the radial temperature gradients impressed by the cold finger. The existing temperature gradients presumably drive a convection current which adequately mixes the solution. Cooled liquid descends at the center of the ampoule, flows over the solid–liquid interface and rises at the peripheral of the ampoule. A typical starting charge is 200 gm and contains 27 wt% Cd. This corresponds to an initial solid–liquid interface temperature of approximately 900°C. Boules ranging in weight from 50 to 150 gm have been grown. A 100 gm boule corresponds to a final solid–liquid interface temperature of approximately

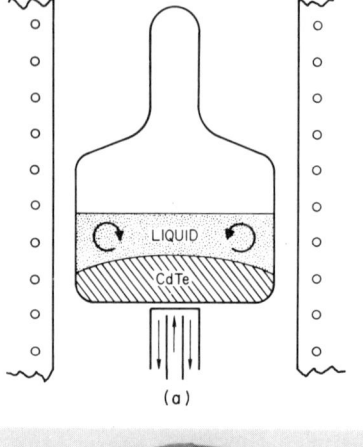

FIG. 1.14. Schematic drawing (a) of a convection assisted solution growth system and a 25 cm³ CdTe boule (b) containing a 20 cm³ single crystal (Zanio, 1974).

650°C. Figure 1.14b shows a 25 cm³ boule of which all but 5 cm³ is a single crystal. The single crystal cleaved on a {110} plane which was perpendicular to the bottom of the ampoule. Typically boules contain single crystals between 1 and 2 cm on a side. Schaub (1976) grows ingots of comparable size from Te-rich solution by lowering the ampoule through a favorable temperature gradient. In order to maintain the solid–liquid interface at the same position, the temperature of the furnace is simultaneously lowered.

An infrared inspection of boules grown by the cold finger technique show single crystalline regions to be nearly free of inclusions. Micron-size inclusions generally decorate the crystal boundaries and twin planes. It has not been resolved whether the smaller inclusions are due to entrapment of liquid during growth or the precipitation of Te-rich solutions from homogeneous CdTe during cooling. Such precipitation is possible if the solidus on the Te-rich side of the phase diagram is indeed retrograde, as shown in Fig. 1.3.

C. Vapor Growth

Single crystals of CdTe have been grown from the vapor by direct synthesis, sublimation, and chemical transport (Hartmann, 1975). Sublimation is the most popular. Variations of the sublimation method result in the largest single crystals, a few cm², and provide facets for the study of growth mechanisms. Thin films are also grown by sublimation, chemical transport, and vacuum deposition. Both bulk single crystals and thin films can be grown by the same techniques (i.e., sublimation and chemical transport). However, the resulting form of the compound depends upon the growth conditions (i.e., single crystal growth represents a closer approach to equilibrium and generally involves higher temperatures; thin film growth represents significant departure from equilibrium and generally lower temperatures). The literature describes vapor growth in more detail than melt growth because vapor growth is more precisely controlled allowing a more quantitative interpretation of the growth process.

1. *Single Crystals*

a. Direct Synthesis

Some of the first crystals of CdTe were formed by the direct synthesis of Cd and Te (Frerichs, 1947). In this method Cd and Te vapors are generated in separate reservoirs and reacted in the presence of H_2 under atmospheric pressure. Lynch (1962) used reservoir partial pressures of from 5 to 10 mm and a reaction temperature of 800°C. Hydrogen is not necessary for the transport of the vapors but is convenient for reducing the components. Resulting crystals are about a millimeter on a side and take the form of dodecahedra or hexagonal rods. By varying the temperature of the resevoir Vanyukov *et al.* (1974) was able to control the conductivity type of crystals grown in an argon stream. This method of growth results in dislocation densities as low as 10^2 cm^2. The synthesis step in forming the CdTe is attractive, since the lower processing temperature reduces contamination. Since the size of these crystals is not of technical importance, this method has not been pursued extensively. However, it is noteworthy to point out that molecular beam epitaxy (Smith and Pickhardt, 1975), a recent development in thin film growth, incorporates the feature of direct synthesis.

b. Sublimation

Crystals are prepared by sublimation techniques over a wide range of conditions. Crystals grown at equilibrium or near-equilibrium conditions result in faceted crystals, permitting optical studies, contact studies, growth mechanism studies, etc. on as-grown planes. The relationship between the partial pressures over the solid phase under equilibrium conditions is given by Eq. (1.6). Sharp temperature gradients across the crystal face result in nonfaceted but larger single crystals. At extreme cases of supersaturation thin films are formed, their quality and crystallite size depending also on the type and degree of preparation of the substrate. The vapor pressures of Cd and Te are quite comparable versus other systems, and consequently the system is more flexible when sublimation is performed under nonequilibrium conditions.

1. Small Faceted Crystals. For growth under near-isothermal conditions, the vapor–solid interfacial energy influences the size and shape of the single crystals and faceted growth occurs. Typically, crystals with volumes up to 100 mm^3 and having well-defined (110), (111), and (100) planes are obtained. Volumes up to $\frac{1}{2}$ cm^3 have been reported. Faceted crystals have been grown by several methods and are classified here by growth in either open or closed systems. Three variations of the open tube technique and a schematic temperature profile are shown in Fig. 1.15. A polycrystalline charge, positioned at S in the hotter region of a quartz ampoule, dissociates, sublimes to the cooler region of the furnace and crystallizes. The static H$_2$ system relies on diffusion and convection to transport the CdTe (Lynch, 1962). The ampoule is connected to an external tee through which H$_2$ flows

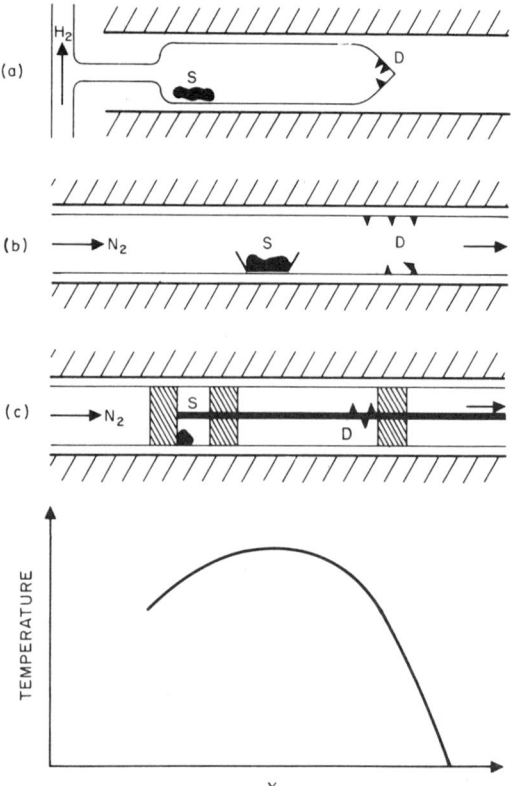

FIG. 1.15. Schematic representation of three open tube systems for growing CdTe from the vapor along with associated temperature profile. Material is transported from the source, S to the deposition regions, D, by (a) diffusion and convection and by (b) and (c) flowing gas.

maintaining a reducing atmosphere at atmospheric pressure (Fig. 1.15a). Good growing conditions occurred when the source was held at 850°C and the growth region was held at 835°C. The rate of crystal growth was 0.01 gm/hr and typical crystal size was less than 100 mm^3.

In the static system, diffusion and convection limits the growth rate. Consequently a carrier gas is introduced to assist in the transport process. However at excessive flow rates the partial pressures of Cd and Te in the carrier gas will be low and the gas will be undersaturated. Supersaturation and deposition will then occur downstream at a low temperature. At a low temperature the deposition rate may be too fast with respect to the kinetics of crystal growth to permit the growth of well-faceted crystals. A balance between sublimation rate and flow rates is therefore important. Sublimation rates depend upon the geometry of the system and increase with the surface area of the source and temperature of the source in accord with Eq. (1.5). The most straightforward way of increasing the sublimation rate is to increase the temperature if this does not compromise the low temperature advantages of vapor growth. Using N_2, Teramoto (1963) optimized the flow rate and sublimation speed. Favorable growth conditions occurred when the source was held between 880°C and 950°C and a flow rate of 100 ml/min was used. A schematic representation of this system is shown in Fig. 1.15b. The crystal habit was found to depend on the degree of supersaturation. At low supersaturation conditions dodecahedra and hexagonal plates are formed while at high supersaturation conditions more dendritic forms are likely (Lynch, 1962; Pashinkin et al., 1960). By enclosing the crystal growth chamber with quartz wool and nucleating crystals on an alumina rod placed on the axis of the crystallization chamber (Fig. 1.15c), well-faceted larger crystals, up to 240 mm^3, have been obtained (Corsini-Mena et al., 1971).

Growth rate is dependent on the crystallographic orientation. In the sublimation of CdTe on single crystal substrates, triangular pyramids form on the flat $\{111\}$ Cd planes (Fig. 1.16a), while on the opposite $\{111\}$ planes less frequent growth hillocks occur (Fig. 1.16b) (Teramoto and Inoue, 1963). The difference in the growth properties between these two surfaces results in an anisotropy in the growth pattern on a surface having only one axis of symmetry. On the $\{1\bar{1}0\}$ surface the pattern is convex to a $\langle\bar{1}00\rangle$ direction and flat to the opposite $\langle 100\rangle$ direction (Teramoto and Inoue, 1963; Akutagawa and Zanio, 1971). This feature, shown in Fig. 1.17, implies the growth rate to be higher in $\langle 111\rangle$ directions than in $\langle\bar{1}\bar{1}\bar{1}\rangle$ directions. Also observed on the $\{111\}$ surfaces were growth spirals which were believed to be due to screw dislocations. The open tube methods, also used to obtain thin films, will be discussed later (Saraie et al., 1972; Fahrenbruch et al., 1974; Vanyukov and Krotov, 1971; Aitkhozhin and Temirov, 1971).

28 1. MATERIALS PREPARATION

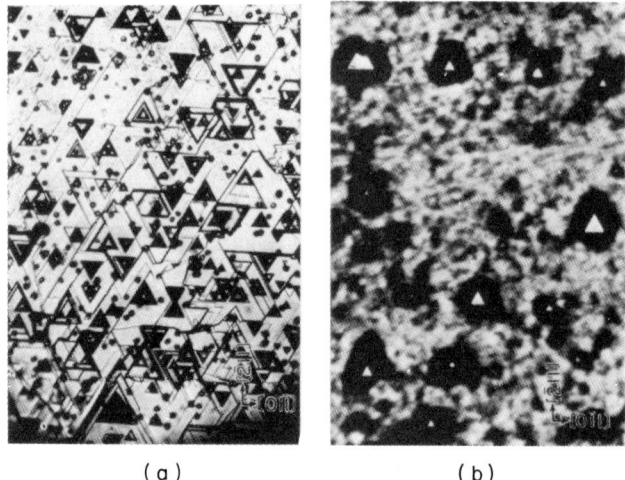

(a) (b)

FIG. 1.16. (a) Growth pyramids on a {111} plane and (b) growth hillocks on {$\bar{1}\bar{1}\bar{1}$} plane at ×308 (Teramoto and Inoue, 1963).

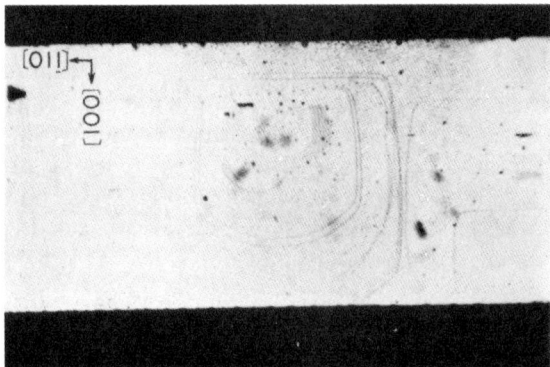

FIG. 1.17. Asymmetrical growth patterns on {110} plane at ×510 due to differences in growth rates along $\langle 111 \rangle$ and $\langle \bar{1}\bar{1}\bar{1} \rangle$ directions (Teramoto and Inoue, 1963).

Using closed tube sublimation techniques, crystals of similar morphology and size have also been grown. Instead of using a carrier gas at atmospheric pressure, the ampoule is evacuated and a Cd vapor pressure is maintained over the crystal by a reservoir held at a lower temperature. By varying the pressure of the Cd it is possible to change the sublimation rate to influence the growth conditions. The sublimation velocity of CdTe and CdSe strongly depends on the vapor pressure of Cd (Höschl and Koňák, 1963a,b, 1965). When diffusion limits mass transport, the sublimation rate is proportional to p_{Cd}^{-3} and is in agreement with theory. Crystal growth occurs at the gradient between the charge and the reservoir. Unsatisfactory results are obtained using low pressures. In the gradient region, crystals are dendritic and have a large number of growth imperfections, characteristic of an excessively supersaturated vapor. By increasing the pressure, $10 \times 3 \times 0.2$ mm^3 platelets and $15 \times 3 \times 3$ mm^3 prisms are obtained. Even larger prisms are found on the surface of the polycrystalline charge.

2. *Large Single Crystals.* When sharp thermal gradients are maintained at the solid–vapor interface, the interfacial energy plays a lesser role in determining the morphology. Reentrant faceted crystal growth is eliminated. Instead large boules having smooth solid–vapor interfaces are obtained. Höschl and Koňák (1963b) obtained preferred nucleation by pointing the end of the ampoule and positioning the ampoule in the furnace so the tip is in a steep temperature gradient. Ingots with large grains grow from the tip. Single crystals with dimensions up to $5 \times 5 \times 15$ mm^3 were obtained. A modification of this growth system has resulted in single crystals up to 3 cm^3 (Akutagawa and Zanio, 1971). With proper temperature gradients it is possible to obtain routinely 1 cm^3 single crystals from the vapor. Instead of pointing the ampoule, a cold finger is spring loaded axially against a flat bottom to attain sharp radial and axial temperature gradients and define the position of initial nucleation (Fig. 1.18a). A steep gradient at the solid–vapor interface encourages radial growth and prevents faceted growth and nucleation on the boule and the walls of the ampoule. Instead of using a Cd overpressure the system is held at p_{min} and a small temperature difference between the source, S, and crystal growth region, D, controls the

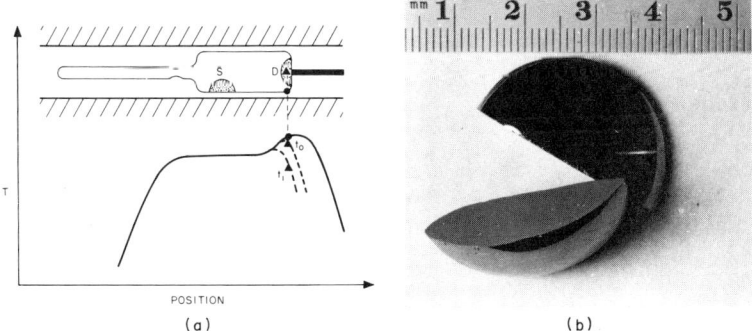

FIG. 1.18. Schematic representation of a closed tube apparatus (a) used to grow single crystals from the vapor under conditions of p_{min} at about 900°C. The temperature of the source S, and the temperature at the periphery of the ampoule near the growth position D, are defined by the solid line. The actual temperature at the growth position before (t_0) and during (t_1) crystal growth are determined from the dashed lines. Sharp axial and radial gradients have resulted in single crystals up to 3 cm^3 (b) with smooth solid–vapor interfaces (Akutagawa and Zanio, 1971).

transport rate. Initially, at time t_0, the temperature at the point of nucleation is above the temperature of the source. The cold finger is then cooled to initiate nucleation at a later time, t_1. The temperature is further decreased until the desired amount of source material is transferred. At 900°C and a transport rate of 20 gm/hr single crystal size is less than 1 mm³. At deposition rates of 0.3 gm/hr 1 cm³ single crystals are routinely obtained from polycrystalline boules. Figure 1.18b shows a boule containing an untwinned 3 cm³ single crystal. The shape of the solid–vapor interface is nearly hemispherical.

Surface diffusion and grain boundary movements are important parameters to consider when growing crystals from the vapor. Höschl and Koňák estimate the mean-free path of an atom due to surface diffusion at 1225°K to be approximately 10^{-5} cm which is smaller but comparable to mean-free paths of 10^{-4} to 10^{-3} cm in a vapor phase of 10^{-2} to 10^{-1} atm. Grain boundary movement occurs especially when temperature gradients are present. At 900°C epitaxial growth was undertaken on {110} single crystal substrates which were orientated such that the temperature gradient was normal to the growth plane (Akutagawa and Zanio, 1971). Although growth was successful on the hot side of the substrate, recrystallization occurred on the cold side of the wafer and grain boundaries propagated through the substrate resulting in no appreciable increase in thickness. Rapid recrystallization and surface diffusion rule out the use of CdTe as a substrate at these temperatures. The use of CdTe as a substrate at lower temperatures appears to be feasible. In H_2 CdS has been epitaxially deposited on CdTe from 620°C to 740°C (Fahrenbruch et al., 1974) and 400°C to 500°C (Yamaguchi et al., 1975).

2. *Thin Films*

a. Sublimation via Carrier Gas

Cadmium telluride in thin film form has been prepared by sublimation in flowing inert gases, gas-transport reactions, and vacuum evaporation. With the flowing gas technique, single crystal films have been deposited on substrates of mica, sapphire, fluorite, and GaAs. Besides acting as a transport agent, the carrier gas prevents substantial reevaporation from the substrate surface when the temperature of the substrate is raised to allow the quality of the film to be improved by surface migration and grain growth. Smooth (111) films were obtained on cleaved (111) CaF_2 crystals (Vanyukov and Krotov, 1971) using a linear velocity of the carrier gas from 1.9 to 2.3 cm/sec, a source temperature of 800°C, and a temperature gradient of from 8 to 10°C/cm in the crystallization zone. Substrate temperatures of from 580°C to 675°C corresponded to growth rates of from 0.5 to 8.0 μm/hr. Aitkhozhin and Temirov (1971) epitaxially deposited (111) CdTe on the shear plane of mica, on the (111) surface of GaAs and on the (0001) surface of Al_2O_3; (110) epitaxy was found on the (100) surfaces of GaAs. At substrate temperatures between 650°C and 700°C film thickness from a few microns to several hundred microns were obtained.

Films have been grown on single crystal wafers of CdTe (Saraie et al., 1972), CdS, and fused sapphire (Fahrenbruch et al., 1974) using a close-spaced vapor technique. Material is transported via H_2 at atmospheric pressure from source to substrate wafers which are held about a millimeter apart by quartz spacers. Smooth epitaxial layers were obtained on (100), (110), and $(111)_{Cd}$ CdTe substrates. Here $(111)_{Cd}$ refers to the polar plane having exposed Cd atoms. On the $(111)_{Te}$ substrate large hillocks grow. Optimum substrate temperatures range between 460°C and 500°C and depend upon source temperature. Excessive temperatures result in pits characteristic of the substrate; inadequate temperatures result in small hillocks. Smooth epitaxial layers are obtained at growth rates of about 10 μm/hr and source temperatures as low as 600°C. A representative rate for CdTe epitaxy on CdS is 100 μm/hr for source and substrate temperatures of 720°C and 610°C, respectively. Grain size depends upon substrate temperature and orientation and quality of the surface. Typically grain size increases from 1 μm to 5 μm for substrate temperatures of from 425°C to 610°C. Best results are obtained for deposition on the (0001) plane of CdS and when the substrate is thermally etched previous to deposition. Epitaxy is favored by higher temperatures and is limited to 670°C by the evaporation of the CdS substrate. No epitaxy occurs on fused sapphire. The resistivity of the films are from three to four orders of magnitude higher than that for deposition on CdS substrates.

b   Chemical Transport.

The halogens have been used in the formation and the transport of CdTe through chemical reactions. Epitaxy was achieved in a Cd:Te:H:Cl open-tube system (Weinstein et al., 1965) and in Cd:Te:I (Alferov et al., 1964) and Cd:Te:N:H:Cl (Paorici et al., 1972, 1973) closed-tube system. In the closed-tube method with I_2, CdI_2, and Te_2, vapors are formed from I_2 and a CdTe source located in a high temperature region. In this step the following reaction equation proceeds to the right:

$$CdTe + I_2 \leftrightarrows CdI_2 + \tfrac{1}{2}Te_2 \tag{1.9}$$

These vapors migrate to the colder region where reaction (1.9) proceeds to the left to form CdTe and I_2. The I_2 then migrates back to the hotter region to complete the cycle. Using this method, smooth monocrystalline films were deposited on a GaAs substrate (Alferov et al., 1964). At a source temperature of 850°C deposition rates of approximately 4 μm/hr were attained. In one variation of the open-tube technique (Weinstein et al., 1965), H_2 and HCl streams were passed over CdS substrates at 500°C. Large area (up to 1 cm^2 and 100 μm thick) metastable hexagonal CdTe single crystalline films were obtained. A closed-tube technique (Paorici et al., 1972, 1973) has also been used to grow epitaxial layers of CdTe on the basal planes of CdS. In this technique the transport agent of the reaction is hydrochloric acid obtained by the thermal dissociation of solid NH_4Cl. When source and deposition temperatures are respectively 800°C and 660°C, 15 to 40 μm layers of CdTe could be deposited in from 7 to 10 hr.

When using chemical transport means care must be taken to introduce the proper amount of halogen and to use the proper source and deposition temperatures (Paorici et al., 1972, 1973, 1974). Inadequate amounts of halogen will result in a transport rate which will be governed by the vapor-phase diffusion of Cd and Te. With further increases in total pressure chemical vapor transport should dominate. Transport rates of about 5×10^{-4} mg/sec over the pressure range 0.5 atm to 1.2 atm were obtained in the Cd:Te:H:I system using a 1 cm diameter and 18 cm long ampoule with source and deposition temperatures of 1093°K and 1023°K, respectively. In theory the transport rates should increase with increasing total pressure. Adding additional amounts of halogen in the form of NH_4Cl or I_2 increases the transport rate, but also increases the Te vapor pressure resulting in the formation of liquid Te. The formation of liquid Te limits the transport and growth rates for these systems. To avoid the formation of liquid Te the Te partial pressure must be lower than the saturated Te pressure in equilibrium with liquid Te. Hydrogen has successfully been introduced into these systems along with the halogen compounds to reduce Te partial pressures and increase the transport rate. Millimeter-size well-faceted crystals have been obtained in the chemical transport system (Ghezzi and Paorici, 1974; Paorici et al., 1972, 1973, 1974). Although the chemical transport process does not result in large single crystals, it is attractive as a purification step if the transport agent can either be removed or does not interfere with the desired properties of the CdTe.

Manasevit and Simpson (1971) used the metal–organics dimethylcadmium and dimethyltellurium to form specular CdTe films on several orientations of Al_2O_3, $MgAl_2O_4$, and BeO at about 500°C. For Al_2O_3 the heteroepitaxy relationship is (111) CdTe on (11$\bar{2}$6) Al_2O_3. The metal–organic approach has several advantages over other chemical vapor deposition processes. For example this process requires only one substrate and it is halide free. Such studies with CdTe and other II–VI compounds have been limited to feasibility studies and require further optimization.

c. *Vacuum Deposition*

1. Evaporation of the Compound. The preparation of thin films in vacuum (10^{-4}–$^{-9}$ Torr) has been reviewed in detail (Maissel and Glang, 1970). The quality of the film depends upon the substrate, preparation and temperature of the substrate, rate of deposition, and to a lesser extent to the composition of the source, quality of the vacuum, and technique of evaporation.

Direct evaporation of the compound rather than the elements is the more popular method of preparing CdTe films. When CdTe is evaporated on glass substrates (Konorov and Shevchenko, 1960; Glang et al., 1963; Goldstein, 1958; Goldstein and Pensak, 1959; Kamiyama et al., 1962), crystallite sizes up to 1μm are obtained. Glang et al. found a definite increase in grain size by increasing substrate temperature. In the temperature range 150°C to 250°C films were found to deposit with a preferential (111) orientation, independent of the angle of deposition. The orientation of the crystals with respect to rotation in the plane of the substrate is random. In contrast to these results, Semiletov (1962) using glass and NaCl substrates and Kamiyama et al. (1962) using glass substrates, found a preferred orientation with respect to the molecular beam. The (111) planes of the cubic phase and the (0001) plane of the hexagonal phase were aligned perpendicular to the molecular beam, The resistivity of these films were in the 10^6 to 10^8 Ω-cm range, the high resistivity being associated with barrier layers at the grain boundaries. Such high resistivity thin films have also been evaporated onto single crystalline CdTe substrates (Bernard et al., 1966).

Improved crystallinity has been attained through subsequent recrystallization. Recrystallization is generally achieved by annealing the substrates under an inert gas and at elevated temperature. Regardless of the degree of recrystallization that may be possible, it is unlikely that large area "single crystal" films comparable in quality to an epitaxial deposit can be prepared on amorphous or polycrystalline substrates.

Studies on single crystal substrates also show that the thin films are two phase, wurtzite and sphalerite, and contain numerous microtwins (Semiletov, 1962; Novik et al., 1963; Rumsh et al., 1962; Semiletov, 1956; Shiojiri and Suito, 1964; Suito and Shiojiri, 1963; Holt, 1966). Consequently electron diffraction patterns of these films all contain satellite spots. The degree of epitaxy is closely related to the temperature of the substrate. Suito and Shiojiri (1963) evaporated films onto mica substrates from 20°C to 350°C; the resulting structures were examined by transmission electron microscopy. As the temperature increased, the grain size increased and the diffraction patterns changed from randomly oriented ring patterns (Fig. 1.19b) to well-oriented spot patterns (Fig. 1.19h). It is difficult to quantitatively assess the degree of epitaxy. The method of Ino et al. (1964) is used to measure the

34   1. MATERIALS PREPARATION

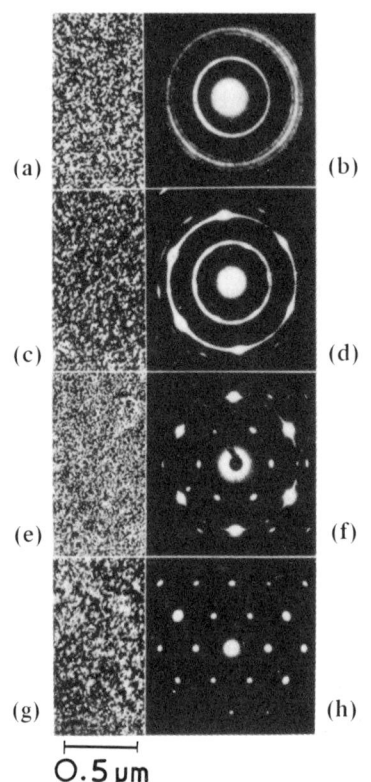

FIG. 1.19. Electron micrographs (left) and corresponding electron diffraction patterns (right) of CdTe films deposited on mica substrates at; 20°C (a) and (b); 150°C (c) and (d); 250°C (e) and (f); and 350°C (g) and (h) (Suito and Shiojiri, 1963).

0.5 μm

degree of epitaxy on Ge (Abdalla and Holt, 1973) and Si (Holt and Abdalla, 1974). At sufficiently low temperatures, films grow with fine grain randomly orientated structures and result in ring electron diffraction patterns. An orientation parameter of $R = 0$ is assigned to the fine grained structure. With increasing temperature an increasing fraction of the volume of the films grow with the epitaxial orientation giving rise to mixed ring and spot diffraction. At higher temperatures the films grow epitaxially resulting in only a spot diffraction pattern. An orientation parameter of $R = 100$ is assigned to the spot pattern with intermediate values of R commensurate with the relative intensities of ring and spot patterns.

For a given substrate temperature the quality of epitaxy is expected to be related to the rate of growth or correspondingly to the temperature of the source. Qualitative relationships determine the maximum and minimum source temperatures on substrate temperature for epitaxy for the II–VI compounds on mica (Muravjeva *et al.*, 1970; Kalinkin *et al.*, 1970). The range of epitaxial growth is shown in the crosshatched regions of Fig. 1.20

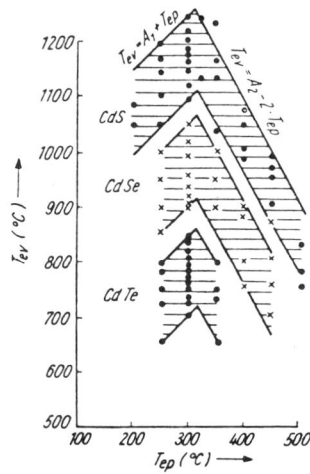

FIG. 1.20. Conditions for epitaxial growth of Cd chalcogenides on mica. The boundaries T_{ev} and T_{ep} respectively represent evaporator and substrate temperatures. The unhatched areas represent regions where the growth of randomly oriented or partially oriented films occur (Kalinkin et al., 1970).

for CdS, CdSe, and CdTe. The left and right hand branches of the crosshatched regions are approximated respectively by $T_{ev} = A_1 + T_{ep}$ and $T_{ev} = A_2 - 2T_{ep}$ where A_1 and A_2 are calculated from the graphs. These relationships were determined for a vacuum of 10^{-4}–10^{-5} Torr, a quartz ampoule as the evaporator of the Knudsen cell type with an inner diameter of about 6 mm and capillary of 0.6 to 0.8 mm and for a source to substrate distance of 85 to 100 mm. Below 310°C the screening effect of both absorbed water and gases from the substrate possibly limits the range of epitaxial growth; above 310°C the range is possibly limited by the recrystallization of the mica.

Both sphalerite and wurtzite are found on mica, NaCl, KCl, and KBr (Holt, 1966), whereas wurtzite is predominant on CdS (Weinstein et al., 1965). Not only the type of substrate but also the orientation of the substrate plays an important role in determining the crystal structure of the film; Abdalla and Holt (1973) find both sphalerite and wurtzite structures on Ge. Sphalerite is formed on (110) and (100) surfaces and wurtzite is formed on a (111) surface. In earlier work using mica, NaCl, KCl, and KBr substrates the films were found to contain twins and included grains of wurtzite material in a sphalerite matrix. The results for wurtzite CdTe on (111) Ge constitutes the first report of the production of epitaxial films free of twins and included grains of the second phase and of high densities of planar defects. Similar work on (111) Si (Holt and Abdalla, 1974) also resulted in films of comparable quality. Cadmium telluride of the sphalerite structure, grown on cubic substrates, has the parallel orientation, that is $(hkl)_{film}$ parallel to $(hkl)_{substrate}$ with $[uvw]_{film}$ parallel to $[uvw]_{substrate}$ (Holt, 1974). For cubic CdTe on Ge the epitaxial relationship is $(100)_{CdTe} \| (100)_{Ge}$ and $[010]_{CdTe} \| [010]_{Ge}$. The quasi-parallel relationship occurs for wurtzite CdTe on cubic structures. For example, for wurtzite CdTe on Si and Ge substrates the orientation is $(0001)_{CdTe}\|(111)_{Si,Ge}$ and $[11\bar{2}0]_{CdTe}\|[1\bar{1}0]_{Si,Ge}$. For nonsphalerite substrates epitaxial relationships have also been established. For cubic CdTe on mica the relationships are $(111)_{CdTe}\|(001)_{mica}$ and $[110]_{CdTe}\|[100]_{mica}$.

There is a tendency for the directional polarity of substrates to be perpetuated for epitaxial II–VI films deposited on them. For CdTe deposited by a vapor-phase technique on the "S face" of CdS, the "Te face" of CdTe was observed as the outermost surface (Weinstein et al., 1965). Igarashi (1971) grew CdS on one face each of GaP and CdTe, two faces of InAs, and six different faces of GaAs and found that the [111] A direction in the CdS films was almost always aligned with one of the [111] A directions in the substrate.

Sublimation under near equilibrium conditions to form single crystals and thin film formation by vacuum deposition are two extreme methods of physical vapor transport. Yezhovsky and Kalinkin (1973) studied the intermediate case of vacuum deposition of CdTe thin films in quasi-closed cells. These studies are interesting in that they evaluate the effect of the departure from thermodynamic equilibrium on the quality of the thin film. As expected, the quality of the thin film improves as equilibrium is approached. The authors report epitaxy on mica at temperatures as low as 50°C.

2. Evaporation of the Elements. Alternative methods have been used to obtain films. The simultaneous evaporation of Cd and Te from separate sources by the three-temperature technique has been successful (Justi et al., 1973). In the coevaporation of the constituents on cleaved NaCl substrates (Sella et al., 1968), the structure was affected by an excess of Cd. Cadmium telluride films showed a cubic structure whereas those with an excess of Cd show a mixture of cubic and hexagonal structures.

Epitaxial films have also been grown on single crystal CdTe substrates from separate Cd and Te sources by molecular beam epitaxy (Smith and Pickhardt, 1975). Molecular beam epitaxy is a closely controlled form of vacuum evaporation in which film constituents are evaporated under ultrahigh vacuum from temperature-controlled Knudsen cells through liquid nitrogen-cooled collimators onto heated single-crystal substrates. Films were grown at 10^{-10} Torr and could be analyzed *in situ* for epitaxy quality by low-energy electron diffraction and for elemental surface composition by Auger spectroscopy. At growth rates of 1 μm/hr highest quality epitaxy was obtained for a substrate temperature of 300°C. Low-energy diffraction patterns showed patterns better than those of the substrates, although surface roughness was greater than 1000 Å. Mirror-smooth surfaces could be produced at lower temperatures but such films were always polycrystalline.

II. CRYSTAL GROWTH

Molecular beam epitaxy contains the element of direct synthesis, the approach used by Fredrichs in 1947 to obtain some of the first crystals of CdTe.

Ueda (1975) obtained the best epitaxy on NaCl when the ratio of the intensity of the Cd beam to the intensity of the Te beam was 10. The hexagonal form of CdTe was preferred to the cubic form at lower temperatures. For cubic CdTe on cleaved surfaces the epitaxial relationship is CdTe(001)[100]||NaCl(001)[100]. For the hexagonal form there are two relationships: CdTe(0001)[100]||NaCl(001)[110] and CdTe(3034)[100]|| NaCl(001)[110]. Epitaxy can be lowered to near room temperature by applying a dc voltage between the source and substrate.

Rather than evaporate the elements, Cusano (1966) introduced Cd and Te powers separately onto an inverted hot wall bell jar which was continuously pumped on (Fig. 1.21). Predried and dispersed powder mixtures of Cd plus CdI_2 and of Te were fed gradually into regions B and A, respectively, from dispensers located above the O-ring seals. The vapors of Cd, CdI_2, and Te_2 diffuse within the bell jar and react at the hot surfaces. A 10 μm film of I-doped CdTe could be obtained in about 30 min on the inside surface of a Mo foil which was precoated with from 0.5 to 1.0 μm of polycrystalline CdS. Often an intermediate thin film layer of CdS is used (Justi et al., 1973; Adirovich et al., 1969, 1971). Earlier CdTe films in a similar system were also deposited directly on metal (Cusano, 1963). Tellurium has also been deposited onto CdS to form CdTe, the formation of CdTe attributed in the presence of excess Cd (Dutton and Muller, 1968).

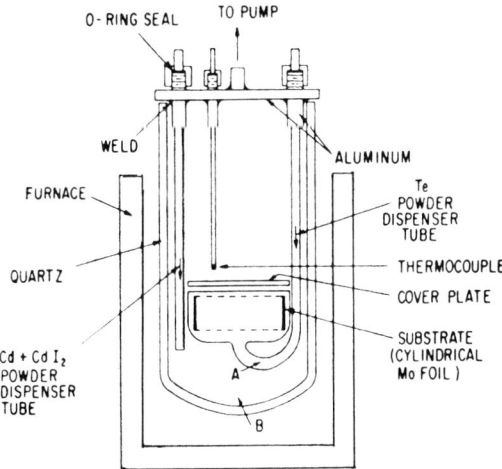

FIG. 1.21. Vapor reaction apparatus for forming CdTe films. Dispersed power mixture of Cd plus CdI_2 and of Te are fed into regions A and B respectively where they react at the hot surfaces to form films. (Reprinted with permission from Cusano, 1966. © 1966, Pergammon Press, Ltd.)

D. Growth from Foreign Hosts

The solubility of CdTe in Sn and Bi at 900°C is respectively 9 and 12 mole%. Rubenstein (1966, 1968) grew crystals from these solvents at this temperature under the driving force of a temperature gradient. Platelets with dimensions of 4 mm × 4 mm × 2 mm and rods with dimensions of 2 mm × 2 mm × 6 mm were obtained. These crystals have not been characterized; however, relatively pure crystals of GaP and GaAs have been obtained by similar solvent growth. Taguchi et al. (1974b) grew CdTe from $CdCl_2$ solutions at 675°C. The solubility of CdTe into liquid $CdCl_2$ at this temperature is about 38 mole%. Crystals were n-type with a mobility of about 1040 cm^2/V sec.

Millimeter-size crystals of CdTe and other II–VI compounds have been grown in $(OH)^-$ solutions of NaOH and KOH under hydrothermal conditions at about 350°C on seeds, with thermal gradients (Kolb et al., 1968). The growth of single crystals of the tellurides of Cd, Zn, and Pb from gels has been unsuccessful, the difficulty being attributed to the instability of the Te ion (Blank and Brenner, 1971).

III. Purification

Cadmium telluride designated as 99.999% pure is commercially available. The physical and chemical properties of material of such quality are not significantly influenced by the residual impurity content. If reasonable cleanliness is observed during processing and subsequent crystal growth, this level of purity can be maintained. However such material is still relatively impure when considered as a semiconductor. For example the electrically active impurities in Si and Ge have been reduced to better than one part per billion. It is unlikely that CdTe will ever be purified to this degree. Even if such a task were accomplished, the semiconducting properties would be limited by the native defects. Rather than strive for the ultimate purity it is important to understand the native defect problem and achieve a happy balance.

Although some purification occurs during the synthesis and subsequent crystal growth of CdTe, a reduction in the impurity content of the compound is most efficiently achieved by purifying the starting materials. Distillation, zone refining, electrolysis, etc. have been used to attain the 99.9999% level and better. The best method or combination of methods to obtain high purity is not straightforward and the degree to which CdTe actually has been purified is probably not known since the reliability and sensitivity of most analytical methods is not much better than 1 part per million atomic (ppma) for most impurities in Cd, Te, and CdTe. A careful mass spectrographic

III. PURIFICATION

analysis has a sensitivity of 0.1 ppma for 90% of the impurities. Consequently transport and optical measurements have also been used to evaluate the purity of the material.

Several examples illustrating the degree of purity that can be achieved for Cd, Te, and CdTe are cited. In evaluating the purity the method of analysis and the interpretation of the results should be carefully considered. Unfortunately, many of the important results regarding the purification of these materials are not available in the open literature. This is especially true for Te, being a component in several compounds used for infrared detectors.

The use of ultrapure starting materials does not guarantee a high purity compound as contamination may occur during processing. For example, Medvedev et al. (1968a) purified both Cd and Te by fractional distillation and zone refining. Electrical measurements on Cd and Te inferred purities of better than 99.9999%. However, the room temperature electron mobility of CdTe indicated electrically active native defects or impurities in concentrations in excess of those characteristic of the starting materials. Although the growth parameters and the annealing characteristics of the material were similar, variations in the mobility were found. The mobility was therefore most likely limited by impurities. In interpreting electrical measurements it is important to realize that the actual concentration of impurities may be below that estimated due to their association with native defects resulting in neutral complexes or second phases. Impurities that are inactive in the elements Cd and Te may or may not be electrically active in the compound.

This section summarizes the methods of purifying Cd and Te and the efforts undertaken to achieve purification during synthesis and crystal growth of the compound.

A. CADMIUM

Cadmium is a relatively rare element and is found primarily in the compound greenockite CdS which is contained in solid solution with zinc blende, ZnS. A secondary mineral also containing Cd is $CdCO_3$. The entire output of primary Cd is obtained as a by-product in the processing of the zinc ores in zinc-retort, copper and lead smelting, and electrolytic plants (Chizhikov, 1966; Kirk-Othmer, 1964).

Intermediate purification is attained by electrolytic refining. Extensive electrolytic refining results in purities up to 99.999%. Final purity depends upon the composition of the electrolytic, the anode metal, the material of the bath, and the surrounding atmosphere. The composition of a typical bath is shown in Table 1.1 (Pakhomova, 1966). A purity of 99.993% is possible for a bath containing a 10:1 ratio of Cd to Zn. Additional electrolytic refining further increases the purity of the Cd. Pakhomova (1966) repeatedly refined electrolytic Cd so as to remove traces of Mg and Al and reduce the

TABLE 1.1

CADMIUM PURIFICATION

Process	Purity of charge		Method of analysis[a]	Reference
	Starting material	Refined material		
Electrolysis	200 g/liter Cd, 17.6 g/liter Zn, 0.184 g/liter Fe 0.0006 g/liter Cu, 2.5 g/liter Mn, 0.048 g/liter Co 0.015 g/liter Cl, 0.123 g/liter Tl, 50 g/liter H_2SO_4	99.993%	—	Pakhomova (1966)
Electrolysis	0.02% Pb, 0.005% Zn, 0.01% Cu 0.002% Fe, 0.015% Tl	~0.01% Cu, Mg-trace, Al-trace	Spectroscopic analysis	Baikova et al. (1966)
	Cu 1 ppm, Ni 0.5 ppm, Pb 4.5 ppm, Zn 6 ppm, Fe 5 ppm, Sb 0.6 ppm, Tl 3.6 ppm	99.998%	—	Zosimovich et al. (1961)
Zone refining	0.011% Pb, 0.006% Cu, 0.004% Ni, Fe trace, possible traces of Tl, Bi, Sb, As	99.99995%	Resistivity	Aleksandrov (1961)
	Technical grade[b]	99.999%	—	Mochalov et al. (1966)
	99.995%	99.99995%	Resistivity	Schaub and Potard (1971)
Distillation	99.98%	99.9994%	Resistivity	Aleksandrov and D'yakov (1962)
	99.98%	99.9999%	—	Aleksandrov and Udovikov (1973)
Ion exchange	0.02% Pb, 0.005% Zn, 0.0036% Cu 0.0026% Tl, <0.001% Fe, <0.001% As <0.001% Sb	0.0001% Cu 0.0001% Pb	Spectroscopic analysis	Chizhikov (1966)

[a] Resistivity measurements refer to electrically active impurities.
[b] Technical Grade—>99.95%, <0.02% Pb, <0.005% Zn, <0.01% Cu, <0.002% Fe and <0.005% Tl. [Taken from Chizhikov (1966).]

III. PURIFICATION

Cu content from a few hundredths of a percent to a trace. Spectroscopic analysis also indicated no traces of other elements.

Higher purity Cd is obtained through distillation processes, zone refining, and ion exchange. Usually combinations of these methods are required to remove impurities to a required limit (Baikova et al., 1966; Zosimovich et al., 1961). Purities better than 99.9999% result when zone refining is used as a final step after electrolysis. The effectiveness of zone refining depends upon crucible design, zone speed, degree of stirring in the zone, and crucible materials. Precise control of the temperature of the zones are required due to the high thermal conductivity and low melting point of Cd. The higher purity quartzes such as supersil, suprasil, etc. are preferred crucible materials. Carbon coating of the quartz should be considered to avoid incorporation of silicon (Schaub and Potard, 1971) and other impurities from the quartz. The segregation coefficients, listed in Table 1.2 for various impurities in Cd, are important in determining the effectiveness of impurity separation during zone refining. Except for Mg and Ag the segregation coefficients for all impurities investigated are less than 1. Unfortunately, the segregation coefficients for most impurities in Cd are either not precisely known or not available on the literature. Zone refining results in different degrees of purification. Using technical grade Cd* a purity of about 99.999% was obtained (Mochalov et al., 1966). However, using comparable starting materials 99.99995% purity was obtained, based upon values of about 25,000 for the ratio of the resistivity of the material at 4.2°K to that at room temperature (Aleksandrov, 1961). Such electrical evaluations are used to determine the concentration of electrical active residual impurities in material better than 99.9999% purity. Various protective atmospheres have been used. Aleksandrov (1961) used a He overpressure to obtain 99.99995% Cd. A reactive overpressure would be expected to assist in the removal of certain volatile impurities. A H_2/N_2 gas mixture was assumed to be helpful in the reduction of P, S, As, and O. The concentration of less volatile impurities such as Ag, Au, Li, Na, and Sb were nearly constant (Schaub and Potard, 1971). Based upon values of 27,000 for the resistance ratio the final material was believed to be of 99.99995% purity. Starting with $6N$ Cd Wernick and Thomas (1960) obtained ratios as high as 38,000.

Comparable results have been obtained by distillation methods. One advantage of the vacuum distillation process with respect to zone refining is that it lends itself more easily to continuous processing. Although Cd is generally distilled in vacuum, the repeated distillation of Cd in H_2 reduces the oxides and prevents CdTe ingots from sticking to quartz crucibles (Lawson et al., 1959, 1960). In general, as with the electrolytic and zone

* See Table 1.1 for a description of technical grade Cd.

TABLE 1.2

Segregation Coefficients for Impurities in Cadmium

Impurity	k	Experimental data	From phase diagrams
Cu	≈0.1	a	d
	<1	b	
Sn	≈0.1	a	
	<1	b	c
	10^{-2}		d
Pb	<0.1	a	
	<1	b	c
	10^{-2}		d
Tl	<0.1	a	
	<1	b	c
	≪1		d
Ga	≪1		d
As	<1	a	
Bi	≪1		d
In	≪1		d
Sb	<1	a, b	c, d
Na	<1		d
Au	0.4		d
Ni	<1	a, b	
Hg	0.5		d
Zn	0.5	a	
	<1	b	
Pt	<1		
Ca	<1		
Al	<1		
Fe	<1		c
Li	≈1.5		d
Ag	>1	b	c
	3		d
Mg	>1	b	c
	≳1		d

[a] Chizhikov (1966, p. 187).
[b] Mochalov et al. in Chizhikov (1966, p. 187).
[c] Vigdorovich and Selin (1971).
[d] Aleksandrov (1961).

refining processes, the final purity is related to the starting material. With relatively impure starting material, purities as high as 99.998% can be obtained (Singh et al., 1968; Taziev et al., 1965; Kotov, 1966). Starting with materials of 99.8% purity so designated as technical grade Cd,* 99.99994% Cd was obtained by distillation at 10^{-4} mm and 450°C to 500°C (Aleksandrov and D'yakov, 1962). Resistance measurements were used to evaluate the purity. Also starting with technical Cd, a single vacuum distillation at 450°C and 5×10^{-5} Torr yielded 99.9999% material (Aleksandrov and Udovikov, 1973) with Pb, Cu, Bi, Tl, and especially Zn and As being the most difficult to remove. Przybyslawski et al. (1967) also found Pb and Bi difficult to remove. These authors were able to obtain purities greater than 99.9999% but found it impossible to reduce the content of Ca, Cu, and Mg below the sensitivity of spectral methods. In some cases better results are attained through vacuum distillation rather than by zone refining (Silvey et al., 1961). When 99.99% material was zone refined trace impurities could not be removed. However, when vacuum fractional distillation in sealed tubes was carried out, the impurities were reduced below the limits of spectroscopic analysis.

Starting with electrolytic Cd, containing 0.02% Pb, 0.005% Zn, 0.0036% Cu, 0.0026% Tl, and less than 0.001% each of Fe, As, and Sb, the ion exchange method (Chizhikov, 1966) has been used to obtain high-purity material having no impurities as determined by spectroscopic analysis except for 0.0001% each of Cu and Pb.

B. Tellurium

There are small concentrations but large quantities of Te in the copper porphyries, copper sulphide, nickel sulfide, and lead sulfide deposits (Kirk-Othmer, 1969). Like Cd there are no deposits that are mined specifically for Te alone and the overall percentage of recovery from these ores is very small. Almost all of the commercial Te is recovered from electrolytic Cu refinery slimes, in which it is present from a trace to 8%. After the slimes are decopperized, the Te is recovered from tellurous acid by several methods the most popular one being by electrolysis. A purity of 99.9999% is possible by electrolytic refinement with a plastic membrane enveloping the cathode (Napolitano and Mantell, 1964). Higher purity requires the assistance of another approach.

Further purification is undertaken by zone refining, distillation, and chemical means (Chizhikov and Shchastliviyi, 1970). Spectroscopically pure Te has been obtained by the decomposition of H_2Te (Weidel, 1954). The yield of this process is only 25%. Bagnall (1966) has reviewed the different chemical purification schemes. The degree of purification that can be attained by chemical methods is much less than by other techniques. Better

* See Table 1.1 for a description of technical grade Cd.

TABLE 1.3
Tellurium Purification

| | Purity of charge | | | |
Process	Starting material	Refined material	Method of analysis[a]	Reference
Electrolysis	—	99.999%	—	Napolitano and Mantell (1964)
	99%	10^{-4}–10^{-5}% each of Al, Fe, and Mg	Spectrographic analysis	Baĭmakov and Petrova (1960)
	—	3×10^{14} cm^{-3}	Hall measurements	Kujawa (1963)
	—	10^{16} cm^{-3} Mg	Mass spectrographic analysis	
	—	$1-3 \times 10^{15}$ cm^{-3}	Hall measurements	Shvartsenau (1960)
	—	0.001% Si and 0.001% Mg	Spectrochemical analysis	
		5×10^{15} cm^{-3}	Hall measurement	Schaub and Potard (1971)
		$\approx 10^{17}$ cm^{-3}	Mass-spectrographic analysis	
Distillation	97.8%	—	Spectroscopically pure	Petrescu et al. (1960)
	—	>99.9999%	—	Przybslawski et al. (1967)
	—	>99.9999%	Spectrographic analysis	Kujawa (1963)

[a] Hall measurements refer to electrically active centers.

than 99.9999% purity has been achieved by both zone refining and distillation. This degree of purity is achieved in zone refining with relatively impure, 99%, starting material (Baĭmakov and Petrova, 1960). These results are summarized in Table 1.3.

Zone refining has been successful due to the favorable values of the segregation coefficients for most impurities. Except for Cd, Se, and Na the segregation coefficients for the 27 impurities listed in Table 1.4 are much less than one. The removal of Se from Te is found to be more effective if H_2 is present during the zone refining (Vigdorovich et al., 1970; Shvartsenau, 1960; Dufresne and Champness, 1973; Schaub and Potard, 1971). Schaub and Potard have found the concentrations of Cr, Na, and Si to be constant during the refining. Although not entirely consistent with Table 1.4 Al, Bi, Fe, and Si have also been found to be difficult to remove by zone refining (Krapukhin et al., 1967). These impurities have been removed from Te by zone refining of $TeCl_4$, followed by reduction of Te. The removal of Fe, Bi,

TABLE 1.4

Segregation Coefficients of Impurities in Tellurium

Element	k	Reference	Element	k	Reference
Se	0.56	a	Mg	10^{-4}	c
	0.3	c		2×10^{-4}	e
	0.2	e		5×10^{-2}	d
	0.5	f	Ca	10^{-3}	c
	0.55	d		8×10^{-4}	e
Cu	9.5×10^{-3}	b	Au	10^{-5}	c
	10^{-6}	c		7×10^{-5}	e
	8×10^{-7}	e		0.1	f
	3×10^{-2}	d	Al	8×10^{-4}	c, e
Ag	2.2×10^{-2}	b		0.38	d
	8×10^{-6}	c	In	5×10^{-3}	c, f
	2×10^{-6}	e		6×10^{-3}	e
	$<10^{-3}$	f	Tl	10^{-3}	c
	4×10^{-2}	d		2×10^{-4}	e
As	8×10^{-4}	c		5×10^{-3}	f
	10^{-3}	e	Fe	5×10^{-3}	c, e
Sb	5×10^{-3}	c		0.2	f
	3×10^{-3}	e		8×10^{-2}	d
	$<6 \times 10^{-2}$	d	Ni	10^{-3}	c
Bi	10^{-3}	c		9×10^{-4}	e
	9×10^{-5}	e	Cr	0.3	c
	2×10^{-2}	f		0.2	e
	<0.2	d	I	5×10^{-3}	c
Ge	10^{-3}	c		<0.1	f
	6×10^{-4}	e	Na	0.25	c
Sn	3×10^{-4}	c	S	<0.1	f
	10^{-3}	e	Cl	$<4 \times 10^{-3}$	f
Pb	10^{-4}	c	Co	$<2 \times 10^{-3}$	f
	10^{-3}	e	Zn	$<2 \times 10^{-2}$	f
	5×10^{-3}	f	Pt	$<10^{-2}$	f
	$>8 \times 10^{-2}$	d	Si	<0.2	d
Cd	2	c			
	>1	e			
Hg	0.1	c, f			
	5×10^{-2}	e			

[a] Movlanov and Kuliev (1963).
[b] Krapukhin et al. (1967).
[c] Kujawa (1964).
[d] Shuleshko and Vigdorovich (1968).
[e] Kujawa (1965).
[f] Zanio, unpublished results.

and Si is also effective by the simultaneous passage of a current through the bar during refining.

In addition to analytical techniques, Hall measurements are used to evaluate the purity of the material. As with the resistivity measurement on Cd, the impurity content as determined by Hall measurement is likely to be less than actually present. This is illustrated in Table 1.3. In high purity Te the concentration of electrically active impurities is typically about 10^{15} cm^{-3}. This is about an order of magnitude less than determined by analytical methods. Carrier concentrations as low as 2×10^{14} cm^{-3} obtained with Czochralski (Dufresne and Champness, 1973) and horizontally grown (Lawson et al., 1959, 1960) Te indicate that higher purities are possible. The source of residual impurities is difficult to determine although it is likely that some contamination occurs from the quartz tube. Electrical measurements on vertical float zone Te and conventional zone refining in a quartz ampoule showed carrier concentrations of 2×10^{14} and $(2-4) \times 10^{15}$ cm^{-3}, respectively (Abdullaev et al., 1965).

Better than 99.9999% material can also be obtained using vapor transport methods. Again, processing Te in the presence of H_2 is effective in reducing the concentration of volatile impurities such as Se, S, As, Hg, Sb, O, and Cl (Vanyukov et al., 1969; Abdullaev et al., 1965). In some cases Se has been reduced below 1 ppm without H_2 (Petrescu et al., 1960; Medvedev et al., 1968b). Distillation of only the first portion of the charge is effective in removing impurities (Au, Fe, Si, Cu, Ag, Bi, Pb, and Sb) having vapor pressures much greater than Te. Spectroscopically pure Te has been produced by distillation from relatively impure 97.8% starting material (Petrescu et al., 1960). Material of better than 99.9999% purity was also obtained by distillation with Pb, Bi, Ca, Cu, and Mg being the most troublesome impurities (Przybslawski et al., 1967). A combination of distillation and zone refining may be more effective than zone refining alone. Using spectrographic analysis, Kujawa (1963) obtained better than 99.9999% material both by vacuum distillation prior to zone refining. By using the complementary approach, Kujawa and others (Petrescu et al., 1960; Vanyokov et al., 1969; Abdullaev et al., 1965; Medvedev et al., 1968b; Kujawa, 1963) obtained material with carrier concentrations in the $(1-4) \times 10^{14}$ cm^{-3} range.

C. CADMIUM TELLURIDE

The preparation of high purity CdTe is not restricted to purifying the starting materials. Zone refining and sublimation of the compound and the crystal growth process are additional purification steps. The electron mobility is often used as a guide to track the purity and, as expected, has increased through the years. However, it is important not to directly relate mobility to purity as large concentrations of native defects also lower the mobility.

Therefore in correlating these parameters it is important to examine growth and post annealing conditions. Electrical measurements are more important in evaluating the compound than in evaluating the elements. Electrically inactive impurities in CdTe may not affect device performance. However electrically inactive impurities in the elements, incorporated into the compound, may become active and be of importance.

Zone refining has been undertaken by the horizontal three-furnace method and the vertical sealed-ingot method. These systems have been described in the previous section. Using the horizontal method, Kroger and de Nobel (1955) found only Si as a residual impurity as determined by mass spectrographic analysis; however at room temperature only 700 cm^2/V-sec was obtained for the electron mobility. Starting with material containing about 0.01% impurities by weight, de Nobel (1959) after forty passes obtained material having a residual impurity content of about 1 ppma as determined by spectrochemical analysis. However the maximum value of the electron mobility was about 850 cm^2/V-sec at room temperature and 1000 cm^2/V-sec at 160°K. With purer (99.999%) starting materials Matveev et al. (1969a) obtained mobilities as high as 990 cm^2/V-sec, the improvement being attributed to the use of graphite rather than quartz crucibles. In these experiments and those of Prokof'ev (1970) mobility generally decreases toward the end of the bar indicating a segregation coefficient of less than one for the electrically active impurities.

Back diffusion of impurities through the hot ingot and vapor are possible transport mechanisms which could offset the purification by the horizontal three-furnace method and are eliminated by the sealed-ingot technique (Lorenz and Halsted, 1963; Cornet et al., 1970b; Woodbury and Lewandowski, 1971; Triboulet and Marfaing, 1973). Because of the sharp temperature gradients in the sealed-ingot method, the main body of the ingot is below 200°C and the molten zone is confined to a small selfsealing region. In this zone refining method from 7 to 22 passes from speeds ranging from 0.25 cm/hr to 3.0 cm/hr are used. Evidence for effective zone refining has been obtained by spectrographic analysis, by luminescent and fluorescent emission spectra, and by electron mobility measurements. Results of spectrographic analysis for the common impurities in II–VI compounds (Lorenz and Halsted, 1963) indicated traces of Cu, Ag, Al, Sb, and Bi in the initial ingot, whereas the center and tip sections show no detectable concentration of these impurities except for one ingot where 0.1 ppm of Cu and Al were found. The presence of these impurities in the tail in concentrations of several ppm indicate that their segregation coefficients are less than one. Fluorescent measurements show either the absence of or at least a ten-fold decrease in the intensity of a Cu band in the center and tip of the ingot relative to the tail. Large values of the electron mobility at room temperature (1050 cm^2/V-sec) and 20°K

TABLE 1.5
SEGREGATION COEFFICIENTS BETWEEN CADMIUM TELLURIDE AND MELT

Element	Value	Note	Reference	Element	Value	Note	Reference
Hg	0.3	a	Zanio (1974)	Cu	0.2	c	Cornet et al. (1970b)
	0.3	b	Zanio (1974)		<1	d	Bell et al. (1970a)
	0.3	c	Ray and Spencer (1967)		<1	c	Lorenz and Halsted (1963)
Zn	6	a	Zanio (1974)		<1	c	de Nobel (1959)
	4	b	Zanio (1974)	Li	0.6	a	Zanio (1974)
	2	c	Steininger et al. (1970)		0.3	b	Zanio (1974)
S	≈10	a	Zanio (1974)	Na	≈0.05	a	Zanio (1974)
	4	b	Zanio (1974)		0.01	b	Zanio (1974)
Se	≈7	a	Zanio (1974)	K	≈0.2	a	Zanio (1974)
	2	b	Zanio (1974)		≈0.01	b	Zanio (1974)
	<1	c	Strauss and Steininger (1970)	Cl	0.005	a	Zanio (1974)
O	≈0.02	a	Zanio (1974)		0.005	b	Zanio (1974)
	≈0.02	b	Zanio (1974)	I	<0.1	a	Zanio (1974)
Al	≈0.1	a	Zanio (1974)		<0.5	b	Zanio (1974)
	≈0.1	b	Zanio (1974)	Sn	0.025	c	Woodbury and Levandowski (1971)
	<1	d	Bell et al. (1970a)	Pb	<0.005	b	Zanio (1974)
	<1	c	Lorenz and Halsted (1963)		<1	c	de Nobel (1959)
	>1	c	de Nobel (1959)	C	0.09	a	Zanio (1974)
B	<1	a	Zanio (1974)		≈0.5	b	Zanio (1974)
	<1	b	Zanio (1974)	N	≈0.005	a	Zanio (1974)
Tl	<0.01	b	Zanio (1974)		≈0.4	b	Zanio (1974)

Element	Value	Note	Reference	Element	Value	Note	Reference
In	0.06	b	Zanio (1974)	Sb	0.2	e	Lorenz and Blum (1966)
	0.11	c	Yokozawa et al. (1965)		0.01	f	Lorenz and Blum (1966)
	0.07	c	Mizuma et al. (1961)		<1	c	Lorenz and Halsted (1963)
	0.5	c	Thomassen et al. (1963)	Bi	<0.001	b	Zanio (1974)
	0.085	e	Lorenz and Blum (1966)		<1	c	Lorenz and Halsted (1963)
	0.49	f	Lorenz and Blum (1966)	Pt	<0.02	a	Zanio (1974)
	<1	c	de Nobel (1959)		<0.01	b	Zanio (1974)
Co	0.03	b	Zanio (1974)	Cr	<1	b	Zanio (1974)
	0.1	c	Slack and Galginaitis (1964)		<1	d	Bell et al. (1970a)
	0.27	c	Woodbury and Lewandowski (1971)	Mg	0.5	a	Zanio (1974)
Fe	0.3	c	Slack and Galginaitis (1964)		<1.5	b	Zanio (1974)
	0.53	c	Woodbury and Lewandowski (1971)		1	c	Woodbury and Lewandowski (1971)
	<1	d	Bell et al. (1970a)		10	c	Lawrenson and Ray (1971)
Mn	0.7	c	Slack and Galginaitis (1964)		<1	c	de Nobel (1959)
	<1	d	Bell et al. (1970a)				
Au	0.1	a	Zanio (1974)				
	0.1	b	Zanio (1974)				
Ag	0.009	a	Zanio (1974)				
	0.009	b	Zanio (1974)				
	<1	d	Bell et al. (1970a)				
	<1	c	Lorenz and Halsted (1963)				
	<1	c	de Nobel (1959)				

[a] Te-rich melt at 725°C.
[b] Te-rich melt at 880°C.
[c] Near-congruent or congruent melt.
[d] Single pass of a Te-rich zone.
[e] At 1035°C with Cd vapor pressure of 5.3 atm.
[f] At 1055°C with Cd vapor pressure of 0.081 atm.

(57,000 cm^2/V-sec) indicate that impurity or native defect scattering was substantially reduced by this method. Continued studies (Woodbury, 1974) have resulted in electron mobilities as high as 140,000 cm^2/V-sec at 20°K. These mobilities are higher than predicted by simple impurity charge scattering models and are thought to be due to a reduction in the concentration of charged scattering centers by pairing at low temperatures. More typically the mobility at 20°K ranges from somewhat less to one-tenth this value.

Woodbury and Lewandowski also used tracer techniques to determine impurity profiles and segregation coefficients. The following segregation coefficients were determined: Fe, 0.53; Co, 0.27; and Sn, 0.025. The profiles for Bi and W, however, were not normal. Since multiple zoning effectively removed these elements, the segregation coefficients are probably less than one. Triboulet and Marfaing (1973) report mobilities as high as 1.46 × 10^5 cm^2/V-sec at 32°K, with the maximum value not measured but probably occurring at a lower temperature. These results should be interpreted carefully, since room temperature mobilities were less than 1000 cm^2/V-sec. This conflict was explained by the presence of microinhomogeneities but may be due to pairing. Mobility and photoluminescence measurements along the length of the ingot were consistent with the displacement of impurities. If the cooling curves and consequently the deviation from stoichiometry were similar at various positions along the ingot, the differences in mobility could be associated with differences in impurity content. The amounts of Si, Ca, and Mg, as determined by emission spectrography, did not vary along the ingots while Cu, Fe, and Ag segregated to the ends of the ingot. Qualitative analysis with an ionic probe revealed the presence of Al, Na, Cl, F, K, Co, Fe, and Si, depending on the samples. About 10 ppm of C and about 1 ppm of S, Na, and K were found by mass spectrometry analysis. Although a broad spectrum of impurities were reported it is likely that the material is quite pure and extreme care was taken in detecting residual impurities. A further indication of the purity of the material prepared by this group is the presence of the 1.605 eV photoluminescence line at 4.2°K which corresponds to the radiative band-to-band transition.

Cornet *et al.* (1970b) have used a split ring rather than a full ring for the graphite susceptor allowing larger diameter ingots, reducing the power requirements, and permitting sharper temperature gradients within the molten zone. The efficiency of the zone refining process was checked by using radioactive Cu. A segregation coefficient of 0.2 was found. Mass spectrographic analysis indicated that impurities in the starting materials were reduced to less than 0.1 ppm.

The effectiveness of purification by crystal growth from Te-rich solutions is also well documented. The passage of a single Te-rich zone through CdTe by the traveling heater method (THM) results in the reduction of Mg, Si,

Ca, Fe, Cu, and Ag (Bell et al., 1970a,b) and B, Na, S, Cl, Y, Zr, Pd, Ag, and In (Triboulet et al., 1974). The concentration of any one particular impurity did not increase in the CdTe after the solvent pass. Higher purity CdTe is even more likely using the sequence of zone refining and THM. It was estimated that the total impurity content excluding carbon was about 10^{15} cm^{-3}. However photoluminescence measurements did not apparently show the radiative band-to-band transition as found for material only zone refined (Triboulet and Marfaing, 1973). Cadmium telluride has been grown from intentionally doped Te-rich solutions by normal freezing to determine the segregation coefficients of many common impurities in CdTe and to determine if there is a significant dependence of the segregation coefficient on temperature of growth. The segregation coefficients of 29 impurities compiled from zone refining studies, pseudobinary phase diagrams, and crystal growth studies are shown in Table 1.5. For crystals grown from Te-rich solutions, the segregation of Se, N, and C becomes more effective when the growth temperature is decreased from 880°C to 725°C. The opposite behavior was found for Na and K. There is no apparent change in the segregation coefficient of Hg for crystals grown from near-congruent melts and Te-rich solutions. However, for Zn, In, and Co, segregation becomes more effective with decreasing temperature. Crystals could be grown from undoped solutions with a residual impurity concentration as low as 1 ppma, as determined by mass spectrographic analysis, except for high concentrations of oxygen. The concentration of the elements, Mn, Fe, and P, and Cu were unknown because their presence is masked by the background of Cd and Te.

Purification by sublimation probably occurs; however, its effectiveness is not well documented in the literature. In one case starting with 99.9999% pure Cd and Te, two sublimations were sufficient so as to result in material having room temperature electron mobilities as high as 997 cm^2/V-sec and less than 1 ppma each each of impurities that could be detected by mass spectrographic and emission analyses (Kyle, 1971). Whether sublimation or crystal growth was more effective as a purification step is presumably not known.

IV. Summary

The crystal growth technology of CdTe is well established. A congruent melting point of only 1090°C and the resulting moderate vapor pressures permit the growth of crystals up to 100 cm^3 by the Bridgman technique in closed systems. Although elemental Cd and Te are available in better than ppma quality, resulting single crystals generally have a higher residual impurity content, which is introduced during processing. Aluminum, oxygen, and carbon are common contaminants. Zone refining by the sealed-ingot method at the congruent melting point improves the transport properties of the material and presumably the purity. Growth at lower temperatures from Te-rich solutions and from the vapor are other approaches to reducing contamination. Modest convection during solution growth restricts dendritic formation and results in single crystals up to 10 cm^3. However, entrapment of Te-rich inclusions at the grain boundary is severe and entrapment within single crystal regions is common. Although the size of single crystals grown by the vapor method are generally in the sub-cm^3 size, their inclusion content is significantly less. Vapor growth is also somewhat unique in that thin films are of both the zinc blende and wurtzite forms.

It is not certain which method of crystal growth is the most appropriate in obtaining high purity material for device applications because of the effects of post annealing. An ingot may contain a high total impurity content. However, depending upon the shape of the cooling curve, the impurities may be distributed in precipitates and inclusions rather than solid solution. The interplay between native defects and impurities, resulting in either neutral complexes or compensated material as discussed in Chapter 3, also determines the electrically and optically active impurity content. Therefore, when characterizing material, more weight should be given to transport and optical measurements (Chapter 2) and device evaluation (Chapter 4) than to analytical techniques which for most impurities are sensitive in the 0.1 to 1 ppma range.

CHAPTER 2

Physics

I. The Zinc Blende Structure

The zinc blende structure is the stable form for bulk single crystals of CdTe at atmospheric pressure (Fig. 1.5). This structure belongs to the cubic space group $F\bar{4}3m$ (T_d^2) (Kittel, 1967) and consists of two interpenetrating face center cubic lattices offset from one another by one-fourth of a body diagonal as shown in Fig. 2.1. There are four molecules per unit cell, the coordinates of the Te atoms being 000, $0\frac{1}{2}\frac{1}{2}$, $\frac{1}{2}0\frac{1}{2}$, and $\frac{1}{2}\frac{1}{2}0$ and those of the Cd atoms being $\frac{1}{4}\frac{1}{4}\frac{1}{4}$, $\frac{1}{4}\frac{3}{4}\frac{3}{4}$, $\frac{3}{4}\frac{1}{4}\frac{3}{4}$, and $\frac{3}{4}\frac{3}{4}\frac{1}{4}$. As a result every atom is surrounded tetrahedrally by four atoms of the other by $\sqrt{3}\,a/4$. The best value of the cubic lattice parameter at room temperature is 6.481 Å (Lawson *et al.*, 1960; Stuckes and Farrell, 1964; Thomassen *et al.*, 1963; Williams *et al.*, 1969). This value is subject to change due to deviations from stoichiometry.

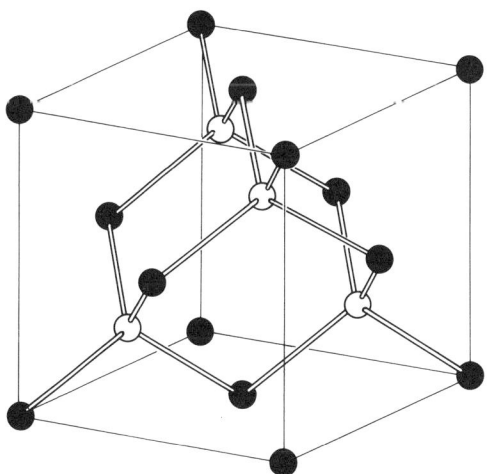

FIG. 2.1. The two equivalent interpenetrating face center cubic lattices of the zinc blende structure.

Medvedev et al. (1972) obtained a value of 6.482 Å (±0.001 Å) for crystals grown from stoichiometric melts. However the lattice constants of crystals grown Cd-rich and Te-rich solutions are 6.480 Å and 6.488 Å, respectively.

Twenty-four symmetry operations leave the zinc blende structure invariant. Besides translation, there are three $\pm\pi$ rotations about the three cubic $\langle 100 \rangle$ axes (C_4^2), six $\pm\frac{1}{2}\pi$ rotations about the same combined with inversion (IC$_4$), eight $\pm\frac{2}{3}\pi$ rotations about the body diagonals $\langle 111 \rangle$ (C_3), and six $\pm\pi$ rotations about the six face diagonals $\langle 110 \rangle$ with inversion (IC$_2$). Such symmetry is evident in faceted single crystals. Forms of faceted CdTe crystals grown from the vapor are shown in Fig. 2.2 (Höschl and Koňák, 1965). The most stable faces are (111), (100), and (110). Solution grown crystals are oriented in Fig. 1.10 so as to show the (100) and (111) faces and illustrate the four-fold and three-fold symmetries.

Inversion symmetry is not present in the zinc blende structure. Defining positions at one-fourth intervals along the body diagonal by atoms, Cd and

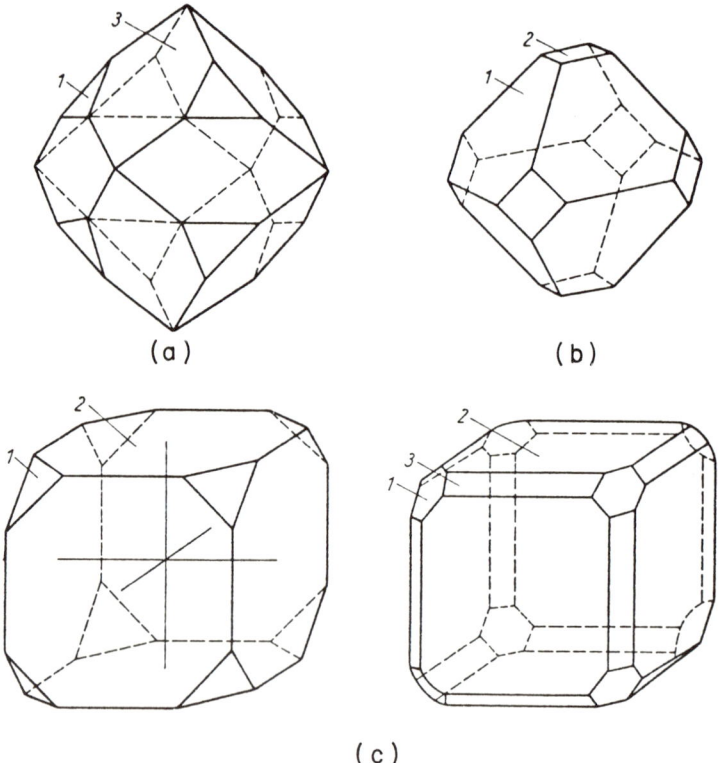

FIG. 2.2. Forms of vapor grown CdTe crystals prepared by Höschl and Koňák (1965): (a) rhombo–dodecaedric, (b) octaedric, (c) hexaedric. 1, (111) plane; 2, (100) plane; 3, (110) plane.

I. THE ZINC BLENDE STRUCTURE

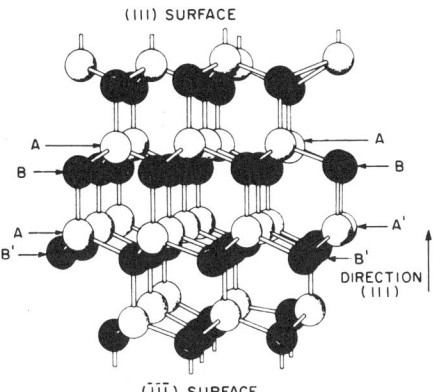

FIG. 2.3. The zinc blende structure viewed nearly normal to the $[\bar{1}\bar{1}\bar{1}]$ direction and illustrating the nonequivalent or polar (111) and $(\bar{1}\bar{1}\bar{1})$ surfaces.

Te, and vacancies V, the sequence in Fig. 2.1 is $\cdots$ TeCdVVTe $\cdots$. This sequence is not invariant since inversion results in the replacement of Te(Cd) atoms by Cd(Te) atoms. If the Te and Cd atoms were indistinguishable, then a center of symmetry would be present and the diamond structure would prevail. This lack of symmetry results in piezoelectric, piezobirefringent, and electrooptic effects which will be discussed later.

The Cd atoms in any {111} plane are bonded to a single neighboring Te atom in a $\langle 111 \rangle$ direction and bonded to three Te atoms on the opposite side of the plane. Viewing the lattice almost normal to a $\langle 111 \rangle$ direction as in Fig. 2.3 makes the double layers of Cd and Te atoms apparent. The separation between Cd and Te planes within the double layer is $0.14a$. The distance between double layers is $0.43a$. As a result of the greater bond density within the double layers and the lack of symmetry, the crystal exhibits crystallographic polarity and the chemical and physical properties on the {111} Cd face differ from those on the opposed $\{\bar{1}\bar{1}\bar{1}\}$ Te face. Since the geometrical structure factor is different in opposite directions along the polar axis, the anomalous dispersion of x rays can identify the Cd or Te-exposed faces (Warekois et al., 1962). Etch pit geometry and etching rates (Inoue et al., 1962) and growth rates and surface texture (Teramoto and Inoue, 1963; Saraie et al., 1972) also differ between faces (Fig. 1.16). By correlation with diffraction results standard etchants can identify either the Te or Cd faces.

Although the zinc blende structure is the stable form for bulk CdTe at atmospheric pressure (Fig. 1.5), thin films in both the sphalerite and wurtzite form are common. Simov et al. (1976) evaluated whiskers to be of the wurtzite structure. The presence of the wurtzite form is not surprising since bulk single crystals of the other II–VI compounds are found in both forms. It is not certain whether the hexagonal structure found in thin films is more stable at lower or higher temperatures as the resulting structure depends upon the substrate material, its orientation, etc. In the process of natural aging of CdTe films, made with an excess of Cd in the vapor, Palatnik *et al.* (1974) observed a phase transition from the metastable hexagonal form to the stable cubic one. When Yezhovsky and Kalinkin (1973) deposited thin films of CdTe on mica by vacuum condensation in a quasi-closed cell, they found that large source/substrate temperature differences promoted wurtzite growth, whereas under conditions close to equilibrium only cubic growth occurred. This is consistent with the work of Ueda (1975). An excess of either Cd or Te in the vapor also promoted the growth of the cubic phase. This is in contrast to earlier results where an excess of Cd was found to promote the growth of the wurtzite structure. (Spinulescu-Carnaru, 1966, Shalimova and Bulatov, 1960, and Sella *et al.*, 1968). Additional thin film work is discussed in Chapter 1, Section II,C.

The common occurrence of both structures is related to the close relationship between the wurtzite and sphalerite forms. Both structures consist of close-packed layers of Cd and Te atoms. The sequence $\cdots a\alpha\, b\beta\, c\gamma\, a\alpha\, b\beta\, c\gamma \cdots$ describes the zinc blende structure, the Greek and italic letters denoting, respectively, the close-packed planes of Cd and Te atoms. In Fig. 2.4 positions A, B, and C define lattice sites, and, therefore, if A refers to Te atoms so do positions B and C. If α is an arbitrary close-packed (111) Te reference plane, then β is an identical Te layer displaced so that each atom lies above a hollow on the α layer. The atoms of the third Te layer γ occupy the alternative set

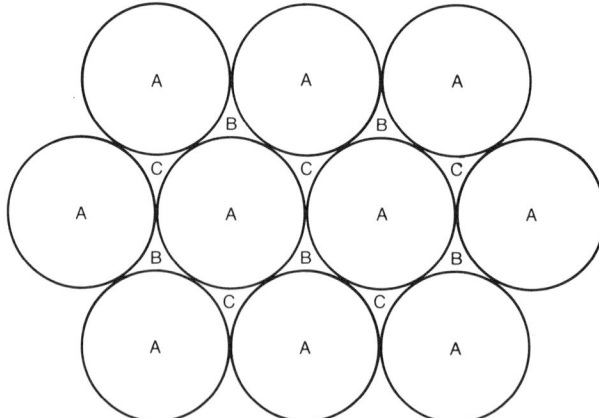

FIG. 2.4. ABCABC stacking sequence of the close-packed face center cubic lattice as viewed from a $\langle 111 \rangle$ direction. If the atoms in the C-layer are positioned over the A-layer atoms the resulting ABABAB stacking sequence is hexagonal close packed.

I. THE ZINC BLENDE STRUCTURE

of hollows above α. The Cd layers are specified in a similar fashion; their position is relative to Te in the double layer shown by Figs. 2.1 and 2.3. In the hexagonal structure the atoms in the third layer are placed over the atoms in the first layer, the sequence being $\cdots a\alpha\ b\beta\ a\alpha\ b\beta\ a\alpha \cdots$. Neglecting the the slight deviations from spherical close packing in wurtzite, each Cd is also tetrahedrally surrounded by four nearest neighbors of Te at the same distance as in zinc blende. The 12 next nearest neighbors are also at the same distance in both structures. Significant differences in the structures can only be found for the third next nearest neighbors. Because of the small difference in bond energies between zinc blende and wurtzite structures, the stacking sequence is very sensitive to growth conditions. The ABCABC cubic stacking sequence can be interrupted by the hexagonal rule ABABAB giving interstratified layers of cubic and hexagonal structures in thin films. Pashinkin et al. (1960) and Semiletov (1962) have identified the formation of a 12-layered hexagonal phase belonging to the 12H polytype. Semiletov (1962) and Novik (1962) have attempted to associate the generation of high photovoltages with these different structural patterns (Goldstein, 1958; Goldstein and Pensak, 1959; Lyubin and Fedorova, 1960).

Twinning in bulk single crystals is common (Kroger and de Nobel, 1955; de Nobel, 1959; Thomas, 1961; Lorenz, 1962c; Lynch, 1962; Teramoto and Inoue, 1963; Akutagawa and Zanio, 1971; Kyle, 1971) and is apparent to the eye in normal laboratory lightning by the change in texture across the twin boundary of a lapped wafer (Fig. 1.7.). The twin plane is parallel with the (111) plane and may be generated by a 180° (60°) rotation about the [111] twin axis (Holt, 1964, 1969). Rotation of separate grains 250° 32 min relative to one another about the [110] tilt axis is an alternative method of generating such a twin (Fig. 2.5). Before rotation the (1$\bar{1}$0) meridian plane defines the two grains. Twinning is common since there are no wrong bonds (i.e., Cd–Cd or Te–Te) across the twin plane. Back reflection Laue normally identifies the orientation of twinned crystals. Twinning is also identified through the orientation of etch pits (Inoue et al., 1962), by the Lang topography method (Ghezzi and Paorici, 1974), and by electron diffraction (Semiletov, 1956, 1962; Shiojiri and Suito, 1964).

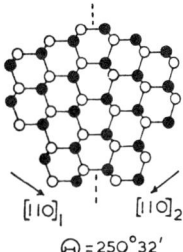

FIG. 2.5. The (111) twin boundary resulting from either a 250° 32′ tilt about the [110] axis or a 60° twist about the [$\bar{1}\bar{1}\bar{1}$] axis (Holt, 1964, 1969).

The transformation to the rocksalt structure at approximately 35 kbar (Fig. 1.5) results in a volume decrease of about 25% with a corresponding decrease on the lattice constant to about 5.86 Å (Mariano and Warekois, 1963; Smith and Martin, 1963; Owen et al., 1963). X-ray measurements at atmospheric pressures on CdTe plate-like samples which had previously made the rocksalt to zinc blende transition also showed extra reflections which were indexed for a rocksalt lattice constant of 5.59 Å (Rooymans, 1963). However, after powdering the samples, the extra reflections disappeared. Smith and Martin (1963) and Owen et al. (1963) found above 90 kbar a transformation to the white tin structure with lattice constants of $a = 5.86$ Å and $c = 2.94$ Å giving a c/a ratio of 0.50. At 300 kbar pressure c/a increased to 0.526 ($a = 5.62$ Å).

Electrical, optical, and thermal properties also reflect these structural changes. From 30 to 35 kbar the electrical conductivity increases about an order of magnitude (Samara and Drickamer, 1962). The resistivity decreases over four orders of magnitude from 85 to 90 kbar at the higher pressure transformation. These resistivity changes are reversible with release of pressure. An abrupt decrease in the absorption edge occurs at about 39 kbar (Edwards and Drickamer, 1961). However optical measurements indicate hysteresis effects. On reduction of the pressure the reverse transition occurs about 20 kbar. Jayaraman et al. (1963) and Cline and Stephens (1965) measured the low pressure transition at about 32 kbar by observing volume discontinuity. Hysteresis effects were also observed. Differential thermal analysis measurements by the same investigators indicate that this transition pressure decreases from 30 kbar at room temperature to 20 kbar at the melting point.

II. Bonding

The zinc blende structure is an immediate result of electron sharing. In fact Garlick et al. (1958) considered the bonding in CdTe to be predominately homopolar. However the ionic and metallic contributions are both significant in CdTe and are reflected in its physical properties. There is an intermediate transition to the NaCl structure with pressure. Transformation to the metallic phase with increasing pressure (Fig. 1.5) may be common to all $A^N B^{8-N}$ structures. However, the presence of d-core levels in CdTe and other structures from this row in the Periodic Table results in the covalent to "white tin" metallic transition at pressures lower than that of the lighter elements and compounds. In fact the diamond to white tin transition for Sn occurs near atmospheric pressure. Structural properties of CdTe along with Sn and its "iso-row" compounds InSb and AgI, are shown in Table 2.1.

TABLE 2.1

Some Properties of Sn, InSb, CdTe, and AgI[a]

	Low pressure structure	High pressure structure	Transition pressure (kbar)	$\Delta V/V$ (%)
Sn	Diamond	White tin	0	21
InSb	Zinc blende	White tin	25	18
CdTe	Zinc blende	NaCl/white tin	30/90	24/3
AgI	Zinc blende	NaCl	3.5	18

[a] Owen et al. (1963).

For this sequence greater pressures are required to attain the metallic phase and less to reach the stability region of a NaCl-type phase as the difference in the electronegativities of the constituent atoms increase.

A. Covalent versus Ionic

The concepts of excited valency state and hybridization in diamond apply to the zinc blende structure. Since the ground state of the carbon atom $1s^2 2s^2 2p^2$ has only two unpaired 2p electrons the corresponding valency state would be expected to result in a covalency of two. However an excited valency state corresponding to $1s^2 2s 2p^3$ obtained by promoting one of the paired 2s electrons into the vacant 2p-orbital results in a covalency of four. This hybridization (s + three p) leads to four orbitals of equivalent bond strength directed toward the corner of a tetrahedron. In CdTe and other II–VI compounds the nearest neighbors have unequal numbers of valence electrons but on the average each atom has four electrons available for formation of the bonds. Two and six electrons are available, respectively, from the kryptonlike-$4d^{10}5s^2$ atomic structure of Cd and the krypton-like-$4d^{10}5s^25p^4$ atomic structure of Te leaving net charges of Cd^{2+} and Te^{6+} on the atomic cores. If the electron charge is shared equally between Cd^{2+} and Te^{6+} as between cores in the diamond lattice, then the effective charges would be Cd^{2-} and Te^{2+}. If the bond is truly ionic, then complete electron transfer results in effective charges of Cd^{2+} and Te^{2-}. Intermediate states between ionic and covalent bonding might be alternatively represented by an electron cloud or bridge whose center of gravity shifts closer to the anion to form the more heteropolar bond and closer to the midpoint to form the more homopolar bond. Unfortunately, x-ray pictures of the electron density are not available for CdTe to assist in this interpretation.

Chelikowsky and Cohen (1976) calculated the electron charge densities as a function of position in the unit cell for CdTe and other semiconductors using wave functions derived from pseudopotential band-structure calculations. The pseudopotential method is discussed in Section IV,B. The charge density for each valence band is given by

$$\rho_n(\mathbf{r}) = \sum_{\mathbf{k}} e|\psi_{n,\mathbf{k}}(\mathbf{r})|^2, \tag{2.1}$$

where the wavefunction $\psi_{n,\mathbf{k}}(\mathbf{r})$ for band n and state $\mathbf{k}$ is summed over all states in the Brillouin zone. Figure 2.6 summarizes these results in contour maps through a {110} plane for the sum of the four valence bands of CdTe and its more covalent neighbors InSb and Sn in the same row of the periodic

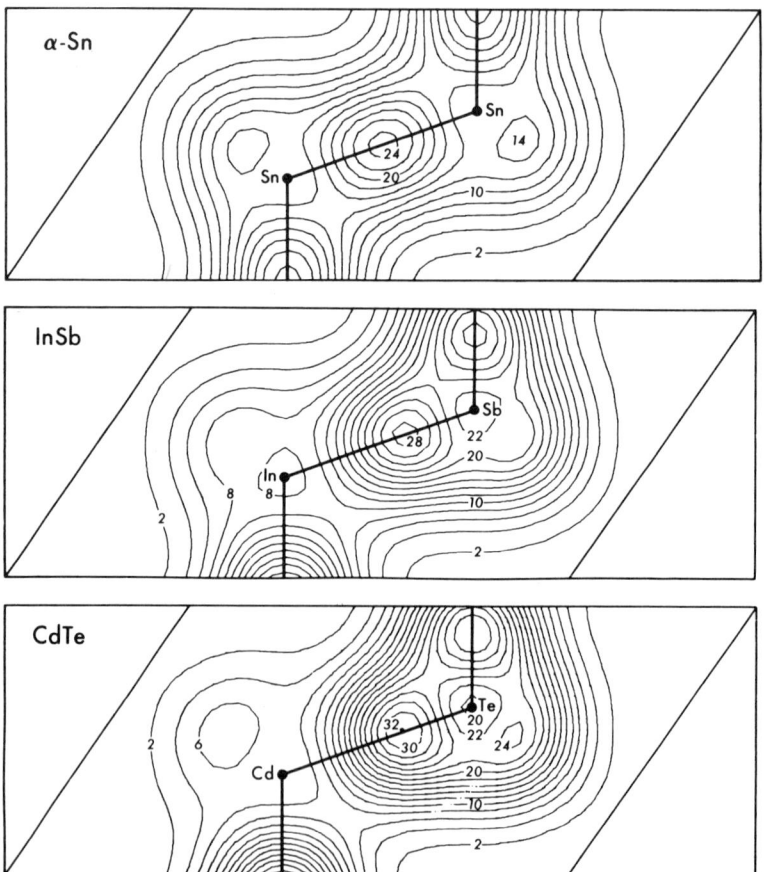

FIG. 2.6. Charge density for the sum of the valence bands 1 through 4 in CdTe, InSb, and Sn as viewed through a {110} plane (Chelikowsky and Cohen, 1976).

table. The tetrahedral covalent bonding is apparent in Sn where the charge density is centered midpoint between the two atoms. For InSb the covalent bonding-charges are displaced towards the Sb atom. For CdTe nearly all the charge is concentrated about the Te atom. Walter and Cohen (1971) define a bonding-charge Z_b by integrating the charge density between the atoms. The values of Z_b in units of e per bond are 0.123 for Sn, 0.091 for InSb, and 0.027 for CdTe. The bonding charge is correlated with the degree of covalency or ionicity of the crystals (Phillips, 1969, 1970; Phillips and Van Vechten, 1969a).

Phillips successfully established an ionicity scale which is capable of predicting, among many other properties, the structure of $A^N B^{8-N}$ compounds. The theory is formulated in terms of the low-frequency electronic dielectric constant of a semiconductor $\varepsilon(0) = n^2$ where n is the index of refraction below the bandgap. The Phillips model uses a bonding energy gap (E_g) which is composed of a homopolar component (E_h) and an ionic component (C) that are related by

$$E_g^2 = E_h^2 + C^2. \tag{2.2}$$

The bonding energy gap (E_h) also represents the isotropic homopolar gap in the Penn model and for CdTe and other materials is based upon extrapolation from the true homopolar materials Si and diamond. The ionic component C is calculated from the electronic contribution to the long wavelength dielectric constant of the solid using

$$\varepsilon(0) = 1 + A(\hbar\omega_p)^2/(E_h^2 + C^2)D, \tag{2.3}$$

where ω_p is the plasma frequency of the valence electrons, A is a constant close to unity which represents a small correction to the bond model due to bonding effects, and D is a parameter introduced by Van Vechten (1969a) to account for d-like core electrons. The plasma frequency squared is proportional to the average number of s–p valence electrons per atom which in true tetrahedral covalent structures is 4. The bonding energy gap represents the difference between the average conduction and valence bands or antibonding and bonding energy levels. It is related to the structural properties and not to be confused with energy difference between the top of the valence band and bottom of the conduction band which determine the electrical properties. Values of the bonding energy gap and the electrical band gap are 5.79 eV and 1.53 eV, respectively. Values of the homopolar and ionic energy gaps in CdTe are 3.08 eV and 4.90 eV, respectively. The fractional degrees of ionic and homopolar bonding are

$$f_i = C/E_g \tag{2.4a}$$

and

$$f_h = E_h/E_g. \tag{2.4b}$$

By this definition CdTe is 72% ionic. Just what this means is difficult to quantitatively access. However, when E_h is plotted versus C in Fig. 2.7 for crystals of the $A^N B^{8-N}$ type having the diamond, zinc blende, wurtzite, or rocksalt structure, the locus of $f_i = 0.785$ separates all the sixfold rocksalt structures from the tetrahedrally coordinated structures. Cadmium telluride is a borderline case. Therefore, it is not surprising that under pressure CdTe transforms from a relatively low density covalent tetrahedral structure to the higher density more ionic octahedral structure (Fig. 1.5). Among the tetrahedrally coordinated crystals CdTe is ionic, ranking as the 35th most ionic out of 42 structures. Among the chalcogenides of Zn and Cd it is the

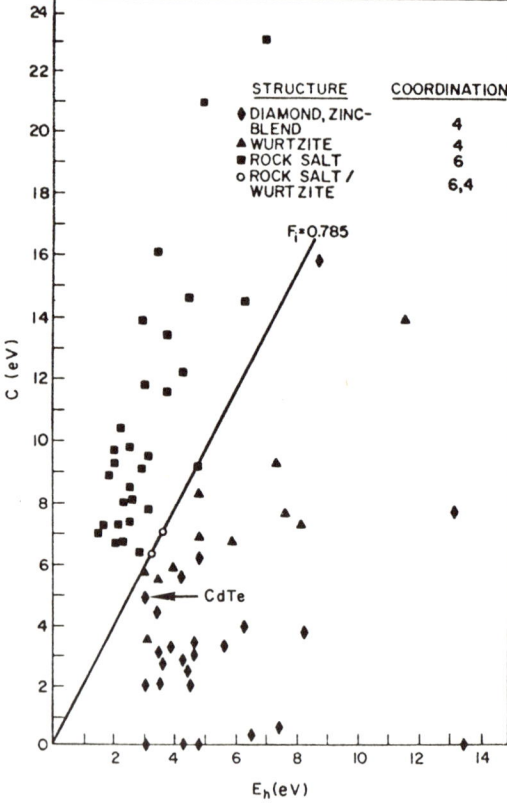

FIG. 2.7. Values of the homopolar, E_h and ionic, C, bonding gap for crystals of the type $A^N B^{8-N}$. The line $f_i = 0.785$ separates all fourfold from all sixfold crystals. (Reprinted from J. C. Phillips, *Rev. Mod. Phys.* **42**, 317, Fig. 10 (1970).)

II. BONDING

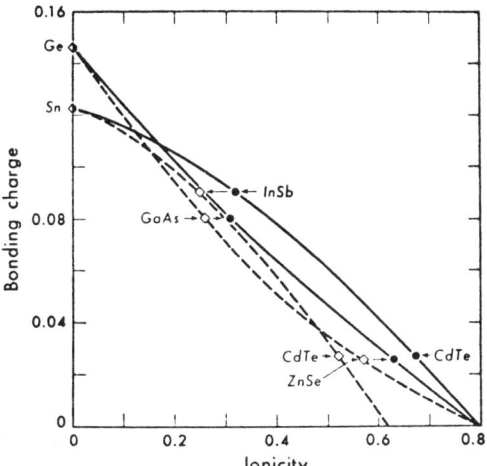

FIG. 2.8. Relationship between degree of ionicity f_i and bonding charge for the iso-rows Sn–InSb–CdTe and Ge–GaAs–ZnSe (Walter and Cohen, 1971). --●--, Phillips (1970); --□--, Pauling (calculated by Phillips (1970)).

most ionic. As a comparison ZnTe with $f_i = 0.609$ is the least ionic. This model correlates closely with numerous physical properties.

Walter and Cohen (1971) correlated this definition of ionicity f_i with the covalent bonding-charge Z_b and found an inverse relationship. Figure 2.8 shows these results along with those of Pauling's. When the results of Phillips are extrapolated to zero for the Sn–InSb–CdTe and Ge–GaAs–ZnSe series, the intercept corresponds to the critical value of ionicity $f_i = 0.785$ which separates the more covalent crystals of fourfold coordination from the more ionic crystals of sixfold coordination. When the amount of charge in the covalent bond approaches zero, the configuration of tetrahedrally directed bonds is no longer stable.

Phillips (1969) relates the cohesive energy for $A^N B^{8-N}$ compounds belonging to a given row of the periodic table to f_i by

$$\Delta G_s = \Delta G_s(0) + k f_i \tag{2.5}$$

where k is a constant. The cohesive energy here is the Gibbs free energy of sublimation of the compound (or the elements for $N = 4$) into neutral atoms. The cohesive energies of rows 1, 2, 3, and 4 are shown in Fig. 2.9 with k constant from row to row.

Interchanging elements from row 3 with those from row 4, such as Zn for Cd to form ZnTe from CdTe, results in the quasi-row 3–4. The linear relationship still holds. The vertical displacement from sequence to sequence

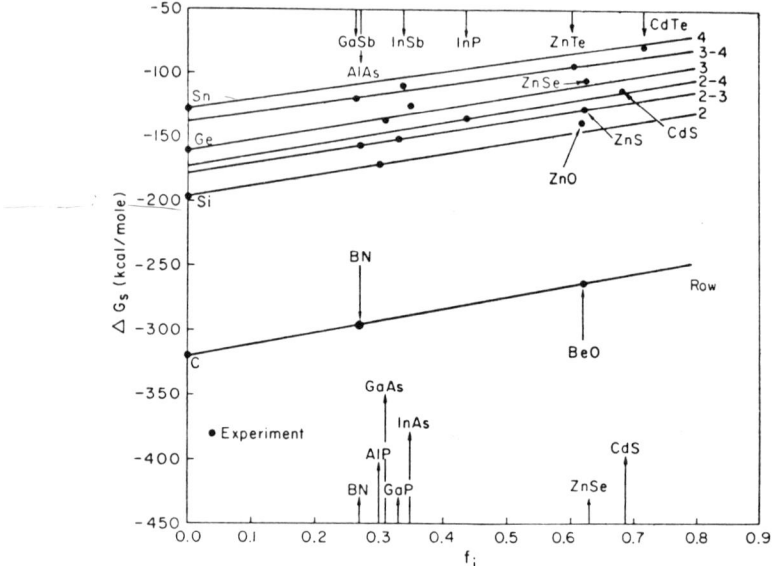

FIG. 2.9. The Gibbs free energy of cohesion versus degree of ionicity for tetrahedral structures. (Reprinted from J. C. Phillips, *Rev. Mod. Phys.* **42**, 317, Fig. 13 (1970).)

is associated with the bond length effect on $\Delta G_s(0)$ and scales approximately as a^{-2}, where a is the bond length. The presence of d-core electrons is responsible for deviations from a constant value of $\Delta G_s(0)/a_2$. It is apparent that in predominately covalent tetrahedral structures, the binding energy decreases with increasing ionicity. Cadmium telluride having a relatively large lattice constant and a high degree of ionicity has a relatively small binding energy.

B. METALLIZATION

Metallization effects are significant in CdTe because of its high atomic number. With increasing atomic number, metallic structures replace covalent ones and heats of formation and cohesive energies decreases. In CdTe and in other structures containing d-core electrons, there is appreciable mixing of the valence and conduction bands with d-levels. Metallization is due to the resulting dehybridization or overlap of the bonding p and antibonding s states and a resulting decrease of the bonding gap E_g. Van Vechten (1969a) associates D in Eq. (2.2) with the degree to which the s-valence electrons penetrate the atomic core. He defines $D = N/4$ where N is the number of valence electrons. For normal valence materials such as Si and diamond there are four valence electrons and no d electrons and $D = 1$. However, for CdTe the effective number of electrons is 5.2 and $D = 1.3$.

Phillips and Van Vechten (1970) include f_i and effects of metallization in a general formula for the heats of formation of $A^N B^{8-N}$ tetrahedrally coordinated structures. With the appropriate scaling factor they arrived at a value of 25.1 kcal/mole for CdTe. This is in good agreement with the experimental value of 22.1 kcal/mole.

C. Elastic Properties

In the cubic system there are three independent elastic stiffness constants C_{11}, C_{12}, and C_{44}. Values for these constants shown in Table 2.2 are related to the type of structure and the specific bonding forces in that structure. The most useful phenomenological description of the short-range valence forces in the tetrahedrally coordinated crystals is the valence–force-field (VFF) approach, in which all interatomic forces are related to bond-bending and bond-stretching forces (Musgrave and Pople, 1962). The bond-bending forces stabilize the loosely packed diamond structure against shear and prevent collapse into a denser structure. Martin (1970) extends this approach to zinc blende structure crystals by adding Coulombic and repulsive forces. Rigid point charges at the atomic sites interact through a screened Coulombic potential, $S = \pm(e^*)^2/(\varepsilon(\infty)r)$. Here e^* is the absolute value of the effective ionic charge of the anion and cation, r is their separation and $\varepsilon(\infty)$ is the optical dielectric constant. In relating the VFF to the distortion energy of the unit cell Martin associates α with the bond-stretching and β with the bond-bending forces. Bond-bending forces are important in the tetrahedral crystals where the sp^3-hybrid orbitals must form tetrahedral angles of $110°$ with each other. Altering this angle requires mixing wave functions of d and f symmetry into the space of the sp^3-valence wave function (Phillips, 1973b). The bond-bending forces in diamond of the first row are especially large because the promotion energy from 2s and 2p valence states to 3d

TABLE 2.2

Elastic and Piezoelectric Constants of CdTe

Experimenters	$10^{10} Nm^{-2}$			C/m^2	T (°K)
	C_{11}	C_{44}	C_{12}	$e_{pol}(\sqrt{3}e_{14})$	
McSkimin and Thomas (1962)	5.351	1.994	3.681		298
Vekilov and Rusakov (1971)	5.33	2.044	3.65		300
	5.57	2.095	3.84		77
Berlincourt et al. (1963)	6.15	1.96	4.30	0.054	77
Greenough and Palmer (1973)	5.38	2.018	3.74		298
	5.62	2.061	3.93		77

and 4f state is large. However, in CdTe, the promotion energy from 5s and 5p to 4f is much smaller. As a result of mixing the wave functions and metallization, the bond-bending forces of CdTe or of Sn of the same row are soft, especially with respect to diamond. Martin's (1970) expressions for the elastic constant in terms of the short range bond-stretching and bond-bending α forces and the long-range bond-stretching Coulombic force S are

$$C_{11} = (\alpha + 3\beta)(\sqrt{3}/4r) - 0.083S, \qquad (2.6)$$

$$C_{12} = (\alpha - \beta)(\sqrt{3}/4r) - 0.136S, \qquad (2.7)$$

$$C_{44} = (\alpha + \beta)/(\sqrt{3}/4r) - 0.136S - \xi^2[(\alpha + \beta)(\sqrt{3}/4r) - 0.27S], \quad (2.8)$$

where ξ is an internal strain parameter responsible for the piezoelectric effect. The ratio β/α is a measure of the ratio of the bond-bending to bond-stretching forces and is also an indication of the degree of ionicity. A plot of β/α versus the degree of ionicity f_i as defined by Phillips for several tetrahedral structures is remarkably linear and confirms this interpretation (Fig. 2.10). Similar plots of the shear moduli C_{44} and $C_{11} - C_{12}$ expressed in terms of

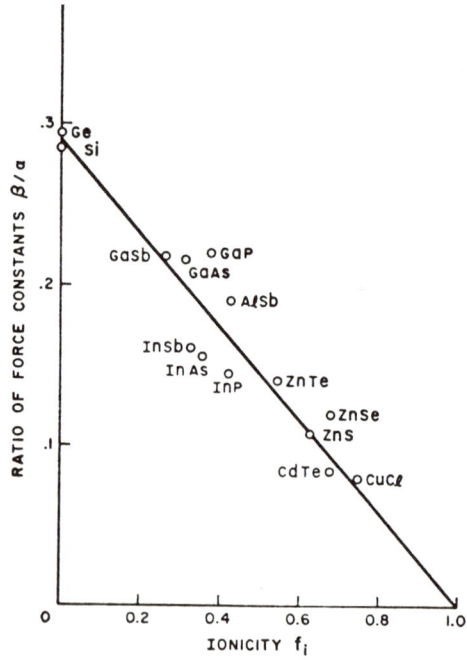

FIG. 2.10. The ratio of the noncentral to the central force constants β/α as a function of Phillips ionicity, f_i (Martin, 1970).

α, β, and S versus f_i show the noncentral or bond-bending force constant β in CdTe to be weak.

Instead of performing an integration over all of the valence bonds, as Walter and Cohen (1971) did in Fig. 2.8 to determine the covalent bonding charge, Rusakov and Anfimov (1976) defined the covalent bond charge for 19 diamond and zinc blende structures in terms of the Phillips ionicity parameter f_i (Eq. 2.4). The bonding charge for CdTe is consistent with the bond-bending and bond-stretching forces as defined by Martin.

Van Vechten (1974) developed a linear relationship between the bond-bending force constants in tetrahedral semiconductors and to the degree of ionicity, valid for $f_i < 0.5$. Competition between tetrahedral and octahedral covalent bonding was suggested as the cause for the failure of this extrapolation to CdTe and more ionic materials. Better agreement for the more ionic materials is found when the force constants are assumed to be proportional to the internal energy between the tetrahedral and pressure-induced rocksalt phases of these compounds.

The adiabatic bulk modulus as calculated from the elastic constants in Table 2.1 is 424 kbar. The experimental value of the isothermal bulk modulus, 252 kbar (Cline and Stephens, 1965), is less than the value of 322 kbar calculated by Nolen and Titer (1973).

D. PIEZOELECTRIC EFFECT

The piezoelectric effect in CdTe is due to internal strains which displace the ionic Cd and Te sublattices nonuniformly relative to one another. The internal displacement of the ions is not the only mechanism causing piezoelectric effects. If the positions of the ionic charges themselves change with strain due to charge redistribution, an additional contribution to the piezoelectric constant occurs which is of opposite sign (Phillips and Van Vechten, 1969b).

Alternate (100) lattice planes of Cd and Te atoms are equally spaced. By symmetry, strain in the ⟨100⟩ direction results in equal spacing before and after strain and therefore no piezoelectric component in that direction. However, in the polar ⟨111⟩ directions the crystal consists on alternating layers of Cd and Te atoms with spacing ratio 1:3. In this case, symmetry arguments cannot argue ⟨111⟩ strain to be uniform before and after displacement. If CdTe is strained longitudinally in the ⟨111⟩ direction, contributions to the dielectric displacement vector are not only in this direction, but also from the other three polar directions; i.e., $[1\bar{1}\bar{1}]$, $[\bar{1}1\bar{1}]$, and $[\bar{1}\bar{1}1]$, because there also is a strain component in these directions. The total dielectric displacement vector for constant electric field in a ⟨111⟩ direction is $D_{\langle 111 \rangle} = \frac{2}{3} e_{\text{pol}} S_{\text{long}\langle 111 \rangle}$, where $S_{\text{long}\langle 111 \rangle}$ is the longitudinal strain in a ⟨111⟩ direction and e_{pol} is the piezoelectric constant (Arlt and Quadflieg, 1968). It is related to e_{14} given in Table 2.2 by $e_{pol} = \sqrt{3}e_{14}$. Mason (1950) relates the piezoelectric constant to various external variables. The two contributions to the piezoelectric effect are contained within

$$e_{pol} = (\xi - d/4)\sqrt{3}(e_C^*/e)(16/a^2). \qquad (2.9)$$

When e_C^* is the Callen ionic charge which determines the splitting of the longitudinal and transverse optic lattice vibration modes, $d(d_{CdTe} = 1.44)$ is a charge redistribution index and $\xi(\xi_{CdTe} = 0.79)$ is the internal strain parameter given by

$$\xi = (2C_{12} - 0.31S)/(C_{11} + C_{12} - 0.31S) \qquad (2.10)$$

Since the II–VI compounds and, in particular, CdTe are ionic, the formal charge on the metal atom would be expected to be positive and associated with the valence $Z = 2$. Likewise, the III–V crystals are more covalent and the negative formal charge on the metal atom in the extreme case would be expected to be one $(4 - z = 1)$. Phillips and Van Vechten assign $e_C^*/e > 0$ and associate the change in sign of e_{pol} on going from the II–VI to the III–V compounds with the relative importance of ξ and $d/4$ in Eq. (2.10). They correlated d with the degree of ionicity f_i and not only found d to decrease with increasing ionicity, but also predicted the sign reversal.

E. Different Approaches in Defining Ionicity

Coulson *et al.* (1962) use the method of the linear combinations of atomic orbitals and assign $\lambda^2/(1 + \lambda)^2$ and $1/(1 + \lambda)^2$ to be the fraction of time that the valence electrons associate with Cd and Te atoms respectively. Using bonding and antibonding orbitals of the form

$$\psi = \lambda\phi_{Cd} + \phi_{Te} \qquad (2.11a)$$

and

$$\chi = \phi_{Cd} - \lambda\phi_{Te}, \qquad (2.11b)$$

$\lambda = 1$ for homopolar bonding and $\lambda = 0$ for ionic bonding. Here $\phi_{Cd}(\phi_{Te})$ is the sp^3-hybrid wavefunction of a Cd (Te) atom. After minimizing the total energy of the system they calculate λ. In reformulating this approach, Phillips (1970) and Van Vechten (1969b) established the Coulson ionicity scale. Although the ionicity increases with N as defined in $A^N B^{8-N}$, the small spread in the values of f_i for compounds within each N-group make a comparison between the compounds in each group difficult. Besides, there is disagreement for the ranking of compounds between the two scales. For example the Coulson definition ranks ZnTe as the most ionic of the Zn and Cd chalcogenides, whereas the Phillips scale ranks it last. Failure of the Coulson-type approximation stems from an underestimation of the covalent energy. Unfortunately, this model does not relate to measured physical parameters.

Pauling (1960) bases his definition of ionicity not on the total energy of the bond but on the empirical heats of formation of diatomic molecules at conditions of standard temperature and pressure. Besides the components being in different standard states than that of the compound, the semiconductor often has a different bonding than that of its components. However the definition of ionicity by Pauling is more meaningful than that of Coulson *et al.*, although his value of −4.4 kcal/mole for CdTe is far from the experimental value.

It is hard to determine the degree of ionicity from either hard sphere theory or interatomic distances. If the CdTe were purely ionic, hard sphere theory would incorrectly predict sixfold coordination. Correlation of the observed structure with the rigid-ion model is poor, even for the predominantly ionic alkali halides. It predicts that several lithium salts should have the zinc blende structure, whereas none of them do. It also predicts that all of the Cs salts will have eight nearest neighbors which is not the case. A comparison of the observed interatomic distances, with that calculated would indicate that CdTe is predominately covalent. The observed interatomic distances 2.80 Å between Cd and Te atoms agree exactly with those calculated for a covalent structure using covalent radii of 1.48 Å for Cd and 1.32 Å for Te (Pauling, 1960). An identical value was obtained more recently by simple quantum-mechanical considerations (Van Vechten and Phillips, 1970). In comparison, the calculated interatomic ionic distance of 3.04 Å for CdTe as well as the rest of the Zn, Cd, and Hg chalcogenides of Te, Se, and S are significantly larger than observed (Roth, 1967). Covalent radii are parameters which are not restricted to homopolar binding and the transition from one type of binding to the next only slightly influences the distance between the atoms.

III. Lattice Dynamics

A. Dispersion Relationship

Dispersion relationships contain information regarding the short-range interatomic forces. However, the strong absorption of thermal neutrons by ^{113}Cd prevents classical neutron inelastic scattering measurements on CdTe. Rowe *et al.* (1974) therefore grew CdTe single crystals from the melt with Cd present as ^{114}Cd to reduce absorption and allow neutron scattering measurements. Results at 300°K for the [100], [111], and [110] directions of propagation are shown in Fig. 2.11. The shape of the dispersion relationship is similar to other zinc blende structures. The transverse optical (TO) and the

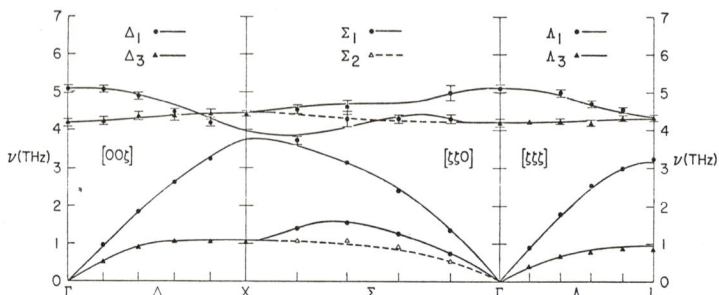

FIG. 2.11. The room temperature phonon dispersion relationship in ^{114}CdTe as determined by neutron inelastic scattering measurements (Rowe et al., 1974).

longitudinal optical (LO) branches at Γ are split into frequencies, 4.20 THz (141 cm^{-1}) and 5.08 THz (168 cm^{-1}) and are in good agreement with results obtained from reflectivity (Fisher and Fan, 1959; Perkowitz and Thorland, 1974; Manabe et al., 1967; Mitsuichi, 1961; Gorska and Nazarewicz, 1974; Vinogradov et al., 1972; Vodopyanov et al., 1972) and compressibility (Moss, 1959) results. Besides the frequencies of the optical modes, the calculated values of the elastic, dielectric, and piezoelectric constants are in good agreement with those listed in Table 2.2.

Similar to other semiconductors not containing first row atoms, the transverse acoustical (TA) mode flattens out. Phillips (1973b) likens the higher Z semiconductors to metals where the large value of the dielectric constant is responsible for screening ion–ion forces at distances large compared with nearest-neighbor distances. Unfortunately, a theoretical model is not available which relates the interatomic forces at the shorter wavelengths to physical parameters. A generalized shell model, incorporating a microscopic theory of dielectric screening and lattice dynamics, is used to interpret the dispersion relationship for α-Sn (Gupta et al., 1974) and Si and Ge (Price et al., 1974). Rowe et al. (1974) use a more simple shell model (Cochran, 1959) to interpret the CdTe data. In this model each atom is regarded as a charged core coupled to an oppositely charged shell. This gives the atom the property of polarizability, not only in an electric field but also under the influence of bonding interactions between adjacent atoms. The continuous lines in Fig. 2.11 represent data fitted to a 14-parameter shell model. These parameters, among others, include the first-neighbor force constants between ion–ion, ion–shell, and shell–shell and general second-neighbor force constants. Unfortunately, these parameters do not provide much physical insight. The shell model is not able to predict an approximate dispersion curve unless critical phonon frequencies are available.

III. LATTICE DYNAMICS

TABLE 2.3

CdTe Phonon Dispersion Models

Model	Author
Shell	Rowe et al. (1974)[a]
	Sennett et al. (1969)
Rigid-ion	Plumelle and Vandevyver (1976)[a]
	Bublik et al. (1972)
	Al'tshuler et al. (1974)
Modified rigid-ion	Gaur et al. (1971)
	Vetelino et al. (1972)
Second neighbor ionic	Talwar and Agrawal (1973)

[a] Fit to neutron data.

The rigid ion model of Plumelle and Vandevyver (1976) also provides a good fit to the neutron data. In this model the ions are rigid and non-polarizable; the model includes nearest-neighbor and next nearest-neighbor interactions and assumes long-range Coulomb interaction. It is not apparent whether either the shell model or the more simplistic rigid ion model is more appropriate. However, it is evident that neutron data is essential to the development of any theory. Earlier workers (see Table 2.3) also constructed the shell model, the rigid ion model, modified rigid ion model, and the second-neighbor ion models based upon optical and x-ray data and the elastic constants. There is a noticeable lack of agreement between these earlier calculations and the two correct fits to neutron data.

Although detailed information regarding the short-range forces cannot be obtained, at least a relative comparison between other structures of the same row is possible as dispersion measurements have also been made on α-Sn and InSb (Price et al., 1971). The isoelectronic semiconductor sequence α-Sn → InSb → CdTe is characterized by increasing band gap and ionicity while the structure remains the same and the lattice spacing and atomic masses change very little. Therefore, the effect of ionicity on lattice dynamics can be examined. Peaks occur because of singularities on the phonon frequency distribution. These singularities arise from critical points in the Brillouin zone where the frequency versus wave vector curves have zero slope. Figure 2.12 shows the frequency distribution for α-Sn, InSb, and CdTe. The splitting of the LO and TO modes at Γ is larger for CdTe than for InSb, as expected from its greater ionicity. A general "softening" of the optic modes (4 to 6 THz) and the LA modes (3 to 4 THz) in the sequence α-Sn → InSb → CdTe is also evident from the figure while the transverse acoustic modes ($\approx$ 1 THz) are much less affected.

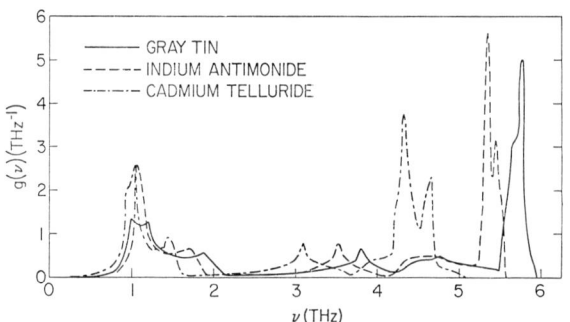

FIG. 2.12. The phonon distribution in the iso-row Sn–InSb–CdTe at room temperature (Rowe et al., 1974).

B. INFRARED RESPONSE

Infrared studies of lattice absorption and reflection and Raman data have been helpful in determining characteristic energies at the zone center as well as determining characteristic phonon frequencies. The usual method to obtain the frequency of the transverse optical mode at the zone center $\omega_{TO(\Gamma)}$ is to analyze the Reststrahlen band by Drude dispersion theory:

$$n^2 - k^2 = \varepsilon(\infty) + 4\pi\rho\{(\omega_{TO(\Gamma)}^2 - \omega^2)/[(\omega_{TO(\Gamma)}^2 - \omega^2)^2 + \gamma\omega^2]\}, \quad (2.12)$$

where n is the refractive index, k is the extinction coefficient, ω is the phonon frequency, $\varepsilon(\infty)$ is the optical dielectric constant, γ is the damping constant, and ρ the oscillator strength. Mitsuishi (1961) analyzed the reflection data with the Kramers–Kronig relationship to obtain the optical constants n and k. Using these constants he found at room temperature a value of 141 cm^{-1} for TO(Γ) to be the best fit of the experimental data to Eq. (2.12). The static dielectric constant $\varepsilon(0)$, the longitudinal optical phonon frequency at the zone center $\omega_{LO(\Gamma)}$, and the Szigeti effective charge e_s^* are calculated from

$$\varepsilon(0) = \varepsilon(\infty) + 4\pi\rho/\omega_{TO(\Gamma)}^2, \quad (2.13a)$$

$$\omega_{LO(\Gamma)}/\omega_{TO(\Gamma)} = (\varepsilon(0)/\varepsilon(\infty))^{1/2}, \quad (2.13b)$$

$$e_s^* = \frac{3}{\varepsilon(\infty) + 2}\left\{(\varepsilon(0) - \varepsilon(\infty))\frac{\pi M}{N}\right\}^{1/2}\omega_{TO(\Gamma)}, \quad (2.13c)$$

where M is the reduced mass and N is the density of ion pairs. The values of these constants are summarized in Table 2.4. Vinogradov et al. (1972) show that the frequencies of the transverse and longitudinal optical phonons correspond to the points at which the second derivative changes sign on the

TABLE 2.4

FUNDAMENTAL OPTICAL DATA FOR CdTe

Investigator	T (°K)	$\omega_{TO(\Gamma)}$ (cm^{-1})	$\omega_{LO(\Gamma)}$ (cm^{-1})	$\varepsilon(\infty)$	$\varepsilon(0)$	ρ (10^{24} sec^2)	γ_{TO}	e_s^*/e
Manabe et al. (1967)	300	141	169	7.1	10.2	4.4	6.6	0.74
	100	144	165	7.1	9.4	4.6	2.2	0.75
Mitsuishi (1961)	300	140					7–8	0.83
	90	144					3	
Moss (1959)		145		7.2				
Yamada (1960)	300				10.3			0.83
Fisher and Fan (1959)	300	147		7.13	10.63			0.76
Halsted et al. (1961)	20		172					
Strzalkowski et al. (1976)	100				10.55			
	300				11.00			
de Nobel (1959)	20				10.9			
	77				11.0			
Lorimor and Spitzer (1965)	300			7.05	10.6			
Górska and Nazarewicz (1974)	300	143	168	7.2				0.70
Berlincourt et al. (1963)	77				9.65			
Marple (1964)				7.2				
Rowe et al. (1974)	300	141	168	6.9	10.2			
Mooradian and Wright (1968)	4.2–300	140	171					
Perkowitz and Thorland (1974)	300	140.7		7.1	10.2		7.0	
Vinogradov et al. (1972)		140	168					
Vodop'yonov et al. (1972)		141	167					
Vodop'yonov et al. (1974)	80	143	171	7.2	10.5		2	0.77
	293	140	167	7.3	10.4		5.6	0.73
Johnson et al. (1969)	300			6.7	9.4			0.73
	8			6.7	9.0			
de Nobel (1959)	300				11.0			

Reststrahlen band. Their values of 168 cm^{-1} and 140 cm^{-1} for the longitudinal and transverse optical branches are in excellent agreement with the above method of calculation and are probably accurate to within 1%. Due to phonon anharmonicity, cooling the sample from room temperature to about 90°K results in a decrease in the damping constants by more than a factor of 2. Vodop'yanov et al. (1974) also interpret γ to increase with surface damage and decrease with phonon frequency.

The static dielectric constant is not as accurately known as values for TO and LO. At 300°K $\varepsilon(0)$ ranges from 10.2 to 11.0 with the largest value measured most recently with capacitance measurements (Strzalkowski et al., 1976). The optical constants are relatively independent of temperature. The static dielectric constant increases by approximately 5% from 100°C to 300°C (Segall et al., 1963; Manabe et al., 1967; Strzalkowski et al., 1976). The increase in $\varepsilon(\infty)$ is much smaller (Rode, 1970). Considering these dependencies and the small decrease in $\omega_{TO(\Gamma)}$ with increasing temperatures (Table 2.4) $\omega_{LO(\Gamma)}$ should be nearly constant by the Lyddane–Sachs–Teller relation.

Corresponding transmission spectra of CdTe are shown in Fig. 2.13. The strong absorption near 140 cm^{-1} is due to the transverse optical mode and is in good agreement with reflection studies. Vodop'yanov et al. (1974) demonstrated that for oblique incidence in s-polarized light only transverse phonons are excited, whereas for p-polarized light both modes are active. The intensity of the LO absorption band also increases linearly with film thickness and the shape of the band is in good agreement with theory. Phonon energies away from the zone edge are usually deduced from a number of weaker absorption bands, interpreted in terms of multiphonon processes in which a photon is absorbed and either two or more phonons are emitted or phonons are both emitted and absorbed. Bottger and Geddes (1967) resolved

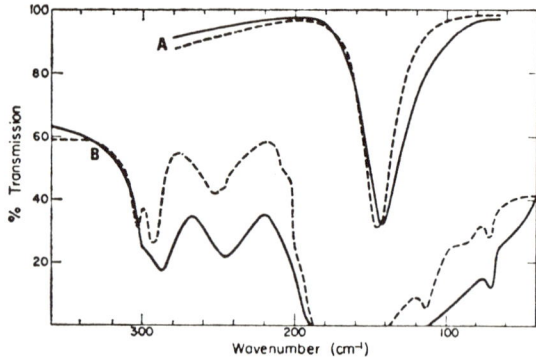

FIG. 2.13. Infrared transmission spectra of CdTe: A, a thin film at 300°K (solid line) and at 140°K (dashed line); B, a 0.3-mm-thick IRTRAN-6 sample at 300°K (solid line) and a 110°K (dashed line) (Bottger and Geddes, 1967).

III. LATTICE DYNAMICS

TABLE 2.5

CHARACTERISTIC PHONON FREQUENCIES

Investigator	T (°K)	Phonon frequency (cm^{-1})			
		LO	TO	LA	TA
Bottger and Geddes (1967)	110	152	142	104	35
	300	148	139	97	35
Vodop'yanov et al. (1974)	300	107	140		
Stafsudd et al. (1967)	88	151	125	50	37
Slack et al. (1966)	4.2	180	140	105	65[a]
Slack et al. (1969)	20			104.5[b]	46[c]
Mitsuishi et al. (1958)			150		
Fisher and Fan (1959)	300		143		
Mooradian and Wright (1968)[d]	4.2–300	150	139	97	35
Rowe et al. (1974)[e]	300	155	142	102	34, 47
Plumelle and Vandevyver (1976)[e]		153	142	115	34, 53

[a] Interpreted later by Slack et al. (1969) to be incorrect assignment.
[b] Transition assigned to L critical point in Brillouin zone.
[c] Transition assigned to X critical point in Brillouin zone.
[d] Assigned to zone boundary from infrared absorption and Raman scattering.
[e] From dispersion models.

seven bands in the infrared absorption spectra from 360 cm^{-1} to 60 cm^{-1} and assigned two phonon processes to them. Characteristic phonon frequencies of calculated and observed TO, LA, TA, and LA modes are in reasonable agreement and are listed in Table 2.5. Mooradian and Wright (1968) also measured the infrared absorption spectrum as well as the one and two phonon Raman spectrum. Raman data at the zone center are in good agreement with reflection data. They interpret their two phonon Raman and absorption data to be characteristic of phonon frequencies at the zone boundaries. These authors, as well as Bottger and Geddes (1967), took care to analyze their data so as to satisfy the requirements of Brout's sum rule. Characteristic phonon frequencies determined from neutron diffraction studies confirm these assignments. In the neutron diffraction studies, two values are assigned to TA. The smaller value, 33 cm^{-1}, is probably associated with transitions near X and L and the larger value with a transition along Σ. A more detailed discussion of absorption mechanisms around 1000 cm^{-1} follows in Chapter 4, Section III,B.

C. THERMAL PROPERTIES

Thermal expansion measurements are in good agreement between polycrystalline material (Novikova, 1961; Browder and Ballard, 1972; Williams et al., 1969) and single crystals (Greenough and Palmer, 1973; Smith and

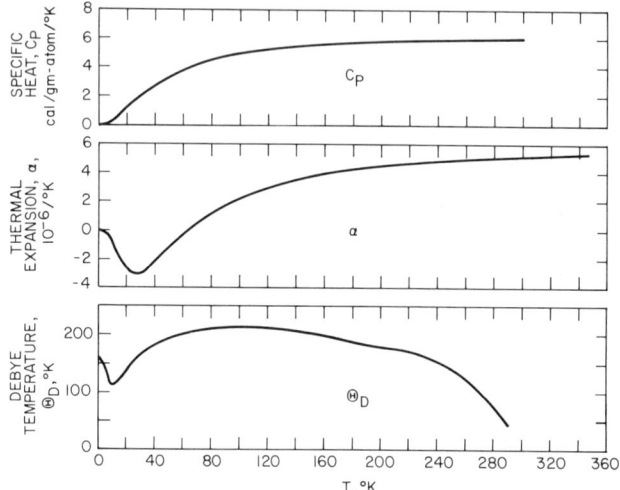

FIG. 2.14. (Top) Heat capacity at constant pressure C_p for CdTe using a smooth curve fit to the data of Birch (1975) below 25°K and of Rusakov et al. (1970) and Demidenko (1969) above 25°K. (Middle) Coefficient of linear thermal expansion α for CdTe using a smooth curve fit to the data of Smith and White (1975) below 35°K, of Browder and Ballard (1969, 1972) between 35°K and 300°K, and of Novikova (1960) and Williams et al. (1969) above 300°K. (Bottom) Debye temperature Θ_D for CdTe using a smooth curve fit of the data of Birch (1975) below 25°K and of Demidenko (1969) above 25°K.

White, 1975). The precise measurements of Smith and White (Fig. 2.14) show that the data below 4°K follow

$$\alpha = -(170 \pm 10)T^3/10^{12} \quad (°K^{-1}). \tag{2.14}$$

Negative thermal expansion is characteristic of the tetrahedrally coordinated compounds at low temperature. Kittel (1967) shows that a positive coefficient of expansion is due to anharmonic terms in the potential energy of a pair of atoms in a classical linear oscillator. For a positive coefficient of thermal expansion the Grüneisen factor

$$\gamma_j = d \ln v_j / d \ln V \tag{2.15}$$

is positive. Here v_j is the frequency of the jth mode and V is the molar volume. However, for a transverse mode in the linear chain, vibration results in an increase in the displacement between the atom and each of its two adjacent atoms. These forces on the adjacent atoms act to reduce their separation. The spring constant of the linear chain and the Grüneisen constant for this mode decreases. The value of the Grüneisen constant depends upon the weight of each of the modes and is given by

$$\gamma = 3\alpha V B_s / C_p, \tag{2.16}$$

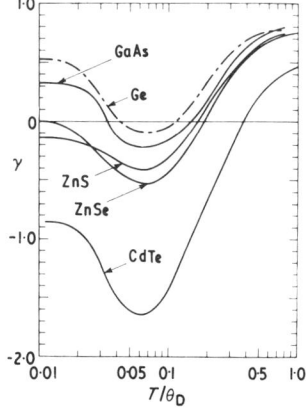

FIG. 2.15. Variation of the Grüneisen's constant γ, with reduced temperature T/Θ_D. The Debye temperature Θ_D has values of 374°K (Ge), 339°K (ZnS), 271°K (ZnSe), 345°K (GaAs), and 160°K (CdTe). (From Smith and White, 1975. Copyright by The Institute of Physics.)

where C_p is the specific heat at constant pressure and B_s is the adiabatic bulk modulus. Compared with other tetrahedral structures the Grüneisen parameter for CdTe (Fig. 2.15) is more negative. Smith and White relate the value at 0°K, γ_0, to the stability of the structure; i.e., the more negative γ_0, the more susceptible is the lattice to shear distortion. They show in a plot of γ_0 versus fractional ionicity as defined by Phillips (1968) that as the degree of ionicity decreases, γ_0 becomes more negative.

The specific heat at constant pressure and the Debye temperature θ_D are also shown in Fig. 2.14 as a function of temperature (Birch, 1975; Rusakov et al., 1970; Demidenko, 1969; Gul'tyaev and Petrov, 1959). The value of the Debye temperature in the zero temperature limit from the calculations of Birch (1975) is 160 ± 2°K and is in good agreement with values of 164 ± 5°K (Vekilov and Rusakov, 1971) and 162.7°K (Greenough and Palmer, 1973) calculated from the elastic constants. Plumelle and Vandevyer (1976) calculated the variation of the Debye temperature from neutron data using the rigid ion model. Their results are also in good agreement with experimental data. From specific heat data, moments of the phonon spectrum have also been calculated (Rusakov et al., 1970). Further calculations using neutron data should serve as a check on these moments as well as other lattice dynamic related phenomena.

IV. Band Structure

A. INTRODUCTION

The general shapes, location of extrema, and the degeneracies and anisotropies of the bands of many semiconductors became clear during the mid-1960's. In fact, until the absorption measurements of Marple (1966), the relative position of the conduction band and valence band extrema in **k**-space for CdTe was not in general agreement. The difficulties are related not only to the considerable calculations that are required after suitable approximation procedures are established, but more importantly to the fact that the crystal potentials are not well known. More recently, the bands have become better defined as a result of the use of high speed computers, refined experimental techniques and, most important of all, because of the use of the pseudopotential method. Extensive theoretical calculations using this approach were first performed for silicon and germanium. Only through these earlier studies has the band structure of the zinc blendes been solved. The purpose of this section is to summarize the methods used to determine the band structure of CdTe, culminating in the use of the pseudopotential method of Cohen and Bergstresser (1966).

Regardless of the approach used to determine the band structure, various approximations are made, the most important of those being the one-electron approximation where the total wave function of the system is represented by a combination of wave functions each of which involves the coordinates of only one electron. The field seen by a specific electron is that of the nuclei plus some average field produced by the charge distribution of all other electrons. The problem reduces to finding an appropriate Hamiltonian which is compatible with the periodic potential of the zinc blende structure. The Brillouin zone of the diamond cubic as well as the zinc blende structure is a truncated octahedron as shown in Fig. 2.16, the primitive translation vectors of the reciprocal lattice being

$$\mathbf{K}_1 = 2\pi \mathbf{a}^{-1}(1, 1, -1), \quad \mathbf{K}_2 = 2\pi \mathbf{a}^{-1}(1, -1, 1),$$

and

$$\mathbf{K}_3 = 2\pi \mathbf{a}^{-1}(-1, 1, 1). \tag{2.17}$$

The most important symmetry points are the center of the zone (Γ), the [111]-axes (Λ) and their interaction with the zone edge (L), and the [100]-

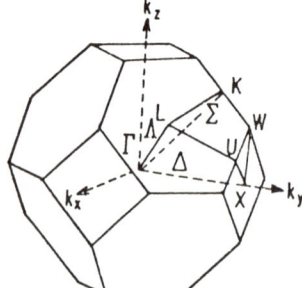

FIG. 2.16. The Brillouin zone for the zinc blende structure with lines and points of special symmetry.

IV. BAND STRUCTURE

axes (Δ) and their intersection (X). The [110]-axes (Σ) and points (U, W, K) along the intersecting planes are also defined.

The band structure may be approached from two extreme limits—the tight binding theory and the nearly free-electron theory. In the tightly bound approximation the potential in the neighborhood of the cores is strong compared with the potential in the intermediate regions, and the isolated uncoupled atoms provide a good first approximation to the solid. The energy of the electron is based upon a linear combination of atomic orbitals (LCAO) and represented by

$$\psi_k(\mathbf{r}) = \sum_j C_j(\mathbf{k})\phi(\mathbf{r} - \mathbf{R}_j), \quad (2.18)$$

where $\phi(\mathbf{r})$ is the wave function of the electron at position $\mathbf{r}$ for a free atom and $\mathbf{R}_j$ is a lattice vector. In this approximation an electron in the vicinity of a particular nucleus j is only slightly influenced by the presence of other atoms. The coefficient C_j is taken to be equal to $\exp(i\mathbf{k} \cdot \mathbf{R}_j)$ since the potential field is periodic and the wave satisfies the characteristics properties of a Bloch wave. For the alkali halides there is very little overlap of the wave function and a summation of $\psi_k(\mathbf{r})$ over nearest neighbors gives good results. However, for the III–V and II–VI compounds having the zinc blende structure, accurate results cannot be obtained without including interactions between quite distant atoms.

In the nearly free-electron model the wave function can be expressed as a Fourier series

$$\psi_k = \sum_g a_g \exp i(\mathbf{k} - \mathbf{g})\mathbf{r}, \quad (2.19)$$

where the electrons are only weakly perturbed by the periodic potential of the ion cores. In this approximation it is assumed that the potential energy $V(\mathbf{r})$ can be expressed by a constant and the first few terms of the expansion whose magnitude is small compared with the relevant kinetic energies. This is not a good approximation since the potential becomes very strong at small r inside the core and the model cannot simultaneously accommodate the atomic-like radial wiggles of ψ_k near the atomic sites and the long wavelength nature of ψ_k between the atomic sites. In principle, the energy $E(\mathbf{k})$ at some $\mathbf{k}$ could be calculated by forming the plane wave representation of the Schrödinger equation (Cohen and Heine, 1970):

$$\det|[\tfrac{1}{2}(\mathbf{k} - \mathbf{g})^2 - E]\delta_{gg'} + V_{gg'}| = 0, \quad (2.20)$$

where $\mathbf{gg'}$ are reciprocal lattice vectors and $V_{gg'}$ the $(\mathbf{g} - \mathbf{g'})$th Fourier component of $V(\mathbf{r})$. However, a secular equation on the order of $10^6 \times 10^6$ would be needed for convergence. The orthogonal plane wave (OPW) and the augmented plane wave (APW) are two methods used to solve the problem of the wiggles near the ion core. The wave function must be orthogonal to all the bound states which lie at a lower energy. In the OPW method this condition is met by the addition of suitable linear combinations of the bound state. The wave function is plane outside the core but now has the right shape on the core to provide a reasonable convergence of the secular equation [Eq. (2.20)]. In the APW method the lattice is separated into nonoverlapping spheres surrounding each atom and interstitial region. Inside the spheres the potential is assumed to be spherically symmetrical so that the radial wave equation satisfies the Schrödinger equation inside the sphere and joins correctly at the radius of the sphere the plane wave on the outside. Each of the APW's satisfy the Schrödinger equation within the sphere. A reasonably small sum of APW's will now smooth the kink at the radius of the sphere, represent the slowly varying wave function outside the sphere, and will again result in convergence of the secular equation with a reasonable number of terms.

B. Pseudopotential Approach

Other potentials or pseudopotentials have been generated which by making the transition smooth between the core and the interstitial region also eliminate the sharp oscillations of ψ_V inside the core and yet preserve the proper shapes of ψ_V outside the core. With these pseudopotentials the shape of ψ_V is now sufficiently simple so that the interaction of the outer electron with the ion and its surrounding of outer electrons is easily described. Replacing $V_{gg'}$ by V_{ps}, the secular Eq. (2.20) will now converge, the degree of convergence determined by the choice of V_{ps}. The potential is normally expressed as (Cohen and Bergstresser, 1966)

$$V(\mathbf{r}) = \sum_{|\mathbf{G}| \leq G_0} (S^S(\mathbf{G})V_G^S + iS^A(\mathbf{G})V_G^A)e^{-i\mathbf{G}\cdot\mathbf{r}}, \quad (2.21)$$

where $\mathbf{G}$ is a reciprocal lattice vector, $S(\mathbf{G})$ is the structure factor, and V_G is a pseudopotential form factor. For the II–VI compounds it is convenient to consider the pseudopotential to consist of a part which is symmetric (S) about the center of the bond joining the Cd and Te atom and a part which is antisymmetrical (A). If we take the origin of the coordinates to be halfway between these atoms the structure factors become

$$S^S(\mathbf{G}) = \cos \mathbf{G} \cdot \boldsymbol{\tau} \quad (2.22a)$$

and

$$S^A(\mathbf{G}) = \sin \mathbf{G} \cdot \boldsymbol{\tau}, \quad (2.22b)$$

where $\pm \boldsymbol{\tau} = \pm(\frac{1}{8}, \frac{1}{8}, \frac{1}{8})a$ and a is the lattice constant. The symmetric and antisymmetric form factors become

$$V_G^S = \tfrac{1}{2}[V_{Cd}(G) + V_{Te}(G)] \quad (2.23a)$$

TABLE 2.6

Comparison of the CdTe Form Factors Used in EPM

Reference	$V^S(3)$	$V^S(8)$	$V^S(11)$	$V^A(3)$	$V^A(4)$	$V^A(11)$
Cohen and Bergstresser (1966)[a]	−0.20	0.00	0.04	0.15	0.09	0.04
Saravia and Casamayou (1972)[a]	−0.202	0.0021	0.04	0.155	0.08	0.04
Chadi et al. (1972)[a]	−0.200	−0.012	0.027	0.168	0.075	0.028
Chelikowsky et al. (1973)[b]	−0.245	−0.015	0.073	0.089	0.084	0.006
Chelikowsky and Cohen (1976)[b]	−0.220	0.00	0.062	0.060	0.050	0.025

[a] Local calculations.
[b] Nonlocal calculations.

and

$$V_G^A = \tfrac{1}{2}[V_{Cd}(G) - V_{Te}(G)]. \quad (2.23b)$$

The appropriate coefficients for the fcc structure are $V^S(111)$, $V^S(220)$, $V^S(311)$, $V^A(111)$, $V^A(220)$, and $V^A(331)$. Form factors determined for the empirical pseudopotential model (EPM) are listed in Table 2.6. As a first approximation to the symmetric form factors, the potential of Sn is used as an average potential of the Cd and Te atoms. Cadmium and Te contain d states in the atomic cores and therefore for a more accurate picture of the electronic structure, it is necessary to include nonlocal effects (Chelikowsky and Cohen, 1976). Table 2.7 summarizes the approaches used to construct the band structure for CdTe.

TABLE 2.7

Approaches Used to Establish the CdTe Band Structure

Method	Author
Local empirical pseudopotential	Cohen and Bergstresser (1966)
	Cohen (1967)
Nonlocal empirical pseudopotential	Chelikowsky et al. (1973)
	Chelikowsky and Cohen (1976)
Orthogonal plane wave	Shay et al. (1967)
	Shay and Spicer (1967)
	Herman et al. (1967)
Korringa–Kohn–Rostocker	Treuch et al. (1967)
	Madelung and Treusch (1968)
	Eckelt (1967)
k · **p** perturbation	Cardona (1963)
	Pollak (1967)

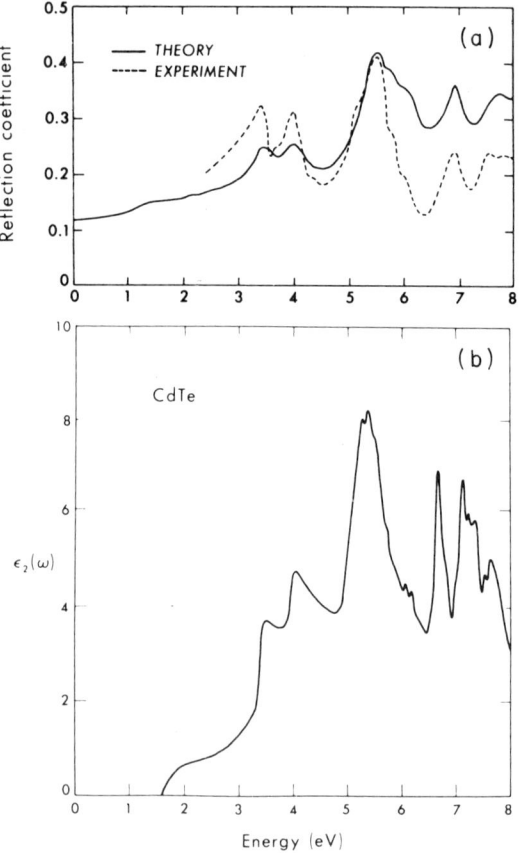

FIG. 2.17. (a) Experimental and theoretical reflectivity spectra (Chelikowsky and Cohen, 1976), and (b) imaginary part of the dielectric function for CdTe (Chadi et al., 1972).

C. Optical Properties

1. Reflectivity

Reflectivity measurements are important to the development of the band structure by the pseudopotential technique. The three antisymmetric form factors [Eq. (2.23b)] are adjusted to give band structure calculations [Eq. (2.20)] consistent with the principal features (i.e., fundamental gap, spin–orbit splitting, etc.) of reflectivity data (Fig. 2.17a). A direct, more detailed comparison between the calculated and experimental reflectivity permits further adjustment in the form factors. The reflection coefficient is given by

$$R(\omega) = \frac{[n(\omega) - 1]^2 + k^2(\omega)}{[n(\omega) + 1]^2 + k^2(\omega)}, \tag{2.24}$$

IV. BAND STRUCTURE

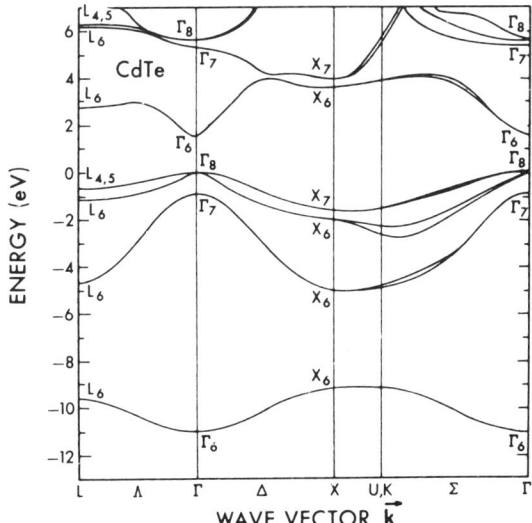

FIG. 2.18. Electronic band structure of CdTe in the principal symmetry directions using the nonlocal empirical pseudopotential method (Chelikowsky and Cohen, 1976).

where $n(\omega)$ and $k(\omega)$ are, respectively, the real and imaginary parts of $[\varepsilon_1(\omega) + i\varepsilon_2(\omega)]^{1/2}$. From the initial band structure the imaginary part of the dielectric constant $\varepsilon_2(\omega)$ can be calculated (Cohen, 1967; Cohen and Heine, 1970):

$$\varepsilon_2(\omega) = \frac{\hbar^2 e^2}{3\pi m^2 \omega^2} \sum_{c,v} \int d^3k |M_{cv}(\mathbf{k})|^2 \, \delta(E_c - E_v - \hbar\omega), \quad (2.25)$$

where $M_{cv}(\mathbf{k})$ is the momentum matrix element. The summation is performed over all conduction (c) and valence (v) bands. Figure 2.17 shows the reflectivity and $\varepsilon_2(\omega)$ using EPM form factors. The Kramers–Kronig analysis is then used to obtain the real part of the complex dielectric constant $\varepsilon_1(\omega)$ from $\varepsilon_2(\omega)$ (Cardona, 1965; Cohen, 1967; Pollak, 1967). The band structure of CdTe calculated by the nonlocal EPM is shown in Fig. 2.18.

The reflectivity method is simple and samples are easily prepared by cleaving crystals. This is in contrast to transmission measurements, which require extremely thin crystals that are difficult to prepare by standard polishing and grinding techniques. Pinhole-free samples for transmission measurements can also be prepared by thin film techniques, although they are not single crystalline. Table 2.8 summarizes the sources of data used in establishing the CdTe band structure.

A broad interest of the band structure about 1966 was timely since earlier optical and electrical studies led to erroneous conclusions as to the location

TABLE 2.8

OPTICAL AND PHOTOEMISSION STUDIES USEFUL IN
THE CONSTRUCTION OF THE CdTe BAND STRUCTURE

Method	Author
Reflectivity	Cardona (1961)
	Cardona and Greenway (1963)
	Chadi et al. (1972)
Electroreflectance	Cardona et al. (1967)
Transmission	Cardona and Harbeke (1962)
Polarimetry	Marple and Ehrenreich (1962)
Photoemission	Shay et al. (1967)
	Shay and Spicer (1967)
	Ley et al. (1974)
	Eastman et al. (1974)
Electrical measurements	Segall et al. (1963)
Faraday rotation	Marple (1963)
Cyclotron resonance	Kanazawa and Brown (1964)
Piezoresistance	Sagar and Rubenstein (1966)

in k space of the absolute extrema of the valence and conduction bands (Yamada, 1962; Davis and Shilladay, 1960; Koňák, 1963; Spitzer and Mead, 1964). The established model shows the conduction band minimum and valence band maximum to occur at $\mathbf{k} = 0$, classifying CdTe as a direct gap material.

Reasonable agreement between the calculated and experimental $R(\omega)$ curves provides a firm basis for interpretation of the band structure of CdTe. The valence band is associated with the 5s and 5p atomic orbitals of Cd and Te. Neglecting spin, the valence band is composed of four subbands. The fourth or lowest valence band state is the Γ_1 level primarily associated with s-orbitals of Te. The three upper bands are degenerate at $\mathbf{k} = 0$ and form the upper edge of the band. The higher valence band states consist largely of the p-electron states of Te and a small admixture of s-electron states of Cd. For $\mathbf{k} \neq 0$ the threefold degeneracy at the top of the band is split into a twofold degenerate and a nondegenerate band. When spin is taken into account, each band is doubled. At $\mathbf{k} = 0$ the $p_{3/2}$ level is fourfold degenerate (Γ_8-symmetry), the four states corresponding to m_j values of $\pm\frac{3}{2}$ and $\pm\frac{1}{2}$. The $p_{1/2}$ level is doubly degenerate (Γ_7-symmetry) with $m_j = \pm\frac{1}{2}$. The $p_{3/2}$ states are higher in energy than the $p_{1/2}$ states by the spin–orbit energy. For $\mathbf{k} \neq 0$ the Γ_8 term splits into two bands, each of the bands having different curvatures and consequently different effective masses for holes. The lowing-lying conduction band states are also associated with s and p orbitals which are heavily concentrated about the cations.

IV. BAND STRUCTURE

Papers of Chadi *et al.* (1972), Saravia and Casamayou (1972), and Chelikowsky and Cohen (1976) include a critical point analysis of the $\varepsilon_2(\omega)$ and $R(\omega)$ curves (Fig. 2.17). The major features of $\varepsilon_2(\omega)$ are a threshold at 1.59 eV and about six major peaks up to 8 eV. The threshold corresponds to the $\Gamma_8 - \Gamma_6$ transition between bands 4 and 5 in Fig. 2.18 and also corresponds to the first peak in theoretical $R(\omega)$. The spin–orbit splitting separates band 2 by 0.91 eV from the two upper valence bands. Structure due to spin–orbit splitting is outside of the range of measurements of Chadi *et al.* and are only weakly resolved in polarimetry (Marple and Ehrenreich, 1962), thermoreflectance (Matatagui *et al.*, 1968), and other reflectivity (Cardona and Greenway, 1963) measurements. However, the splitting is beautifully demonstrated in the electroreflectance measurements (Cardona *et al.*, 1967; Tyagai *et al.*, 1973). The next two peaks in $R(\omega)$ and the first two peaks in $\varepsilon_2(\omega)$ are due to spin–orbit split transitions at L or along Λ between bands 4 and 5 and 3 and 5. The calculated values of 3.42 eV and 3.97 eV agree well with the experimental values of 3.49 eV and 4.04 eV, respectively. Splitting near L ($\Delta_1 = 0.57$ eV) is in good agreement with earlier data (Cardona *et al.*, 1967; Cardona and Greenway, 1963; Marple and Ehrenreich, 1962; Matatagui *et al.*, 1968).

Details of these and other transitions are itemized in Table 2.9. A similar analysis has been done by Saravia and Casamayou (1972). A comparison

TABLE 2.9

Theoretical and Experimental Reflectivity Structure for CdTe and Their Identifications, Including the Location in the Brillouin Zone, Energy, and Symmetry of the Calculated Critical Points[a]

CdTe reflectivity structure (eV)		Associated critical points location in zone	Symmetry	Critical-point energy (eV)
Theory	Experiment			
1.65	1.59	$\Gamma_8^v - \Gamma_6^c$ (0.0, 0.0, 0.0)	M_0	1.59
3.49	3.46	$L_{4,5}^v - L_6^c$ (0.5, 0.5, 0.5)	M_1	3.47
4.04	4.03	$L_6^v - L_6^c$	M_1	4.00
5.16	5.18	$\Delta_5^v - \Delta_5^c$ (0.5, 0.0, 0.0)	M_1	5.14
5.50	5.53	Plateau near (0.75, 0.25, 0.25)	—	
5.68	5.68	$\Delta_5^v - \Delta_5^c$ (0.75, 0.0, 0.0)	M_1	5.58
6.00	5.95	$\Delta_5^V - \Delta_5^c$	M_1	5.96
6.91	6.82	$L_{4,5}^v - L_6^c$	M_1	6.83
—	7.44			
7.79	7.6	$L_6^v - L_{4,5}^c$ (0.5, 0.5, 0.5)	M_1	7.53

[a] Chelikowsky and Cohen (1976).

of these critical point analyses shows that the transition energies and their location within the Brillouin zone differ. The enhancement of the experimental peaks on the reflectivity spectrum are attributed to excitons (Phillips, 1966; Kane, 1969). Since excitons were not considered in the theory, a more detailed analysis may not be justifiable. Excitons are discussed in a later section.

2. *Photoemission*

Absolute assignment of all transitions interpreted from reflectivity measurements is impossible. Photoemission measurements (Shay et al., 1967; Shay and Spicer, 1967; Eastman et al., 1974; Ley et al., 1974) are very helpful in that they determine the absolute energies between levels; optical measurements determine only energy differences. The extra degrees of freedom gained by varying the work function and measurement of the energy distribution allows one to fix the levels of the conduction band on an absolute energy scale as well as to determine energy differences between the conduction and valence bands. The bandwidth and critical point positions with respect to the valence band edge for the valence band have been determined using photoemission densities of states (PED), derived from photoemission spectra, and are shown in Table 2.10. The photoemission density of states is in good

TABLE 2.10

Experimental and Theoretical Valence-Band Positions (in eV below the Valence-Band Edge) for CdTe[a]

Approach	L_3	$\sum_{1\,min}$	X_3	X_1	Γ_1	Reference
Photoemission ($h\nu = 24$ eV)	0.7 ± 0.2	2.8 ± 0.2	4.7 ± 0.2	8.8 ± 0.3	—	Eastman et al. (1974)
Photoemission (X-ray)	0.9 ± 0.3	2.7 ± 0.3	5.1 ± 0.2	—	—	Ley et al. (1974)
Photoemission ($h\nu < 11.6$ eV)	0.7	1.9	3.1 (L_1)	—	—	Shay et al. (1967)
Theory (EPM)[b]	0.4	1.9	2.9	11.9	12.4	Cohen and Bergstresser (1966)
Theory (EPM)[b]	0.4	2.5	3.15	—	—	Chadi et al. (1972)
Theory (EPM)[c]	0.9	2.7	5.2	9.1	11.1	Chelikowsky and Cohen (1976)
Theory (ROPW)	0.75	2.3 (3/2 X_5)	3.9	9.6	10.4	J. P. Van Dyke and F. Herman, ARL Rep. No. 69–0080, Sec. 19

[a] Eastman et al. (1974).
[b] Local.
[c] Nonlocal.

IV. BAND STRUCTURE

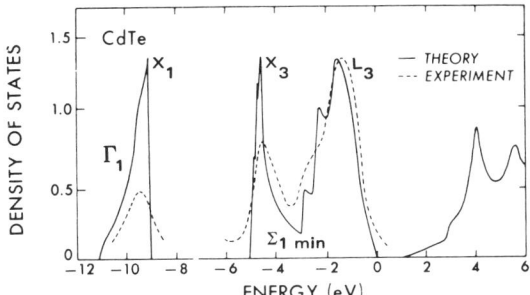

FIG. 2.19. Comparison of the photoemission density of states by Eastman *et al.* (1974) (dashed lines) with the corresponding empirical pseudopotential method calculation density of states with nonlocal terms added to the pseudopotential by Chelikowsky and Cohen (1976) (solid lines).

agreement with the density of states calculated by the nonlocal EPM (Fig. 2.19). Ultraviolet photoemission spectroscopy (UPS) and x-ray photoemission spectroscopy (XPS) confirm that nonlocal terms must be added to the pseudopotential. Comparable fits to the PED data are obtained by OPW calculations and nonlocal EPM calculations. Photoemission studies on CdTe were obtained with 24 eV photons for UPS and 1487 eV photons for XPS. They give an overview of the bandwidth and are not based upon any model. Previous photoemission studies (Shay *et al.*, 1967), limited to lower photon energies, i.e., $hv < 11.6$ eV, were concerned with states near the gap and were analyzed in the light of the work of earlier optical data.

The shape of the density curves relate to the ionicity of the crystal and chemical bonding. As one goes from the covalent homopolar group IV compounds to the heteropolar III–V compounds and then to the more ionic II–VI and I–VII compounds, the gap widens between the two lowest valence bands $(X_3 - X_1)$ and the width of bands two through four $(E_v - E_3)$ decreases. This is also evident in photoemission results (Eastman *et al.*, 1974) in Fig. 2.20 where $(E_v - \sum_{1\min})$ and $(X_3 - X_1)$ are plotted as a function of increasing ionicity for the Ge, GaAs, and ZnSe row in the periodic table and for Sn, InSb, CdTe, and AgI in the row below it. The parameter $X_3 - X_1$ increases with ionicity and is approximately equal to C (Fig. 2.7) as defined in Eq. (2.3).

3. *Pressure Effects*

Measurements of the shift of the transition energies with hydrostatic pressure are helpful in identifying the symmetries of the lowest conduction bands (Paul, 1968). Similar coefficients support the argument that transitions are due to a particular point in **k**-space. The pressure coefficients for CdTe have been measured by absorption (Babonas *et al.*, 1965), reflection (Langer,

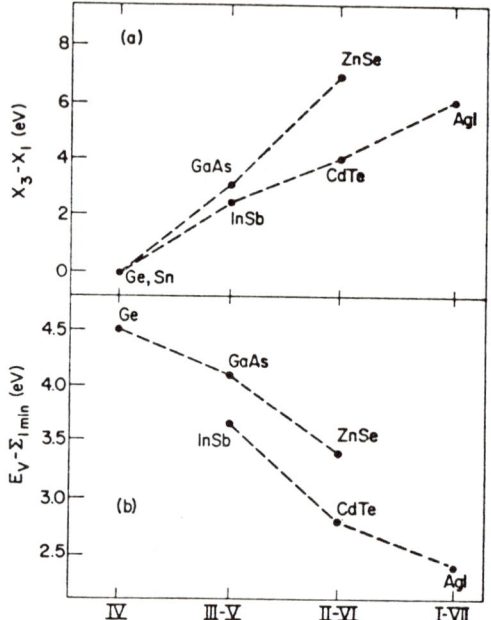

FIG. 2.20. (a) the ionicity gap $(X_3 - X_1)$ and (b) width of the upper valence bands $(E_v - \Sigma_{1\,\mathrm{min}})$ as a function of valence difference for the constituents of compounds in rows 4 and 5 of the periodic table (Eastman *et al.*, 1974).

1964), and electroreflection (Babonas *et al.*, 1971) and are shown in Table 2.11. Good agreement is found for values at the fundamental band gap. Calculated pressure coefficients using the dielectric theory of the chemical bond are in good agreement with experiments (Camphausen *et al.*, 1971). Values of 0.8×10^{-6} eV/bar and 0.7×10^{-6} eV/bar have also been measured by Babonas *et al.* (1971) and Mertz (1969), respectively, for the pressure coefficient of the spin-orbit splitting at Γ. Using a volume compressibility of 2.38×10^{-12} cm²/dyn (McSkimin and Thomas, 1962; Cline and Stephens, 1965), Langer (1964) and Babonas *et al.* (1971) both calculated values of -3.4 eV for the dilation coefficients of the fundamental band gap. Langer also calculated coefficients of -2.7 eV and -2.8 eV for the 3.4 eV and 4.0 eV peaks, respectively, and assigned them to transitions at L. His assignment is different from that by others (Babonas *et al.*, 1971; Chadi *et al.*, 1972) along Δ. However, similar coefficients for these peaks support the argument that the peaks are due to transitions between the split-valence bands and the conduction band at only one point in **k**-space. Both the pressure coefficient and the dilatation coefficients for the 3.4 eV peak measured by Babonas *et al.* are different from those of Langer.

TABLE 2.11

Hydrostatic Pressure Coefficients dE/dP (10^6 eV/bar) in CdTe

	Transition energy				
	1.5 eV	2.4 eV	3.4 eV		4.0 eV
Investigator	(Γ)	(Γ)	(Λ)	(L)	(L)
Edwards and Drickamer (1961)	4.4[a]	—	—	—	—
Langer (1964)	8.0	—	—	6.4	6.7
Babonas et al. (1965)	7.8	—	—	—	—
Babonas et al. (1971)	7.9	8.7	4.5	—	—
Camphausen et al. (1971)[b]	8.1	—	3.9	—	—

[a] Possible nonhydrostatic loading of sample.
[b] Theory.

4. *Free Carrier Absorption*

At longer wavelengths and for moderate to large carrier concentrations the absorption spectra are characteristics of free carrier absorption (FCA). In this region the absorption coefficient is nearly proportional to the free electron concentration and increases with λ^n, where n characterizes the absorption mechanism and is discussed in more detail in Section VI of this chapter under Transport Properties. Figure 2.21 is a plot of the absorption coefficient versus wavelength at room temperature for various carrier concentrations (Strauss and Iseler, 1974). For wavelengths greater than 1.4 μm absorption is characteristic of intraband FCA. Transitions from the valence to the conduction band at $\mathbf{k} = 0$ determine the absorption coefficient for wavelengths less than 0.95 μm. Absorption characteristic of the intrinsic process is shown for the sample having the lowest carrier concentration. In this sample FCA is negligible even for longer wavelengths. Between 0.95 μm and 1.4 μm the measured absorption in the higher carrier concentration sample is significantly higher than the intraband FCA extrapolated from longer wavelengths (Strauss and Iseler, 1974). This additional absorption is believed to be partly associated with interband electron transitions from the Γ minimum to a higher conduction band minimum. An analysis of FCA in III–V compounds shows this to be a reasonable interpretation. A comparison of the shapes of the absorption coefficients versus photon energy between CdTe and InP indicates that the separation between the Γ minimum and the next higher conduction band is about 0.9 to 1.0 eV. This is about 0.4 eV less than predicted by Fig. 2.18. Vul *et al.* (1970) analyzed the infrared absorption spectra of p-type CdTe at 300°K and attributed other than λ^n behavior due transitions from the heavy- to the light-hole subband by the selective absorption of free carriers.

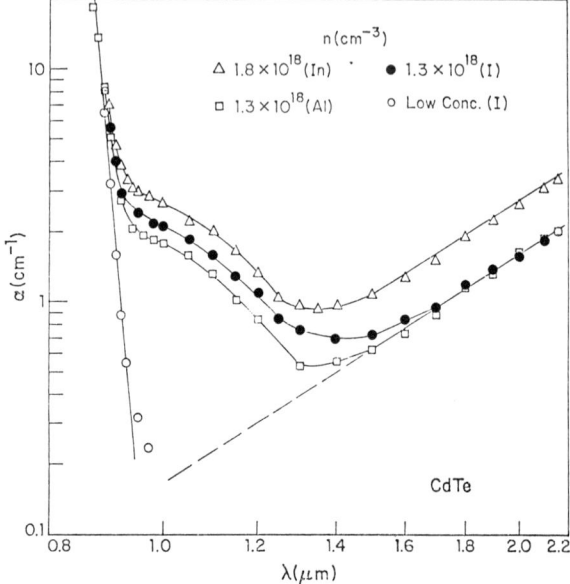

FIG. 2.21. Absorption mechanisms for donor-doped CdTe. The absorption coefficient is characteristic of free carrier absorption (FCA) at the longer wavelengths, of interband FCA the intermediate wavelengths and intrinsic absorption at the shorter wavelengths (Strauss and Iseler, 1974).

Reflectivity, photoemission, intraband free carrier absorption, and pressure-dependent transitions are all phenomena which help describe the electronic band structure. A second group of optical properties, luminescence, absorption, the Faraday effect, and piezobirefringence, besides being helpful in describing band structures, strongly relate to excitons and lattice defects and will consequently be used in their description. The following section will discuss excitons; Chapter 3 will discuss lattice defects.

V. Excitons

A. Introduction

In discussing interband transitions in the last section, little reference was made to exciton effects. Wannier (1937) treats the exciton as electron and hole wave packets taken out of states from the conduction and valence bands. The exciton may also be considered as a neutral electron–hole pair which is bound by a Coulombic potential and is free to migrate through the lattice. It may be formed by exciting a valence electron to a state just below the conduction band by the absorption of a photon. The II–VI compounds

V. EXCITONS

are rich in exciton-associated transitions because they are direct gap materials; consequently, such transitions do not require extensive phonon cooperation. Because the physical separation between the electron and the effective positive charge is large relative to atomic distances, the electron "sees" the positive charge and the average lattice potential rather than the positive charge and the details of the lattice potential. The ionization energy for such a system is

$$E_x = (-m_r^* q^4/2h^2 \varepsilon(0)^2)(1/n^2), \tag{2.26}$$

where q is the electronic charge, and n is an integer greater than or equal to one, depending upon whether the exciton system is in the ground state or in an excited state. The term m_r^* is the reduced mass given by

$$(1/m_r^*) = (1/m_e^*) + (1/m_h^*). \tag{2.27}$$

Here m_e^* and m_h^* are, respectively, the electron and hole effective masses. For CdTe the electron (Marple, 1963; Kanazawa and Brown, 1964) and hole (Segall and Marple, 1967) effective masses are $0.1m$ and about $0.4m$. Because the reduced mass is less than the effective mass of an electron the binding energies of the exciton is lower than the hydrogenic donor or acceptor binding energies. It is customary to use the conduction band edge as a reference energy level for the completely disassociated ($n = \infty$) exciton.

B. REFLECTIVITY MEASUREMENTS

A satisfactory understanding of excitons in the II–VI compounds is due to the measurements on the wurtzite compounds, CdS (Thomas and Hopfield, 1959), and ZnO (Thomas, 1960). Reflectivity measurements were first used to study the optical properties of excitons in CdTe (Thomas, 1961). More precise reflectivity data were obtained by Marple (1967) and are shown in Fig. 2.22. Peaks at 1.596 eV and 1.603 eV correspond to the ground state and first excited state of the exciton, respectively. Applying Eq. (2.26) to the difference of these energies and using a value of 9.8 for the static dielectric constant results in a value of 0.071 for the reduced mass. The sum of the binding energy of the ground state, 0.010 eV and the energy of the ground state yields a band gap energy of 1.606 eV at 1.6°K. The binding energy in Eq. (2.26) refers to crystals having simple valence and conduction bands. However, including spin, the valence band is sixfold degenerate and is split by spin–orbit interaction into an upper fourfold and a lower twofold degenerate band separated by the spin–orbit splitting. Considering the entire upper band Balderschi and Lipari (1970) calculated the binding energy of the exciton to be 0.0117 eV. They neglected the effects of the split-off band because the spin–orbit splitting energy is much larger than the exciton binding energy.

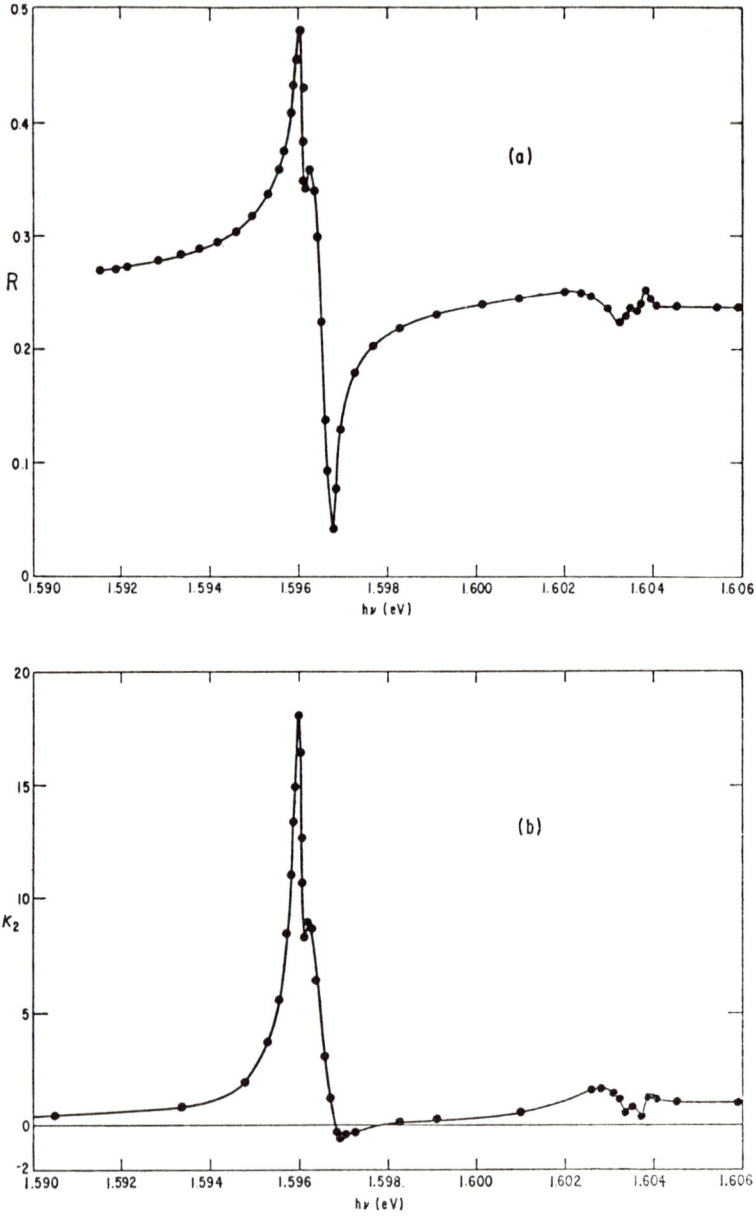

FIG. 2.22. (a) Reflectance for near-normal incidence R as a function of photon energy $h\nu$ as obtained at 2°K by a photoelectric method from the cleaved face of a zone-refined CdTe single crystal. (b) Imaginary part of the dielectric constant $\varepsilon_2(\omega)$ obtained by Kramers–Kronig inversion of reflectance data (Segall and Marple, 1967).

V. EXCITONS

The ground ($n = 1$) and excited ($n = 2$) states of excitons have also been observed in electroreflection at 77°K at energies of 1.584 eV and 1.593 eV (Vavilov et al., 1969). These results are in reasonable agreement with the exciton positions reported by Thomas (1961) and Marple (1967). Cardona et al. (1967) associate the structure of their lowest energy electroreflectance peak with a transition at the fundamental band edge; it is likely that the structure of Cardona et al. is associated with excitons. The role of excitons in reflectivity is important and should be considered in the determination of the band structure.

Experimental oscillator strengths of 6×10^{-4} (Segall and Marple, 1967) and 8.5×10^{-4} (Thomas, 1961) per molecule for the ground state agree with the calculated value of 6.5×10^{-4} (Segall and Marple, 1967). The oscillator strength for the ground state is about an order of magnitude larger than that of the first excited state.

Thomas applied uniaxial compressive stress to single crystals of CdTe and examined the exciton spectrum in reflection. The uniaxial stress splits the degenerate $J = \frac{3}{2}$ valence band at $k = 0$ into two bands, one with $M_j = \pm\frac{3}{2}$ and the other with $M_j = \pm\frac{1}{2}$. From the shifts and splittings of the excitons with stress and from the elastic constants the deformation potentials were determined. The splitting was identical for stress applied in any direction in the (110) plane. Therefore, one rather than two deformation potentials can describe the splitting, although the material is elastically anisotropic.

Excitons would be expected to be found at $\mathbf{k} = 0$ and other positions where the critical point relation

$$\nabla_\mathbf{k} E_c(\mathbf{k}) = \nabla_\mathbf{k} E_v(\mathbf{k}) \tag{2.28}$$

is satisfied. Here $E_c(\mathbf{k})$ refers to the conduction band and $E_v(\mathbf{k})$ refers to the valence band. Cardona and Harbeke (1962) made transmission measurements and found good agreement between their peaks and reflectivity peaks at 3.53 eV and 4.07 eV (Cardona, 1961). Observing the peaks to sharpen with decreasing temperature they associated this structure at L or along Λ as in Fig. 2.17, with excitons rather than interband transitions. Correlating precise reflectivity measurements with a rather special band structure, Marple and Ehrenreich (1962) associated this structure with interband transitions. This conflicting interpretation stimulated further controversy. In a review of fundamental optical spectra Phillips (1966) emphasizes that line shape constitutes the best test for resonance behavior. The sharp upper edge at 3.5 eV followed by the dip is characteristic of hyperbolic or "saddle point" excitons. Duke and Segall (1966) contest this interpretation. Similar but more precise calculations by Kane (1969) supports the interpretation that the structure is indeed due to Coulombic effects.

C. Absorption at the Fundamental Gap

The importance of excitons is also demonstrated in absorption. Figure 2.23 shows $\log I_0/I$ over the energy range 1.5–4.5 eV in a 0.25 μm-thick thin film at 80°K. Here I_0 and I are the incident and transmitted intensities, respectively. Detailed measurements at 2.1°K just below the fundamental band edge with a 74 μm thick single crystal are included as an insert. The absorption coefficient rises rapidly at 1.590 eV to a peak value near 1.596 eV which is associated with the ground state ($n = 1$) for direct exciton absorption. The shape of the curve in this region could not be determined because of the effectiveness of the absorption process. Photon energies for direct exciton absorption by Thomas (1961)(a) and Spitzer and Kleinman (1961)(b) are indicated. A transmission band at about 1.600 eV corresponding to photon energies between the ground and excited states is apparent. The absorption coefficient rises rapidly from about 600 cm^{-1} in the valley to about 10^5 cm^{-1} at 2.5 eV.

The shape of the absorption curve, especially near the band edge, is extremely sensitive to surface preparation and the quality of the crystal. Data for the insert of Fig. 2.23 were taken with a chemically polished sample cut from the purest section of a zone-refined ingot. From about 1.591 eV to 1.594 eV the shoulders have been associated with the creation of excitons

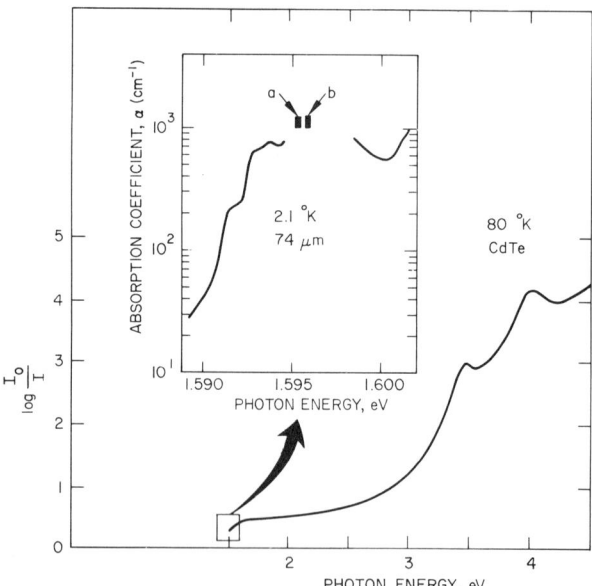

FIG. 2.23. Logarithm of the ratio of incident to transmitted intensities in CdTe thin films at 80°C (Cardona and Harbeke, 1963). The insert shows single crystal absorption measurements just below the fundamental band edge at 2.1°K (Marple, 1966).

bound to lattice defects or chemical impurities and have been found to coincide in energy with fluorescence measurements (Halsted et al., 1961). Hall measurements indicated these samples to be n-type with donor and acceptor concentrations of 1.05×10^{15} cm^{-3} and 6.7×10^{14} cm^{-3}, respectively. Converting this sample to p-type by thermal annealing results in a higher absorption coefficient and an absorption coefficient that is less steep with photon energy and exhibits new shoulders at 1.55 eV and 1.57 eV. Lorenz and Segall (1963) suggest that the structure is associated with transitions from a shallow acceptor state at $(E_v + 0.05)$ eV to the conduction band. Agrinskaya et al. (1970a) associate peaks at lower energies also with defects. The 1.57 eV acceptor to conduction band transition is assisted by the emission of a longitudinal optical phonon. Studies with intentionally doped samples and earlier studies (Davis and Shilliday, 1960; Koňák, 1963) show higher values of α for a fixed photon energy and a less steep rise in α with photon energy. However, these authors did not take into account effects caused by impurities or phonon assisted transitions. Brodin et al. (1970) show that donor impurities shift the absorption spectra towards longer wavelengths. Besides chemical purity of the samples, Marple suggests that surface preparation may play a role in the difference between his results and the results of earlier workers.

Segall (1968) also measured the thermal width of the ground state exciton peak. The width is about 1 meV up to 50°K, with a slight increase over this temperature interval due to interaction with acoustic phonons. Above this, temperature broadening is exponential with temperature and is associated with the longitudinal optical phonons. The interaction between excitons and lattice vibrations is important in the interpretation of optical data.

At higher temperatures phonons play an especially important part in the energy transfer process and along with excitons quantitatively explain the shape of the absorption edge in CdTe. Taking advantage of the availability of the careful and near intrinsic measurements of Marple, Segall (1966) demonstrated that optical absorption in the edge region for photon energies just below the first exciton peak is due to the creation of direct excitons with the simultaneous absorption of one and two longitudinal optical phonons. This process is illustrated in Fig. 2.24.

Perturbation theory quantitatively accounts for the magnitude and the temperature and energy dependence of the absorption coefficient. The method of calculating the absorption coefficient is described in Segall's paper. Basically it requires the total Hamiltonian of the system to consist of unperturbed (H_0) and perturbed (H') parts:

$$H_0 = H_x + H_L + H_R, \tag{2.29}$$

$$H' = H_{eL} + H_{eR}, \tag{2.30}$$

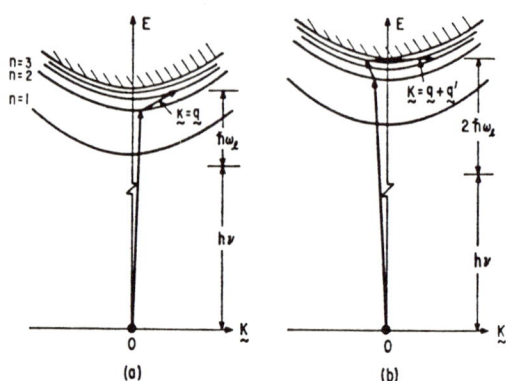

FIG. 2.24. Schematic representation of the one-phonon (a) and two-phonon (b) assisted "direct" exciton optical absorption processes. The dot at $E = 0$ represents the ground state, the parabolas the discrete exciton bands, and the hatched area the continuous spectrum. The smaller arrows between the bands symbolize the "scattering" of the intermediate state excitons by the LO phonons. (Marple, 1966).

where each of the terms are Hamiltonians describing the exciton H_x, the lattice vibrations H_L, the radiation field H_R, the electron–radiation coupling H_{eR}, and the electron–lattice coupling H_{eL}. The density of states are determined and transition probabilities then calculated. The schematic representation of this process (Fig. 2.24) consists of the crystal being excited from the ground state to an intermediate state at $K \approx 0$ by the annihilation of the photon. The exciton is then scattered into the second intermediate state ($K = q$) by the absorption of the phonon of wave vector q; it is subsequently scattered into the final state ($K = q + q'$) by the absorption of the phonon q'. Data and theory for the two phonon process are plotted in Fig. 2.25 as α/N^2 versus $(E_{x1} - h\nu)$ to remove the temperature dependence. Here $N^{-1} = \exp[(h\nu/kT)^{-1}]$. Agreement between theory and experiment is good using a hole mass of 0.4m. The exact expression for α is complicated and is given in Segall's paper. These results are strong evidence of direct band gap transitions.

In semiconductors there is generally an overall phonon-associated shift in the position of the band edge with temperature. It is generally observed that exciton absorption lines shift toward lower energies with increasing temperature and this is certainly true of CdTe. Segall observed that the absorption coefficient has a nearly exponential dependence on $(E_{x1} - h\nu)$ for $170°K > T > 130°K$ in agreement with that predicted by Urbach's rule:

$$\alpha = \alpha_0 \exp[-(E_0 - h\nu)/\gamma kT], \qquad (2.31)$$

where α_0 and γ are constants and E_0 is a constant of the order of magnitude

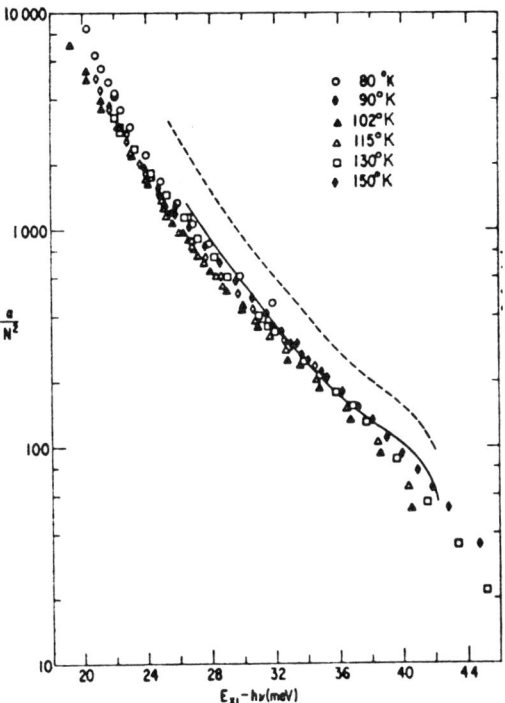

FIG. 2.25. A comparison of data and the theory for absorption by the two-phonon process near the band edge. Data are plotted as α/N^2 versus $E_{x1} - h\nu$ where α is the absorption coefficient, $N = \exp[(h\nu/kT) - 1]$ and E_{x1} is the ground state energy of the exciton. The solid curve represents the calculated result for $m_h = 0.4$ while the dashed curve that for $m_h = 0.5$ (Segall, 1966).

of the peak absorption energy. Segall suggests that for the partially polar compounds such as CdTe, Urbach's rule is intimately related to the two-phonon process described above.

Williams and Schnatterly (1975) looked further into the exponential broadening of the absorption edge by inducing magnetic circular dichroism in CdTe. The difference in the absorption of right and left circularly polarized light ($\Delta\alpha$) was examined at different temperatures on the long wavelength side of the ground state exciton energy. For lower energies $\Delta\alpha/\alpha$ displays rigid-shift behavior increasing with increasing photon energy. However near the exciton energy $\Delta\alpha/\alpha$ decreases and even changes sign. This suggests that the Urbach edge in CdTe is the sum of two exponential edges having opposite magnetic circular polarizations. The role of excitons in the interband Faraday effect has been discussed (Ebina *et al.*, 1965; Karmazin and Miloslavskii, 1971; Kireev *et al.*, 1972a–c; Karmazin *et al.*, 1973). Sign reversal of the

Faraday rotation has been found (Ebina et al., 1965; Karmazin et al., 1973) and is associated with a negative g value for the exciton. On the basis of those results Williams and Schnatterly associate the Urbach edge with comparable broadening of both the exciton peak and the interband edge. Alternatively, they suggest their behavior to be due to competition between light- and heavy-hole masses each having a different broadening mechanism. Segall's model favors the heavy-hole mass and a dominant right magnetic circular polarization, the electric microfield mechanism favors the light-hole mass and a left magnetic circular polarization.

D. TEMPERATURE DEPENDENCE OF FUNDAMENTAL GAP

Because of the difficulty in precisely defining the band edge, it is not straightforward to determine its temperature dependence. Analyzing the ground state of the exciton is the best method of arriving at the band gap. Since the exciton binding energy is much smaller than the band gap, the temperature shift of the exciton should result primarily from the change in the band gap with temperature. Although the value of 1.606 eV at liquid helium is well established, a wide range of values is found at 300°K. The techniques used to characterize the band edge, its value at 300°K, and the temperature coefficient are all summarized in Table 2.12. Many of the measurements are made below room temperature. An extrapolation is used to obtain the band gap at 300°K.

With increasing temperatures excitons are difficult to observe by conventional measurements because of phonon broadening and screening by an increased free carrier density and lattice defects. Absorption, conductivity and photoconductivity measurements are not precise methods to determine the band gap since results must be interpreted indirectly through the interaction of optical phonons with the exciting radiation or even more remotely through the additional interaction of optical phonons with charge carriers. Modulation techniques have been useful in achieving greater resolution. Electroreflectance (Cardona et al., 1967; Babonas et al., 1971), piezoreflectance (Camassel et al., 1973), thermoreflectance (Matatagui et al., 1968), electroabsorption (Babonas et al., 1968), photoreflection (Lisitsa et al., 1974), and the Faraday effect (Zvara et al., 1966; Ebina et al., 1965; Karmazin and Miloslavskii, 1971; Karmazin et al., 1973; Kireev et al., 1972b) are all modulation techniques that have been used to interpret optical data at room temperature. With the interband Faraday effect (Karmazin and Miloslavskii, 1971; Karmazin et al., 1973) excitons have been analyzed at room temperature to determine the forbidden band width of CdTe (1.547 eV), the position (1.538 eV) and half width (0.04 eV) of the exciton peak, and the g factor of the Zeeman splitting of the exciton level (-0.81 eV). Reliable values of the band edge at 300°K, as determined by analysis of the $n = 1$ peak in piezo-

V. EXCITONS

TABLE 2.12

TEMPERATURE DEPENDENCE AND 300°K VALUE OF THE CdTe BAND GAP

Reference	Technique	E_g(eV) (300°K)	$-dE_g/dT$ (10^{-4} eV/°K)
Miyasawa and Sugaike (1954)	Photoconductivity	1.47	4.5
Bube (1955)	Photoconductivity	1.41	3.6[b]
Van Doorn and de Nobel (1956) de Nobel (1959)	Absorption and photovoltaic effect	1.5	2.3 to 5.4[c]
Davis and Shilliday (1960)	Absorption	1.5	5.6
Yamada (1962)	Absorption	1.44	4.4
Konak (1963)	Absorption	1.39	4.1
Spitzer and Mead (1964)	Photovoltaic effect	1.505	5
Segall (1966)	Reflectance	1.528	3.0[d]
Ludeke and Paul (1967)	Electroreflectance	1.5[a]	2.8
Matatagui et al. (1968)	Thermoreflectance	1.5[a]	3.5
Smith (1970)	Hall effect		1.9[e]
Babonas et al. (1971)	Electroreflectance	1.495[a]	4.2
Karmazin and Miloslavskii (1971)	Absorption	1.547	1.97
Camassel et al. (1973)	Piezoreflectance	1.529	3.1
Tsay et al. (1973)	EPM calculation	1.48	4.2

[a] Assignment of peak position and not necessarily E_g.
[b] Measured below 320°C.
[c] Upper value measured at 800°K.
[d] Extrapolation.
[e] Measured over the temperature range 475°C–670°C.

reflectance (Camassel et al., 1973) and in reflectance (Segall, 1966) are, respectively, 1.529 eV and 1.528 eV. In these analyses the temperature dependence of the exciton was determined from 70 to 200°C, its position extrapolated to 300°K, and its 0.010 eV binding energy added to this value. The nonlinearity of the band gap above 78°K may be significant resulting in further uncertainty in its value above room temperature if determined by extrapolation. Absorption measurements show $|dE/dT|$ to increase from 2.3×10^{-4} eV/°K at 77°K to 5.4×10^{-4} eV/°K at 800°K (Van Doorn and de Nobel, 1956; de Nobel, 1959). The value of the band gap at approximately 700°C (1.18 eV) is in good agreement with that found by Hall measurements (1.13 eV) (Smith, 1970). High temperature conductivity measurements appear to overestimate the value of the band gap (Höschl, 1966; Zanio, 1969b; Rud' and Sanin, 1969b).

The temperature dependence of the Γ and Λ transitions has been further separated into electron–phonon and thermal expansion components (Babonas et al., 1971). The electron–phonon interaction dominates the expansion process for both transitions. Using Debye–Waller (DW) factors

(Vetelino et al., 1972) within the framework of the emperical pseudopotential method, Tsay et al. (1973) confirmed these results. The DW factor, $|\mathbf{G}|^2 \langle U_j^2 \rangle / 6$, is included in the structure factor (Eq. 2.21) in the following way:

$$S_j(\mathbf{G}, T) = \exp(i\mathbf{G} \cdot \tau_j) \exp(-|\mathbf{G}|^2 \langle U_j^2 \rangle / 6). \tag{2.32}$$

Here $\langle U_j^2 \rangle$ is the total mean square displacement of the jth atom. They found the contribution of the spin–orbit interaction to be small. They also calculated the width of the band gap from 0°K to 300°K. Although their values at the lower temperature were within the experimental results, their value for the room temperature band gap (1.48 eV) was low.

E. Luminescence Measurements

Free excitons (Noblanc et al., 1969, 1970; Halsted et al., 1961, 1965; Suga et al., 1974; Hiesinger et al., 1975; Triboulet et al., 1970, 1974; Barnes and Zanio, 1975; Camassel et al., 1973; Siffert et al., 1975) and in one case band-to-band radiative recombination (Triboulet and Marfaing, 1973) have been observed in photoluminescence. Generally, photoluminescence studies have been directed toward the identification of native defects and impurities, either through direct transitions associated with band to defect-type transitions or the recombination of excitons associated with defect centers. A strong free exciton line is generally associated with a low concentration of defects and a high mobility. The photoluminescence of material exhibiting a strong free exciton line and also band-to-band recombination is shown in Fig. 2.26. Along with the free exciton line (1.596 eV) the one and two LO phonons at 1.575 eV and 1.554 eV are observed. The line at 1.584 eV probably corresponds to the first phonon replica of the band-to-band transition at 1.606 eV. Hiesinger et al. (1975) associate weak lines at 1.603 eV to the $n = 2$ excited state of the free exciton. The remaining lines are associated with defects and will be discussed in more depth in the next chapter.

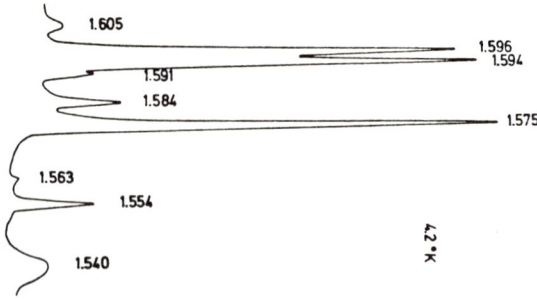

Fig. 2.26. Photoluminescence spectrum of high purity n-type CdTe at liquid helium temperature. (Triboulet and Marfaing, 1973).

FIG. 2.27. Excitation spectra of exciton luminescence of the emission line at 1.596 eV in the energy region higher than the band gap of CdTe. (a) A pure n-type sample at 1.8°K, and (b) the same sample at 20.6°K. (Hiesinger et al., 1975).

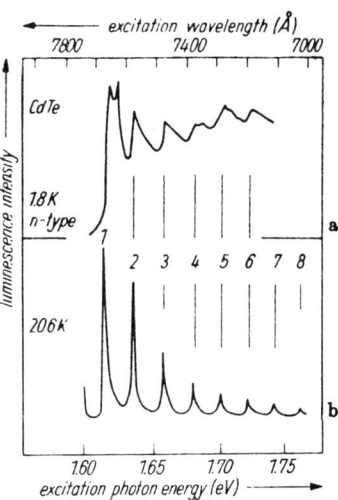

The relaxation mechanisms of excitons in CdTe have been better understood through the interpretation of excitation spectra of luminescence (ESL). The ESL technique consists of recording the change in the luminescence intensity at one fixed energy at a function of exciting photon energy. Figure 2.27 shows the ESL spectrum for a higher purity n-type sample at 1.8°K (Hiesinger et al., 1975). A characteristic of these exciton spectra is equidistant oscillatory peaks superimposed upon the continuous background. The spacing between the peaks is 0.021 eV, the longitudinal optical phonon energy. At 20.6°K, although the spacing between is similar, the peak-to-valley ratios are significantly better. (The abscissa is shifted 0.008 eV to account for the energy shift caused by temperature change). The exciton spectrum is said to relax more readily at the higher temperature by correlated phonon scattering. The strong background at 1.8°K is attributed to excitons not exclusively scattered by the fast LO process. Therefore, these excitons require a longer time to reach the final observed free or bound exciton state. Besides providing physical insight as to the exciton relaxation process, these measurements illustrate that the shape of the spectrum depends upon excitation energy. Using modulation techniques Norris et al. (1976) also studied the effects of injection intensity on cathodoluminescence.

Noblanc et al. (1969, 1970) observed free exciton lines in cathodoluminescence at 77°K. Excessive sample heating and increased self absorption occurs in cathodoluminescence measurements and therefore sharper line structure is obtained by photoluminescence measurements. Also self-absorption shifts peak positions in cathodoluminescence to lower energies with increasing electron energy. Because of the self-absorption process Noblanc et al. also maintain that the cathodoluminescence spectrum cannot unambiguously interpret the different exciton lines and that clarity requires a comparison of cathodoluminescence, reflectivity data, and detailed balanced arguments.

F. Piezobirefringence

Cadmium telluride is birefringent under the action of a uniaxial stress caused by the splitting of the valence band at $\mathbf{k} = 0$. Under uniaxial stress Thomas (1961) observed the splitting of the exciton line in reflection and measured the deformation potentials. Wardzyński (1970) observed the piezobirefringence from 2 μm to the absorption edge and measured the spectral dependence of piezo-optic coefficients (Fig. 2.28). The difference in the refractive indices $n_\parallel$ and $n_\perp$ for polarization parallel and perpendicular to the stress axis, respectively, according to Nye (1960) is

$$\Delta n = n_\parallel - n_\perp = -\tfrac{1}{2}n_0^3 \pi p, \qquad (2.33)$$

where n_0 is the zero stress refractive index, p is the compressive stress, and π is a linear combination of the piezooptic constant. For stress along [100] and [111] the piezooptic constants are, respectively, $\pi_{11} - \pi_{12}$ and π_{44}. Yu and Cardona (1973) obtained similar results over comparable wavelengths. Using an interferometric technique at 10.6 μm Weil and Sun (1971) calculated values of -8.13×10^{-13} cm^2/dyn and -2.85×10^{-13} cm^2/dyn for $\pi_{11} - \pi_{12}$ and π_{44}. After a measurement of the phase retardation at 10.6 μm Pitha and Friedman (1975) calculated 1.04×10^{-13} cm^2/dyn for $\pi_{11} - \pi_{12}$.

Piezobirefringence changes sign near the absorption edge. Yu and Cardona show this to be consistent with similar studies on other cubic semiconductors. For wider band gap materials like CdTe, Δn is in positive at longer wave-

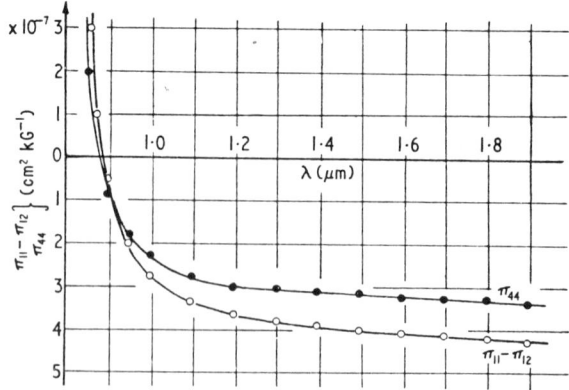

Fig. 2.28. Spectral dependence of piezo-optic coefficients $\pi_{11} - \pi_{12}$ and π_{44} in CdTe. (From Wardzynski, 1970. Copyright by The Institute of Physics.)

lengths and reverses sign when approaching the fundamental absorption edge. Yu and Cardona discuss piezobirefringence in the light of Phillips theory of the dielectric constant. The long wavelength contribution is due to that of the centroid of oscillator strength or average gap. The shorter wavelength contribution of opposite sign is determined by the splitting of the valence band under stress. In fact utilizing a parabolic band model, they were able to determine the shear deformation potentials for several III–V compounds. However Yu and Cardona were not able to calculate the deformation potentials in CdTe and other II–VI compounds due to strong response from excitons. However, using deformation potential data of Thomas (1961) and Gavini and Cardona (1970) they were able to obtain a good fit between their experimental and theoretical results.

VI. Transport Properties

A. Electrical Properties

In a review of the transport properties of II–VI compounds Devlin (1967) shows that the interaction of electrons and holes with optical phonons is the primary intrinsic scattering mechanism which defines charge transport in CdTe. Recently, however, acoustical deformation potential scattering has been suggested to be more important than earlier considered (Rode, 1970; Kranzer, 1974). Scattering studies have been primarily interpreted through Hall mobility data. Measurement of the drift mobility for electrons and holes with the time of flight technique permits an independent check of the mobility in high resistivity material and insight into new material parameters and transport phenomena.

1. Electrons

The interpretation of Hall data (Fig. 2.29) by Segall et al. (1963) limits the intrinsic scattering process of electrons exclusively to longitudinal optical phonons. In this scattering process, optical phonons induce an electrostatic potential which is proportional to the difference of the effective charge on the Cd and Te atoms. The resulting polarization effectively couples the electrons to the phonons. When the coupling is weak the mobility can be investigated on the basis of a "bare" electron interacting with the phonon field by a weak coupling treatment. When the interaction is strong the electron induces a polarization cloud about itself which lowers the electron energy. For the latter case must consider the mobility of the "polaron" whose mass is higher than that of the free electron.

The coupling constant α measures the strength or degree of coupling between the electron and phonon field and is given by

$$\alpha = (m^*/m)^{1/2}(\text{Ry}/\hbar\omega_l)^{1/2}[\varepsilon(\infty)^{-1} - \varepsilon(0)^{-1}], \quad (2.34)$$

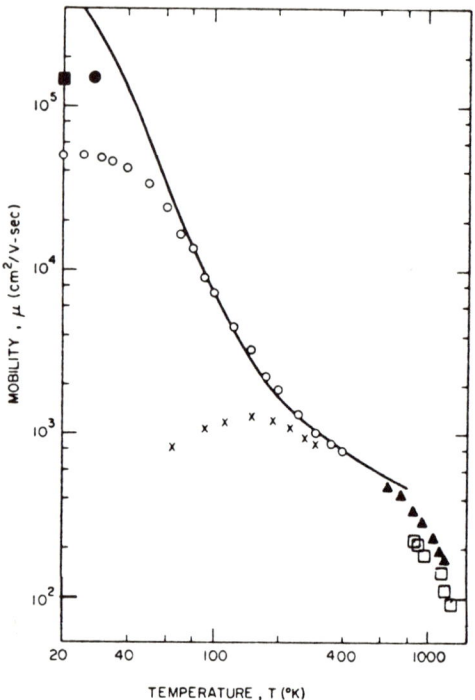

FIG. 2.29. Electron Hall mobility in CdTe by —, Rhode (1970); ○, Segall *et al.* (1963); ×, Inoue (1969); ▲, Smith (1970); □, Chern *et al.* (1975); ■, Woodbury (1974); ●, Triboulet and Marfaing (1973). Impurity scattering is significant below 60°K for the data of Segall *et al.* (1963) and below 200°K for the data of Inoue (1969).

where $\varepsilon(0)$ and $\varepsilon(\infty)$ are the low and high frequency dielectric constants, ω is the frequency of the longitudinal optical phonon, Ry is the rydberg, and m^* is the effective mass of the electron. Using the measured values of these quantiaties, Segall *et al.* find α to be 0.39. Since the interaction is weak (i.e., $\alpha < 1$) second order perturbation theory (Ziman, 1960) can be used to calculate transition probabilities of the carriers from one state to another. Waldman *et al.* (1969) made a quantitative check on the validity of the Fröhlich model of electron–LO–phonon interaction by measuring the cyclotron–resonance–absorption fields in *n*-type CdTe at a number of fixed-incident photon frequencies below the LO–phonon energy.

Characterizing the scattering mechanism by the relaxation time approach makes a theoretical description of charge transport convenient. From the relaxation time one can calculate the perturbation of an equilibrium electron distribution by a small electric field and hence the mobility. Unfortunately, the scattering process by longitudinal optical phonons is inelastic and the absorption or emission of such a phonon ($\hbar\omega_l = 0.021$ eV) alters the energy of a carrier appreciably, relative to its average thermal energy. Consequently, a simple relaxation time cannot be defined for polar scattering and the transport coefficients are calculated by a variational technique (Devlin, 1967). The phonon-limited mobility for electrons in CdTe calculated in such a way is (Segall *et al.*, 1963)

$$\mu_{po} = \frac{0.870}{\alpha \hbar \omega_l}\left(\frac{m}{m^*}\right)\left[\frac{\exp Z - 1}{Z^{1/2}}\right]G(Z)e^{-\zeta} \quad (\text{cm}^2/\text{V-sec}). \quad (2.35)$$

Here $Z = \Theta/T$, where Θ is the Debye temperature defined by $\hbar\omega_l = k\Theta$ and $G(Z)e^{-\zeta}$ is a function which includes screening effects. When $\hbar\omega_p/\hbar\omega_l \ll 1$, where ω_p is the plasma frequency, screening effects are negligible.

Segall *et al.* interpreted deformation potential scattering to be insignificant. Deformation potential scattering is due to changes in the crystal potential. These changes in the crystal potential are introduced by the local strains associated with the acoustic modes. According to Segall *et al.* the mobility for deformation potential scattering is

$$\mu_{DP} = 3.0 \times 10^{-5}(m^*/m)^{5/2}C_l T^{-3/2}E_c^{-2} \quad (\text{cm}^2/\text{V-sec}), \quad (2.36)$$

where E_c is the deformation potential for the conduction band in electron volts and $C_l = \rho \langle u_d^2 \rangle$ in dynes per square centimeter, with ρ the density and $\langle u_d^2 \rangle$ the square of the longitudinal sound velocity. A value of 2.5 eV was used for the deformation potential.

Rode (1970) developed an iterative solution of the Boltzmann equation for lattice scattering. His results, which include polar optical mode scattering and acoustic deformation potential scattering, indicate that the latter mechanism is more important than previously expected. His value of 9.5 eV for the acoustic deformation potential used in Eq. (2.36) results in a mobility at least an order of magnitude lower than before. The solid line on Fig. 2.29 shows the theoretical mobility applying Matheissen's rule to polar optical and deformation potential acoustic scattering. From 60 to 300°K the agreement is good with the data of Segall *et al.* Below 60°K defects limit the mobility. For the data of Inoue (1969), the mobility is limited by defects below 300°K. At temperatures greater than 800°K the data of Smith (1970) fall well below the extrapolated theoretical mobility. Chern *et al.* (1975) find even lower value for the mobility at these high temperatures. Figure 2.29 shows the result for undoped material. Additional results not shown here for material containing 2.7×10^{17} In cm^{-3} are identical verifying that intrinsic processes are responsible for the decrease in mobility. Rode associates the decrease in mobility with simultaneous conduction in higher-lying $\langle 111 \rangle$ minima which he assumes to be a few tenths of an electron volt above the central valley. Segall *et al.* and Rode consider piezoelectric scattering to contribute insignificantly to the scattering. However, Crois (1969) calculates mobilities of approximately 10^4 cm^2/V-sec for piezoelectric scattering and concludes that for sufficiently pure samples at low temperatures and weak electric fields this mechanism may dominate. Pairing of charged defects (Woodbury and Aven, 1974) and microinhomogeneities (Triboulet, 1971; Alekseenko et al., 1970) may make a good fit to the calculated mobility over a wide temperature range difficult.

Free carrier absorption in n-type CdTe crystals also show that electrons are scattered by optical phonons (Yamada 1960; Strauss and Iseler, 1974; Perkowitz and Thorland, 1974; Vul et al., 1968; Lisitsa et al., 1970; Planker and Kauer, 1970). The absorption coefficient due to the interaction of carriers in CdTe is proportional to λ^n where n ranges from 2.3 to 3.5. Free carrier absorption due to classical absorption is proportional to λ^2. From theory n is 2.5 when FCA is caused by scattering by optical phonons (Visvanathan, 1960) and is 3.5 when caused by ionized impurities (Fan et al., 1956). Making use of the quantum theory of free carrier absorption, Jensen (1973) also associates the λ^3 behavior in uncompensated samples having less than 10^{17} cm^{-3} electrons with optical phonons. Mobility measurements support the interpretation that higher values of n imply ionized impurity scattering whereas lower values imply optical phonon scattering. Free carrier absorption measurements only suggest whether polar optical or ionized impurity scattering are operative and they are not as precise as electrical measurements. Segall et al. (1963), Alekseenko et al. (1970), and Chapnin (1969), among others, use the Brookes–Herring formula to interpret the mobility. For N_D singly ionized donors,

$$\mu_I = \frac{4(2/\pi)^{3/2}(kT)^{3/2}(4\pi\varepsilon(0)\varepsilon(\infty)^2)}{q^3 m^{*1/2} N_D [\ln(b) - 1]} \quad \text{(cm}^2\text{/V-sec)}, \tag{2.37}$$

where

$$b = \frac{6m^*(kT)^2(4\pi\varepsilon(0)\varepsilon(\infty)}{\pi q^2 h^2 N_N}.$$

Segall et al. found semiquantitative agreement with experimental results at lower temperatures ($<35°$K).

Care must be taken in interpreting mobility data at lower temperatures. Woodbury (1974) reports a mobility as high as 140,000 cm^2/V-sec at 20°K. However the mobility is higher than that predicted by simple charge scattering models. This anomaly is associated with the pairing of charged defects and an effective decrease in the concentration of scattering centers.

Electrical measurements on n-type CdTe have been helpful in describing the properties of the conduction band. Polar optical mode scattering provides quantitative agreement with Hall measurements, assuming $\mathbf{k} = 0$ and

$m^* = 0.11m$ (Segall et al., 1963). Kanazawa and Brown (1964) observed cyclotron resonance for electrons and obtained a value of $0.096m$ for the effective mass. No anisotropy in the effective mass was found, substantiating the conclusion of Segall et al. that the magnetoresistance results of Yamada (1962) are difficult to reproduce. Marple (1963) measured the infrared Faraday rotation and infrared reflectance. By combining these results and assuming a spherical energy surface for the conduction band, he obtained a value of $0.11m$ for the effective mass. There is a good correlation between the bandgap and the electron effective mass for the II–VI compounds (Kurik, 1967; Devlin, 1967) and this value fits nicely within this correlation. Piezoresistance measurements (Sagar and Rubenstein, 1966) are also consistent with the conduction band energy minimum at $\mathbf{k} = 0$.

Iseler et al (1972) tentatively associate electrons in n-type CdTe with impurities that are tied to non-Γ donors. Localized and quasilocalized impurity states may have wave functions from high-energy band extremes that are degenerate in energy with band continuum at Γ. At lower pressures and especially at lower temperatures Sagar and Rubenstein found the resistivity of their n-type CdTe to increase with increasing pressure and associated the change with the mobility. At higher pressures, changes are clearly associated with changes in the carrier concentration. Relative movement occurs between the conduction band and a non-Γ band located at atmospheric pressure above the Γ minimum (Foyt et al., 1966; Paul, 1968). With an increase in pressure, the separation between the Γ minimum and the energy level (and also between the Γ minimum and the higher minimum) decreases, resulting in a transfer of electrons from the Γ conduction band minimum into the donor levels associated with Ga, In, Cl, and Br. The relative change is about 20 meV/kbar. At higher pressures and at lower temperatures, it requires an unusually long time for the carrier concentration to reach the steady state. Photoconductive decay experiments show that there is a potential barrier of approximately 0.5 eV to the transfer of electrons between the Γ conduction band and the non-Γ donor levels. The nature of these non-Γ donors is not clear but they presumably have either X or L symmetry. Based upon experience with other semiconductors, Paul (1968) considers these levels to be associated with the minimum at X rather than the minimum at L or the valence band. Unfortunately, little is known about the relative energy shift between the central and satellite minima with pressure. Table 2.11 shows, however, that the relative energy shift between the $\Gamma_8 \to \Gamma_6$ (7.9 eV/kbar) and $\Lambda_{4.5} \to \Lambda_6$ (4.5 eV/kbar) transitions with pressure is 3.4 meV/kbar. Evidence in Chapter 3, Section II, A, 1 associates non-Γ donor phenomena with atomic movements.

The drift velocity characteristics for electrons (Fig. 2.30) in semiinsulating CdTe have been measured by the time of flight technique from 77 to 370°K

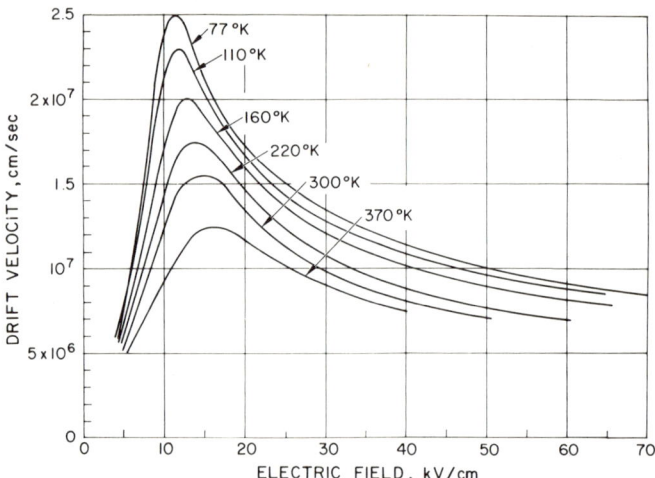

FIG. 2.30. Drift velocity for electrons in semiinsulating CdTe using the time of flight technique (Canali et al., 1971b).

for electric fields up to 70 kV/cm (Canali et al., 1970, 1971a,b, 1974; Martini et al., 1972a). A negative differential mobility occurs beyond the threshold field which increases from 11 kV/cm at 77°K to 16 kV/cm at 370°C. The position of the maximum drift velocity is slightly higher than the Gunn thresholds determined by current voltage measurements on lower resistivity material (Oliver et al., 1967; Oliver and Foyt, 1967; Ludwig, 1967; Picus et al., 1968a).

A theoretical analysis of high field transport in CdTe has been performed with the Monte Carlo technique (Jacoboni and Reggiani, 1970; Ruch, 1972; Borsari and Jacoboni, 1972). Polar optical, deformation potential acoustic, and ionized impurity scattering in both the central and satellite valleys were considered. At the lowest electric fields polar optical scattering dominates the scattering process. At higher electric fields polar optical scattering is no longer able to dissipate the energy gained by the electrons from the field. Intervalley scattering becomes more effective and the electrons populate the region of negative differential mobility having a higher effective mass. The sudden increase in the electron temperature at ~ 8 kV/cm is confirmed by the sudden increase in the trapping times (Canali et al., 1971b). Thermoelectric measurements (Alekseenko and Veigner, 1971) also show the electron temperature to increase with increasing electric field. A best fit to the drift velocity characteristic by Ruch (1972) results in values of 0.15 m_0 for the satellite effective mass, 10^9 eV/cm for the optical deformation potential, and 0.5 eV for the energy difference between the satellite and central valleys.

VI. TRANSPORT PROPERTIES

Coulombic interaction between electrons and ionized centers reduces the drift velocity in high resistivity material (Canali et al., 1973, and 1975b). The effect of ionized impurity scattering is appreciable at lower temperatures and at electric fields below the threshold value where ionized centers would be expected to have a negligible effect on electron transport. Scattering and trapping lowers the total drift velocity even at high mean electron energies by acting only on the low energy electrons contained within the low energy of the distribution function (Bosari and Jacoboni, 1972). In lower resistivity heavily compensated material, the mobility and drift velocity also decrease and current oscillations normally observed at higher electric fields do not occur. Instead, recombination radiation as a result of impact ionization occurs. Picus et al. (1968b) and Van Atta et al. (1968) associate this phenomenon with a decrease in the negative differential mobility. This interpretation is in qualitative agreement with drift velocity measurements on semiinsulating material containing large concentrations of indium and presumably a higher ionized impurity content. With increasing impurity content the peak-to-valley ratio decreases. In effect, the region of negative differential mobility is restricted to higher electric fields with increasing ionized impurity content.

2. Holes

Hall mobility data for holes are available from 100 to 320°K (Yamada, 1960) and from 600 to 900°K (Smith, 1970). In the low temperature range the data are fit to the polar optical scattering theory of Low and Pines

$$\mu_{PO} = \frac{1}{2\alpha\omega} \frac{e}{m} \left(\frac{m_p}{m}\right)^3 f(\alpha)\{\exp(\Theta/T) - 1\}, \tag{2.38}$$

where m and m_p are the effective mass of the electron and the polaron, α is the strength of the interaction of the electron with the lattice, ω is related to the Reststrahl wavelength, and Θ is the Debye temperature. Using a Reststrahl wavelength of 66.7 μm (Mitsuishi et al., 1958, and Table 2.4) a Debye temperature of 252°K was calculated. The following expression for the mobility results when 0.57 is used for the coupling constant and 0.63 is used for the effective mass:

$$\mu_{PO} = 57\{\exp(252/T) - 1\} \quad (\text{cm}^2/\text{V-sec}). \tag{2.39}$$

Smith (1970) obtains a value of 0.66 m for the effective mass. The fit of Eq. (2.39) to the Hall data for holes is shown in Fig. 2.31. This model assumes a single valence band having s-like symmetry. Further theoretical analyses are more complex because of the degeneracy of the valence band and the p-like symmetry of the hole–wave function. In determining the concentration of holes (p) on their Hall measurements Vul *et al.* (1970) considered both the heavy and light holes. Assuming the valence subbands to be parabolic and the relaxation times of the heavy and light holes to be equal to the Hall coefficient is:

$$R = \frac{1}{ep} \frac{[1 + (m_1/m_2)^{1/2}][1 + (m_2/m_1)^{3/2}]}{[1 + (m_2/m_1)^{1/2}]^2} = \frac{\eta}{ep}. \tag{2.40}$$

From an analysis of optical transitions between light and heavy hole subbands in the same material a value of from 5 to 8 was found for the ratio of the heavy to light hole mass (m_1/m_2). Assuming $m_1/m_2 = 8$, they found $\eta \approx 2.2$. A Monte Carlo technique (Costato *et al.*, 1972) extends the theoretical interpretation by considering scattering between the two valence bands and including the p-like symmetry. These results are also in agreement with Hall data and are shown in Fig. 2.31. Values of $1.0m$ and $0.1m$ were used,

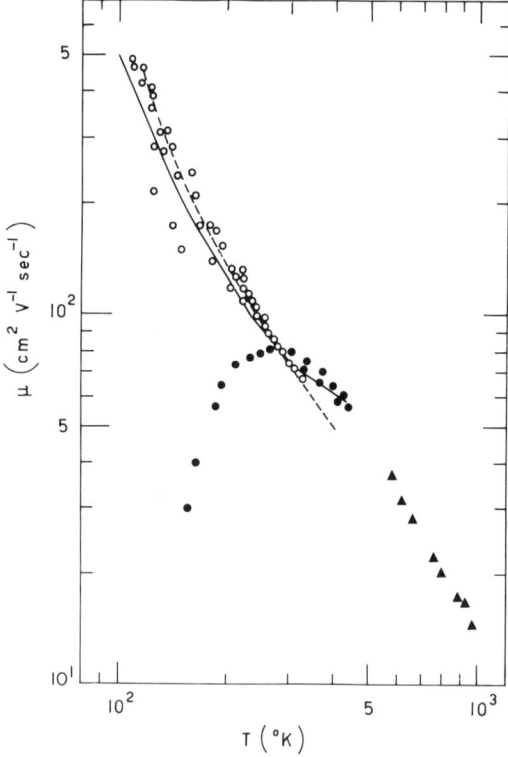

FIG. 2.31. Hall mobility of holes in CdTe. ○, Yamada (1960); ▲, Smith (1970); ---, Yamada (1960); —, Costato *et al.* (1972). The drift mobility (●, Ottaviani *et al.* (1973)), measured by the time of flight technique, is reduced by trapping and detrapping below 300°C.

respectively, for the heavy and light hole masses. The value of $0.1m$ is in good agreement with the values for the average light hole mass calculated by Lawaetz (1971) using the **k** · **p** approach ($0.103m_0$), by Brodin et al. (1968) including spin–orbital interaction ($0.125m_0$), and by Čápek et al. (1973) using the **k** · **π** approximation ($0.19m_0$). The value for the heavy hole mass is somewhat high but is less than the value of $1.38m_0$ for the heavy hole mass using the **k** · **p** approach, the value of $2.1m_0$ including spin–orbital interaction, and the value of $1.72m_0$ using the **k** · **π** approach. Substantially lower values for the effective mass have been used to fit other experimental data. From an investigation of the Seebeck effect a polaron effective mass of $0.41m_0$ was found (de Nobel, 1959). Considering a reduced mass of 0.071 from exciton absorption and an effective electron mass of $0.096m$ an average hole mass of 0.4 (± 0.2) is deduced (Segall and Marple, 1967).

Kranzer (1973a,b, 1974) uses a difference equation method to solve the Boltzmann equation for polar optical scattering. He shows acoustical deformation potential scattering to be important at temperatures below 100°K. In order to obtain the best fit to the mobility data, Kranzer in his latest analysis (1974) uses a polaron effective mass of $0.8m_0$ and 5.2 eV for the effective acoustic deformation potential. The spread in the valence band parameters is apparent and more precise experimental data are necessary.

The hole drift mobility has been measured at room temperature by the time of flight technique in the space-charge-limited mode (Canali et al., 1971c). A value of 70 cm^2/V-sec was obtained for electric fields up to 35 kV/cm. This value is slightly less than the Hall mobility values of Yamada (1960). The theoretical analyses of the drift velocity versus electric field considering polar optical scattering results in a drift mobility of 75 cm^2/V-sec which is independent of electric field at room temperature (Ottaviani et al., 1973). More recently room temperature values as high as 90 cm^2/V-sec were obtained (Canali et al., 1974; Bell et al., 1974a). This is inconsistent when considering that the Hall mobility is expected to be higher than the drift mobility by the factor r_H which is about 1.3 for electrons using the single band model (Segall et al., 1963). A comparable value of r_H is expected for holes for the degenerate valence band structure (Kranzer, 1973b; Beer, 1963). Unfortunately, r_H for holes cannot be precisely determined because large Hall angles cannot be attained with such a low mobility.

B. THERMAL PROPERTIES

The thermal conductivity of CdTe at room temperature is about 0.06 W/cm°K (Stuckes, 1961; Devyatkova and Smirnov, 1962; Ioffe and Ioffe, 1960; Chasmar et al., 1960; Slack and Galginaitis, 1964), and increases with decreasing temperature; the increase is due to a reduction in phonon scattering by the umklapp process (Fig. 2.32a). At lower temperatures it

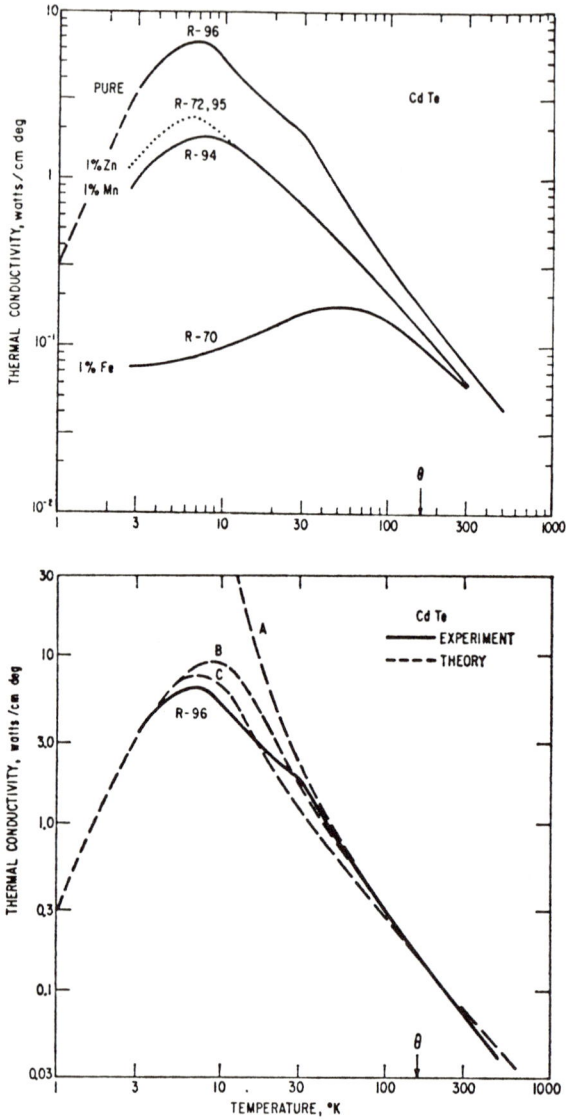

FIG. 2.32. (Top) The thermal conductivity versus temperature for pure, as-grown CdTe, and for three doped, heat-treated samples. (Bottom) Theoretical and experimental curves of the thermal conductivity of pure CdTe versus temperature. Curve A assumes only umklapp scattering is present, and that the exponential factor $b = 3.0$. Curve B assumes that boundary, isotope, and umklapp scattering are present and $b = 3.0$. Curve C assumes the same three mechanisms as in curve B, but now $b = 4.5$ (Slack and Galginaitis, 1964).

decreases, the decrease being due to boundary and isotope scattering. Slack and Galginaitis fit their experimental data for a "pure" sample to

$$K(T) = \frac{k}{2\pi^2 v}\left(\frac{kT}{\hbar}\right)^3 \int_0^{\Theta/T} \tau_c \frac{x^4 e^4 \, dx}{(e^x - 1)^2}, \qquad (2.41)$$

where k is Boltzmann's constant, v is the average sound velocity, Θ is the Debye temperature, $x = \hbar\omega/kT$, and ω is the phonon frequency. The phonon scattering relaxation time (τ_c) consists of the umklapp scattering relaxation time (τ_u), the crystal boundary scattering time (τ_b), and the isotope scattering time (τ_i). The umklapp scattering time is expressed as

$$\tau_u^{-1} = \frac{a\hbar\gamma^2 \omega^2 T}{m_a v^2 \Theta} \exp(-\Theta/bT), \qquad (2.42)$$

where a and b are constants, γ is the Grüneisen constant, and m_a is the average mass of a single atom. Curve A in Fig. 2.31b is a theoretical fit to the experimental data of "pure" CdTe considering only umklapp scattering. Data were fitted, assuming a to be close to unity, b equal to 3.0, and the Grüneisen constant to be 2.0 and independent of temperature. A good empirical fit to the data is found. However, detailed measurements of Grüneisen's constant (Fig. 2.15) show that γ is a strong (Smith and White, 1975) function of temperature. At a Debye temperature of 160°K it has a value of 0.46 and it decreases to a minimum value of -0.65 at 10°K. Curves B and C include isotope and boundary scattering with values of 3.0 and 4.5, respectively, for b in Eq. (2.42).

The thermal conductivity decreases with increasing impurity content (Ioffe and Ioffe, 1960; Stuckes, 1961; Devyatkova and Smirnov, 1962; Slack and Galginaitis, 1964). Point defect scattering, combined with the umklapp and boundary scattering that was used for pure CdTe, give a good fit to the Zn-doped curve (Slack and Galginaitis, 1964). Similar additions of the magnetic impurities Mn and Fe also result in a decrease in the conductivity. For Mn-doped material the decrease is comparable to that of Zn, whereas for Fe-doped material the decrease is significant. For the Fe-doped material additional magnetic scattering effects exists. The extra magnetic scattering is null or absent for the Mn-doped material because the energy levels of the d-shell in Mn are not comparable with the phonon energies in the temperature range studied.

VII. Summary

The Phillips and Van Vechten dielectric theory of ionicity classifies CdTe as being ionic relative to the other zinc blende and wurtzite compounds. Many of the physical properties and constants (e.g., free energy of sublimation, force constants, piezoelectricity, specific heat) scale well with this theory. However, lattice dynamic models for the simple diamond structure are in their infancy. Therefore, an understanding of bonding mechanisms for the more complex zinc blende structures is even more difficult to realize. However, more physical insight is available when CdTe is analyzed within the Sn–InSb–CdTe–AgI isorow of the periodic table.

The empirical pseudopotential model of the band structure provides a good interpretation of reflectivity and photoemission data. The width and separation of the valence bands are consistent with the above definition of ionicity. Being a direct gap material, CdTe is rich in exciton transitions. Excitons are important in luminescence and reflection. Absorption at the fundamental band edge and the transport properties also involve extensive longitudinal phonon cooperation.

Cadmium telluride has not been investigated as thoroughly as some of the other tetrahedral structures such as Si, Ge, and GaAs. However, among the wide band gap II–VI compounds it is probably the most extensively studied. This has been due to a combination of factors: single crystals are readily available; material is both n- and p-type; the technological importance of the material is based upon single crystals rather than thin films for phosphors, etc. The resulting information on CdTe is a valuable contribution to understanding the physics of semiconductors.

CHAPTER 3

Defects

Control of the electrical and optical properties in CdTe as well as in any other compound semiconductor requires an understanding of the native defect structure. Unraveling the defect structure is not a trivial problem. The relatively open zinc blende structure should easily accommodate interstitial atoms. In fact the atomic spacing in CdTe is larger than the spacing of most tetrahedral structures. Correspondingly the cohesive strength of CdTe is smaller than that of most tetrahedral structures (Fig. 2.9) suggesting the energy of vacancy formation to be smaller and the concentration of vacancies to be relatively larger. For the zinc blende structures greater concentrations of native defects are more likely in the II–VI compounds where the more ionic bonding requires alternating positive and negative charges than in the III–V compounds where the more covalent bonding requires a tetrahedral relationship between atoms. These factors contribute to the general disorder in CdTe. Van Vechten (1975) calculates similar values for the energy of formation of neutral vacancies in the II–VI and III–V compounds. However it is likely that many of the defects are charged in CdTe and their concentration is larger.

Fortunately, equilibrium between the various defects in the metal chalcogenides is readily attained by heating the crystal above 500°C and allowing diffusion to occur. Making appropriate measurements while varying the deviation from stoichiometry, and consequently the various defect concentrations, results in considerable insight into the high temperature defect structure. Unfortunately, the kinetics of defect migration are not slow enough to assume that the high temperature disorder can be frozen in while rapidly cooling samples to room temperature. Consequently, the high temperature and room temperature atomic defect structures are quite different and the interpretation of more convenient room temperature electrical and optical data based upon the higher anneal temperature can be misinterpreted.

3. DEFECTS

I. Defects at High Temperature

A. HIGH TEMPERATURE EQUILIBRIUM STRUCTURE

de Nobel (1959) laid the groundwork for defect structure calculations in CdTe. However his interpretation of the high temperature equilibrium was indirectly based upon room temperature electrical measurements. Unfortunately the high temperature defect structure cannot be frozen in.

Conductivity and Hall *measurements at high temperature* are a direct approach in determining the concentration of electrically active majority native defects. By changing the component pressure over samples held at elevated temperatures, the native defect concentrations and hence the free carrier concentration adjust. The conductivity and electron concentration for pure samples in Cd overpressures over the temperature range 600 to 900°C are nearly proportional to $p_{Cd}^{1/3}$ (Whelan and Shaw, 1968; Zanio, 1969b, 1970a; Smith, 1970; Rud' and Sanin, 1971a; Chern *et al.*, 1975). Contrary to this Matveev *et al.* (1969b) found a $p_{Cd}^{1/2}$ behavior. The first high temperature Hall effect measurements on CdTe were undertaken by Smith and are shown in Fig. 3.1. A doubly ionized majority donor defect, either a Cd interstitial (Cd_i^{2+}) or a Te vacancy (V_{Te}^{2+}), provides a consistent explanation of the results. The incorporation of an atom from the vapor Cd(v) into the lattice as a doubly ionized interstitial defect is described by

$$Cd(v) \leftrightarrows Cd_i^{2+} + 2e. \tag{3.1}$$

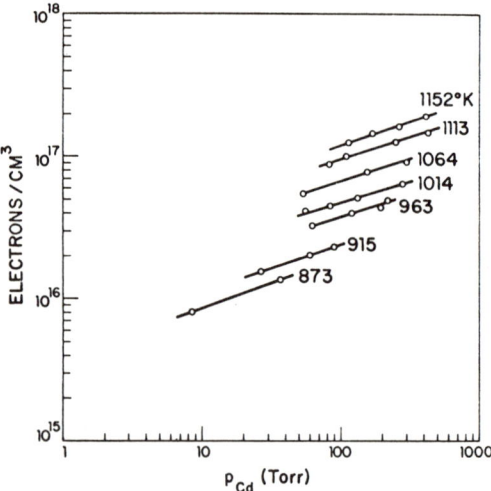

FIG. 3.1. The equilibrium electron concentration in CdTe as a function of Cd pressure (p_{Cd}). A $p_{Cd}^{1/3}$ dependence is found to be consistent with a doubly ionized donor native defect (Smith, 1970).

Applying the laws of mass action to this reaction equation results in

$$[Cd_i^{2+}][e]^2/p_{Cd} = A \exp(-E_i/kT), \qquad (3.2)$$

where A is a constant and E_i is the enthalpy of formation of the doubly ionized defect. A similar relationship results for the doubly ionized Te vacancy. For accurate calculations degeneracy should be considered. If the doubly ionized donor, assumed here to be interstitials, is the majority electrically active native defect, then the electroneutrality condition is

$$[e] = 2[Cd_i^{2+}].$$

The resulting electron concentration is

$$n = (2A)^{1/3} \exp(-E_i/3kT) p_{Cd}^{1/3}. \qquad (3.3)$$

The actual electron concentration measured over the range of temperatures and Cd pressures shown in Fig. 3.1 is

$$n = 6.9 \times 10^{16} \exp(-1.70/3kT) p_{Cd}^{1/3} \quad \text{Torr}^{1/3}/\text{cm}^3. \qquad (3.4)$$

Smith calculated a value of 1.70 ± 0.1 eV for E_i which was independent of defect concentration and temperature. This value is in good agreement with a value of 1.65 ± 0.15 eV earlier obtained by Whelan and Shaw (1968). The depth of this doubly ionized center is 0.24 eV or less from the conduction band. A similar analysis, also based upon conductivity measurements, results in values of 1.35 ± 0.15 eV for the enthalpy of formation and ≤ 0.21 eV for the ionization energy of the center (Rud' and Sanin, 1971a). The Brouwer plot of Fig. 3.2 (Brouwer, 1954) shows doubly ionized donors to be the majority defect at high Cd pressures. It assumes residual impurities to be absent.

High temperature Hall effect (Smith, 1970) and conductivity (Zanio, 1969b) measurements on undoped samples are nearly independent of Te pressure. Therefore either an intrinsic phenomenon or foreign atoms are responsible for the electrical activity on the Te-saturated side of the phase diagram. At lower temperatures the hole concentration varies between samples, and the material is p-type because of the presence of residual acceptor impurities. Smith places an upper limit of 2×10^{16} cm^{-3} native defects at 700°C near Te saturation conditions. Upon increasing the temperature the intrinsic carrier concentration increases and the type changes from p to n because of the higher mobility of electrons as compared to that of holes. The intrinsic concentration according to Smith is

$$n_i = 1.8 \times 10^{15} T^{3/2} [\exp\{-(1.32 + 1.9T \times 10^{-4})\} \text{ eV}/kT] \quad \text{cm}^{-3} \qquad (3.5)$$

and is designated as the carrier concentration over the electroneutrality range $[e] = [h]$ in Fig. 3.2. Simultaneous solution of Eqs. (3.3) and (3.4)

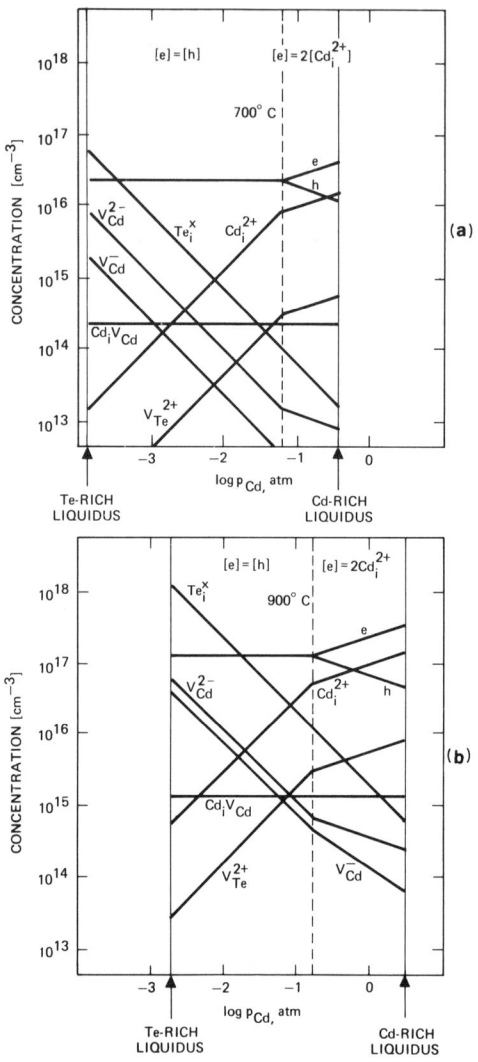

FIG. 3.2. Brouwer plots for CdTe at 700 and 900°C.

determines the transition pressure from the region $[e] = [h]$ to $[e] = 2[Cd_i^{2+}]$. If native acceptors rather than native donors are the electrically active majority defect on the Te side of the solid region in pure material, then the Cd vacancy is favored because of the large ionic radius for Te. For doubly charged Te the radius is 2.22 Å (Pauling, 1960). Consideration should also be given to the possibility that donors in the form of Cd interstitials are the majority electrically active native defect across the entire solid region. Considering the contributions of only ionized vacancies to the conductivity in undoped material, Van Vechten (1975) predicts the zinc blende structure to be n-type (p-type) if the tetrahedral covalent radius of the anion is smaller (larger) than that of the cation. This correlation is in good agreement for the II–VI compounds and is based upon a direct relationship between the radius and the enthalpy of formation of the neutral vacancy. An estimate of 2.75 eV for the formation of both types of vacancies also agrees with the fact that CdTe is both n-type and p-type. (On the assumption that covalent binding occurs between Te neighbors, Bailly (1968) calculates a value of 0.85 eV for the energy of formation of Cd vacancies.) However, no evidence of Cd vacancies is found in high temperature electrical measurements. This may be due to either a significantly larger enthalpy necessary to form charged cation vacancies than charged anion vacancies or the predominance of interstitial defects. If the latter case is true then charged Cd interstitials would dominate because of the large-radius of the charged Te interstitial.

Chern et al. (1975) propose a higher concentration of electrically active acceptor defects than that shown in Fig. 3.2. Consequently, the native defects rather than residual acceptor impurities would be responsible for influencing the electrical properties at low Cd overpressures and the concentration of holes would decrease with increasing p_{Cd}. This is in conflict with the experimental results.

Tracer diffusion studies are also helpful in understanding the defect structure since the tracer self-diffusion coefficient is proportional to the concentration of a particular defect which promotes diffusion. Results are not as straightforward as Hall effect measurements since the diffusion coefficient is determined by the native defect having the largest concentration–mobility product and not necessarily the largest defect concentration. Woodbury and Hall (1967) and Borsenberger and Stevenson (1968) interpret the self-diffusion of Te over most of the solidus to be a result of a fast diffusing, neutral, interstitial Te species. The Te self-diffusion coefficient for Te-saturated conditions is

$$D^*_{Te} = 1.66 \times 10^{-4} \exp(-1.38 \text{ eV}/kT) \quad \text{cm}^2/\text{sec} \quad (3.6)$$

and is proportional to the Te pressure. This proportional relationship is also consistent with the diffusion of selenium and Te in other zinc and Cd chalcogenides. Woodbury and Hall estimate these chalcogenides to contain approximately 10^{19} cm^{-3} chalcogen interstitials at about 900°C under chalcogen saturated conditions. The Brouwer plot shows Te interstitials to be the majority native defect over a large portion of the solidus region.

Under Cd saturation the diffusion coefficient is more independent of component pressure, suggesting the onset of diffusion by Te vacancies (Woodbury and Hall, 1967; Chern and Kröger, 1975).

With the exception of the preliminary results of de Nobel (1959), the Cd tracer diffusion coefficient in undoped material is nearly independent of component pressure (Whelan and Shaw, 1967; Borsenberger and Stevenson, 1968; Chern and Kröger, 1975). Whelan and Shaw find the diffusion coefficient to be

$$D_{Cd}^* = 1.25 \exp[(-2.07 \pm 0.08)/kT] \quad cm^2/sec, \tag{3.7}$$

which corresponds to about 10^{-10} cm²/sec at 800°C. Tracer diffusion in undoped material may be due to either a pressure independent ring or exchange mechanism which is independent of lattice defects or by a mechanism involving a neutral defect complex possibly the neutral association of Cd vacancy and a Cd interstitial. Chern and Kröger find a small but evident increase in D_{Cd}^* for undoped crystals at higher Cd pressures which they attribute to ionized Cd interstitials. Although he did not include data, de Nobel (1959) suggested Cd interstitials.

Foreign atoms alter the native defect structure. The incorporation of indium, gallium, or aluminum on a Cd site results in a singly ionized donor and an extra electron. If the energy of formation of a Cd vacancy is comparable to or less than the difference in energy between the conduction band and the first acceptor level of the vacancy, then reduced Cd pressures are conducive to the formation of compensating acceptors (Mandel, 1964). The energy of formation of a neutral Cd vacancy is probably comparable to that of a neutral Te interstitial, another potential acceptor defect. From size considerations, singly ionized Cd vacancies are preferred to Te interstitials. However, if the first charged level of the vacancy lies significantly closer to the valence band than that of the interstitial, the net energy required to form V_{Cd}^- would be considerably smaller than that needed to form Te_i^-. The energy to form indium complexes, $(In_{Cd}V_{Cd})^-$ or $(In_{Cd}Te_i)^-$, might even be smaller.

At higher Cd pressures where the concentration of native acceptors is reduced and when the indium concentration is somewhat greater than the concentration of Cd interstitials, the electroneutrality condition becomes $[In_{Cd}^+] = [e]$. The concentration of ionized native donors and acceptors, respectively, are proportional to and inversely proportional to the Cd pressure in this region.

For CdTe containing an unspecified amount of indium, Rud' and Sanin (1972) find at 650°C and at a Cd pressure of approximately 4×10^{-2} atm a definite change in the slope of the conductivity. At higher pressures the conductivity is independent of p_{Cd} corresponding to $[n] = [In_{Cd}^+]$ and at

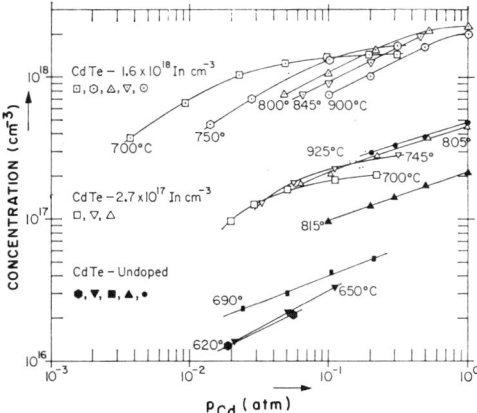

FIG. 3.3. Electron concentration isotherms versus p_{Cd} for undoped CdTe and CdTe doped with 2.7×10^{17} and 1.6×10^{18} In cm^{-3} (Chern et al., 1975).

lower pressures the conductivity is proportional to $p_{Cd}^{1/2}$ corresponding to $2[V_{Cd}^{2-}] = [In_{Cd}^{+}]$. Such an abrupt change in slope is probably idealistic. Figure 3.3 shows the general behavior of the electron concentration with p_{Cd} for undoped material and CdTe doped with 2.7×10^{17} and 1.6×10^{18} cm^{-3} In. The dependence of the electron concentration on temperature and pressure for undoped material is similar to that in Fig. 3.1. In the lightly doped material the temperature dependence is weak. At 700°C the electron concentration is comparable to the indium concentration and weakly dependent upon p_{Cd}. With increasing temperature the concentration of native donor defects increases resulting in an increase in the electron concentration and a more distinct pressure dependence. A significant increase in the carrier concentration occurs for material containing 1.6×10^{18} In cm^{-3}.

An important anomaly to consider in more heavily doped material, especially at lower pressures, is the *decrease* in the electron concentration with *increasing* temperature (Strauss, 1971). With increasing dopant concentration a greater fraction of the dopant is in the form of neutral donor–native defect complexes. An increase in temperature reduces the concentration of these complexes. Although the total dopant concentration is constant the concentration of compensating acceptor complexes, such as V_{Cd}^{2-} and V_{Cd}^{-}, increase thus decreasing the net electron concentration. Also one cannot rule out an increased concentration of neutral $In_{Cd}^{+}Te_{i}^{-}$ complexes caused by an overwhelming increase in Te_i with increasing temperature. The result again would be a decrease in the concentration of isolated indium and likewise free electrons.

When Borsenberger and Stevenson (1968) added aluminum to undoped CdTe they found an increase in D_{Cd}^{*}. The enhanced diffusion suggests that the concentration of the native-acceptor defect (V_{Cd}^-) is increased and that the charge neutrality condition is $[V_{Cd}^-] = [Al_{Cd}^+] \cong [Al]_{total}$. This phenomenon becomes more operative at low temperatures where the concentration of intrinsic defects is low. Chern and Kröger also observe enhanced diffusion in In-doped material. For heavily In-doped material there is a significant increase in D_{Cd}^* at low p_{Cd} and a small increase for high p_{Cd} when compared to undoped material. The annealing of donor-doped CdTe crystals at lower p_{Cd} establishes more favorable conditions for self-compensation, i.e., the introduction of acceptor defect (V_{Cd}^{2-}) in concentrations comparable to the donor impurity concentration. The pressure dependence of the self-diffusion process is consistent with the compensation mechanism.

Phosphorus (Smith, 1970), copper (Rud' and Sanin, 1972), and gold (Borsenberger and Stevenson, 1968) are the p-type impurities studied at elevated temperatures. Phosphorus substitutes for Te and in Te atmospheres increases the conductivity. Borsenberger and Stevenson observed an increase in D_{Cd}^* when Au is incorporated into the lattice. The enhanced diffusion suggests that the concentration of Cd interstitials is increased by the presence of Au_{Cd}^-. Since the addition of both Au and Al enhances D_{Cd}^*, they ruled out diffusion associated with a bound neutral Frenkel pair. They conclude that self-diffusion of Cd occurs by the motion of both ionized Cd vacancy acceptors and interstitial Cd donors and that an ionized Frenkel disorder on the Cd sublattice is the dominant high temperature defect structure. This interpretation is not consistent with the defect structure in Fig. 3.2. Whelan and Shaw (1967) and Chern and Kröger (1975) propose a separate pressure-independent ring mechanism involving Cd_iV_{Cd} to be the predominant mechanism for diffusion in undoped material. After doping, additional diffusion occurs through enhanced concentrations of either Cd_i or V_{Cd}.

Using ion channeling techniques with 2 MeV helium ions Akutagawa et al. (1974, 1975) determined that Au predominantly occupies substitutional Cd lattice sites at 800°C. More importantly, these results strengthen the model advocating the major electrically active native defect to be doubly ionized. The total concentration of Au is determined from the random backscattering spectrum of CdTe, previously annealed at elevated temperatures in Cd and Te overpressures under Au-saturated conditions. The total concentration or solubility of Au varies approximately as $p_{Cd}^{-1/3}$. Channeling spectra along $\langle 110 \rangle$ determine the fraction of Au off-substitutional sites. The difference between the random and aligned spectra determines the substitutional concentration. For an anneal at 900°C with $p_{Cd} = 2.5$ atm the solubility is 3.2×10^{19} cm^{-3}, the off-substitutional concentration is 8×10^{18} cm^{-3} and the substitutional concentration is 2.3×10^{19} cm^{-3}.

The $p_{Cd}^{-1/3}$ dependence at 800°C and appropriate defect relationships result in a charge state of -1 for substitutional Au and $+2$ for the native donor assumed here to be Cd interstitials. This corresponds to a charge state of between 0 and -1 for interstitial Au. Although the solubility is less dependent upon pressure at 900°C doubly ionized Cd interstitials still provide the best fit to the data. Most of the Au is retained in the dissolved state after quenching to room temperature as a result of the stabilizing effect of Cd_i^{2+}. Infrared inspection showed all Au-doped samples on the Cd-rich side of the phase field to be virtually free of precipitates. However, increased opacity occurs toward the Te solidus. The nature of this absorption was not determined. However, analysis of precipitates of samples showing the greatest absorption show them to contain mainly Te with approximately 6% Au. The presence of Au in the precipitates did not alter the results since the Au content is significantly less than the total.

Cadmium interstitials are stable at room temperature only in the presence of Au or another suitable acceptor. In undoped material a decrease in the temperature results in excess neutral Cd which most likely forms precipitates.

Room temperature Hall measurements on undoped samples previously annealed at elevated temperatures in Cd overpressures and quenched to room temperature (Fig. 3.4) show the carrier concentration to have a pressure dependence stronger than $p_{Cd}^{1/3}$ and to *decrease* with *increasing* anneal temperature (de Nobel, 1959; Kröger, 1974; Selim *et al.*, 1975). These results differ from the high temperature results. The room temperature carrier concentration is also expected to follow the $p_{Cd}^{1/3}$ pressure dependence and to the first approximation is also expected to be comparable to the concentration of net electrically active majority defects. Also the *p–n* transitions would be expected to occur close to the intersection of $[Cd_i^{2+}]$ and $[V_{Cd}^{2-}]$, which is about 3×10^{-3} atm at 700°C and just above 10^{-2} atm at 900°C. The room temperature *p–n* transition point increases with increasing temperature but to a much greater extent than predicted by the high temperature defect structures. Chemical diffusion, association of native defects with one another and with residual impurities, and subsequent precipitation explain these differences.

B. CHEMICAL DIFFUSION

A typical interval for quenching a sample from elevated to near room temperature is a few seconds. Rapid atomic rearrangement occurs within this period as a result of chemical diffusion. The chemical diffusion coefficient for CdTe and other binary systems is

$$D_{12} = [N_2 D_1^* + N_1 D_2^*][d(\ln a_1)/d(\ln N_1)], \qquad (3.8)$$

where N_1 and N_2 are the mole fractions of components 1 and 2, D_1^* and D_2^*

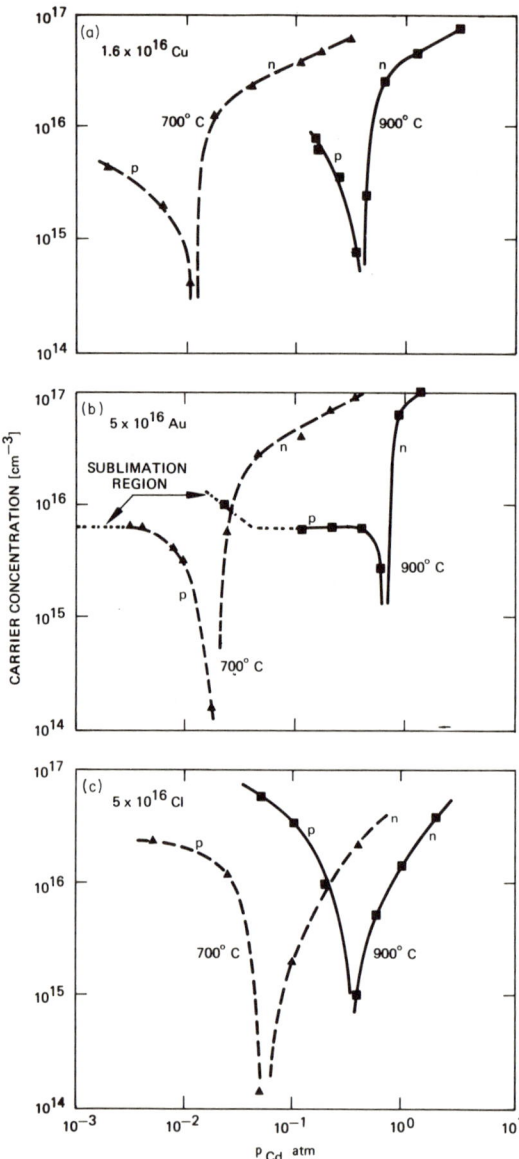

FIG. 3.4. Carrier concentration versus p_{Cd} at 700 and 900°C for CdTe containing (a) 1.6×10^{16} cm^{-3} residual Cu (de Nobel, 1959), (b) 5×10^{16} cm^{-3} Au (de Nobel, 1959), (c) 5×10^{16} cm^{-3} Cl (Selim et al., 1975).

are the tracer diffusion coefficients, a_1 is the activity of component 1, and $d(\ln a_1)/d(\ln N_1)$ is the thermodynamic factor and is the driving force for rapid compositional changes. The thermodynamic factor in CdTe is about 10^5, resulting in a significant increase in diffusion over that expected by tracer diffusion alone (Zanio, 1970a). The chemical diffusion coefficient for CdTe measured over the temperature range 550 to 800°C is

$$D_{CdTe} = 4\exp(-1.15 \pm 0.1 \text{ eV}/kT) \text{ cm}^2/\text{sec}. \qquad (3.9)$$

At 800°C, D_{CdTe} is about 10^{-5} cm²/sec and is in reasonable agreement with the theoretical value. Rud' and Sanin (1972) obtained an activation energy of 1.1 eV and a diffusion coefficient of about a factor of two higher at the same temperature. On the basis of a defect model and assuming the ambipolar diffusion of Cd_i^{2+} and e^-, Chern and Kröger (1975) also calculate a theoretical D_{CdTe} and also obtain reasonable agreement. Both these theoretical calculations require that D_{Cd}^* be characteristic of an interstitial mechanism and increase with increasing p_{Cd}. The theories are not entirely consistent with experimental data which show D_{Cd}^* to be nearly independent of p_{Cd}.

The chemical diffusion coefficient was measured from relaxation times characteristic of conductivity changes which were induced by stepwise changes in the Cd partial pressure (Δp_{Cd}). In contrast to the results of Smith (1970), Rud' and Sanin (1972), Zanio (1970a), and Chern and Kröger (1975) find the relaxation time (τ) to depend upon the sign of Δp_{Cd}. The relaxation time is larger for Δp_{Cd} negative (τ_-) than the relaxation time for Δp_{Cd} positive (τ_+) and in Eq. (3.9) corresponds to $\Delta p_{Cd} > 0$. In the process of chemical diffusion a particular specie may be created, and another annihilated. Since the process is an approach to equilibrium, the reaction rate proceeds favorably in one direction and results in asymmetrical relaxation times.

In In-doped samples and for material not following the $p_{Cd}^{1/3}$ dependence in Eq. (3.3) and presumably containing residual impurities, Rud' and Sanin found lower chemical diffusion coefficients, smaller activation energies, and larger τ_-/τ_+ ratios than in pure material. For impure samples with $\Delta p_{Cd} < 0$, complex relaxation behavior often occurs and it is not possible to characterize the decay process by a single value of τ. Rud' and Sanin associate the complex behavior with a transition from one electroneutrality region to another resulting in additional interactions between the mobile species and impurities. Chern and Kröger (1975) associate secondary relaxation processes with nucleation effects.

C. Precipitation

The Brouwer plot shows the equilibrium concentration of Cd interstitials at the Cd-rich solidus to be about an order of magnitude larger at 900°C than that at 700°C. The excess concentration is significantly less at lower temperatures. In fact, Strauss (1971) extrapolates the equilibrium solubility of excess Cd at room temperature to be only 10^6 cm^{-3}. Assuming one second is required to rapidly cool a sample from 900 to 600°C and using the value of D_{CdTe} at the lower temperature ($D_{CdTe} \approx 10^{-6}$ cm^2/sec), the minimum value for the diffusion length is 10 μm, a distance comparable to the separation between dislocations or other potential nucleation sites. If precipitation occurs subsequent to such rapid diffusion, the native defect concentration and correspondingly the carrier concentration should significantly decrease upon cooling to room temperatures. Cooling the crystal through the retrograde solidus of Fig. 1.3 should result in precipitation, and indeed random gross defects ranging from a few hundred angstroms to about a micron in size or larger are present in CdTe. Micron-size and larger inclusions are more commonly found along crystal boundaries and twin planes with an infrared microscope.

Lorenz and Segall (1963) find that slow cooling avoids the formation of finally divided precipitates always evident in Te-fired and quenched samples. Slow cooling permits the excess to diffuse toward precipitates and grain boundaries leaving the intermediate regions clear. Transmission electron microscopy (TEM) shows Cd precipitates and/or intrinsic dislocation loops for CdTe fired in Cd, suggesting interstitial Cd to be the precipitating species (Selim *et al.*, 1975). The microstructure of a crystal cooled after annealing at 900°C in Cd overpressure of 2.0 atm shows small precipitates and prismatic interstitial-type dislocation loops. Nonequilibrium deviations from stoichiometry are not limited to inclusions. Akutagawa and Zanio (1971) observed macroscopic voids in vapor grown crystals. Transmission electron microscopy studies confirm, in addition to inclusions, the presence of voids for material grown from the vapor by either physical vapor transport or by the deposition of Cd and Te using separate carrier gases (Hall *et al.*, 1975). The precipitates are rods with an average diameter of 50 Å and an average length of 200 Å. The voids tend to be spherical and range in diameter from 40 to 120 Å. Both defects are in the $2-5 \times 10^{16}$ cm^{-3} range. The defect microstructure also shows stacking faults. Gallium-doping has little effect on the degree of disorder in these studies.

In the case of material grown from the melt it is not easy to determine to what extent inclusions are associated with the precipitation phenomenon since constitutional supercooling also occurs. Growth from Te- or Cd-rich solutions are more likely to result in constitutional supercooling and inclusions than growth from congruent melts. However, the subsequent annealing of inclusion-laden crystals at elevated temperatures usually results in cleaner material (Zanio, 1971). This is especially true for the solution-grown material. Unfortunately, the effects of annealing on precipitation in undoped material has only been studied with TEM in detail using

I. DEFECTS AT HIGH TEMPERATURE   **127**

Cd overpressures. Further studies using Te atmospheres and more extensive studies in Cd atmospheres should result in further insight into the precipitation process.

Additional annealing and precipitation studies have been undertaken in doped material. Magee *et al.* (1974, 1975) observed In_2Te_3 and $CdCl_2$ precipitate platelets in CdTe intentionally doped with In and Cl and annealed from temperatures of from 100 to 900°C from 1 to 48 hr in Cd atmospheres. This observation is not consistent with measurements that report In to be soluble in CdTe up to 40 mole % (Woolley and Ray, 1960). Transmission electron microscopy (Fig. 3.5) also shows polycrystalline Te rings ≈ 60 Å in diameter in concentrations from 10^{14} to 10^{16} cm^{-3} to be present in

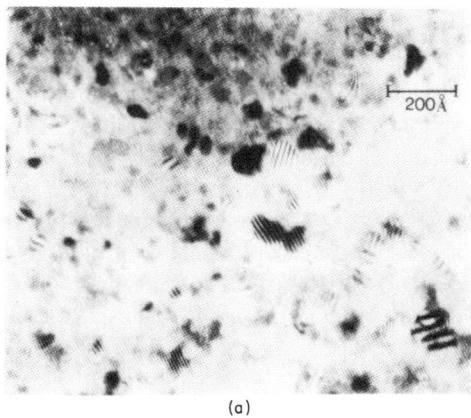

(a)

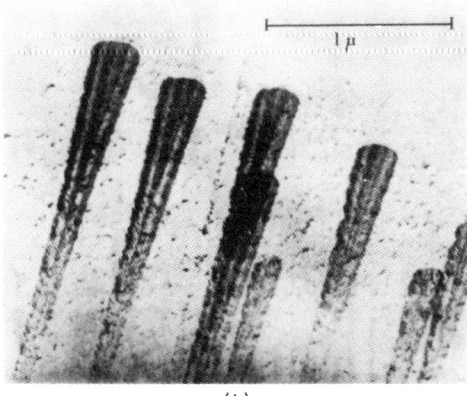

(b)

FIG. 3.5. Transmission electron micrographs of (a) In-doped CdTe containing In_2Te_3 platelets and (b) Cl-doped material containing stacking faults (Magee *et al.*, 1975).

In-doped samples. However, when samples were annealed at temperatures at or above 500°C no evidence of Te precipitation was found in TEM or electron diffraction patterns. Electron micrographs also show that localized regions of high dislocation density are associated with the presence of excess Te.

Anomalous quenching effects occur. Selim *et al.* (1975) reported that the room temperature carrier concentration depends upon the sequence of cooling the reservoir and the sample. For CdTe containing 1.2×10^{17} cm^{-3} In and fired at 700°C and 900°C, rapidly cooling the reservoir first results in electron concentrations that are comparable to the In content in agreement with the result of de Nobel (1959). However, cooling the crystal first results in a significant decrease in the carrier concentration. For CdTe containing 2×10^{18} cm^{-3} In cooling the crystal first after annealing at 700°C also results in a carrier reduction below the In level. This is not the case for a 900°C anneal where the higher carrier concentration is retained. Crystals having a significant decrease in the room temperature electrical activity are expected to have a large precipitate content. Ambiguous results are found. No difference is found in the microstructure between pure CdTe and CdTe containing 1.2×10^{17} cm^{-3} In when the samples are annealed at 700°C and the crystal is quenched first. However, when similarly doped samples are annealed at 800°C, 100 to 500 Å diameter precipitates in concentrations of $\sim 10^{13}$ cm^{-3} are found. This occurs when the sample is quenched first, but not in the opposite sequence. Selim *et al.* suggest these precipitates to be caused by In. When the sample is quenched first for CdTe containing 2×10^{18} In cm^{-3} and annealed at 900°C with $p_{Cd} = 2.0$ atm, the micrograph shows the precipitate content to be significantly less but to contain dislocation loops. Selim *et al.* assume that the excess Cd condenses as dislocation loops rather than being involved in the precipitation of In. They suggest that the absence of heavy precipitation is due to the difficulty in initiating new nucleation sites in the presence of large excess of In. Heating to the higher temperatures of 900°C removes all nuclei present in the sample as a result of its previous history. Unfortunately, it was not possible to determine positively the nature of the precipitating phases.

Because the degree of precipitation and thus the native defect concentration is not well defined it is not realistic to extend Brouwer plots to room temperature using the firing temperature to be the characteristic temperature. Rapid defect migration continues to occur even at room temperature. From an extrapolation of the chemical diffusion coefficient to room temperature [Eq. (3.9)], local changes in composition over distances on the order of a micron are expected within a period of a day. Therefore, it is not unusual to expect marked changes in the conductivity near room temperature. At 150°C, Triboulet (1971) observed an increase in the resistivity from 500 ohm-cm to approximately 10^8 ohm-cm in minutes. At room temperature Lorenz et al. (1964) found 10% increases per day in as-grown samples and even more drastic changes in quenched samples. For material containing about 10^{14} cm^{-3} electrically active centers, Woodbury and Lewandowski (1971) associate changes with about 10^{13} cm^{-3} centers.

II. Defects at Low Temperature

High temperature electrical measurements show that impurities determine the conductivity on the Te-rich side of the stoichiometric composition. They should play even a greater role in the electrical properties near room temperature where even a lower concentration of native defects is expected. On the Cd-rich side of the solidus high temperature electrical measurements establish native defects in the form of Te vacancies or Cd interstitials to be simple donors which dominate the electrical properties. As a result of the high mobility of native defects and the ease with which they precipitate, their concentration decreases upon cooling to room temperature. The strongest evidence of native defects at room temperature is the change in type from p to n with increasing p_{Cd} (Fig. 3.4). To what extent native defects on the n-type side determine the room temperature electrical properties is not clear. Regardless, an excess of the residual foreign acceptors stabilizes the native defect concentration through self-compensation. However, when the residual impurities are donors the mechanism for retention may not be as effective. Isolated native defects are present but only as minority species. Because of the interaction of native defects and impurities, simple isolated native defects are more an exception than a rule.

A. Donors

1. Hydrogenlike Donors

de Nobel (1959) and Kröger (1974) assign a donor level 0.02 eV below the conduction band to isolated ionized Cd interstitials. Lorenz et al. (1964) suggest that this level is due to impurities. Additional optical and electrical measurements strengthen the latter viewpoint. Agrinskaya et al. (1975) find the electron concentration over the 10^{16}–10^{17} cm^{-3} range to be proportional to the weight of $CdCl_2$. An extrapolation of the ionization energy of Cl (Agrinskaya et al., 1975) and residual donors (Woodbury and Aven, 1974) to infinite dilution (E_0) using $E_d = E_0 - \alpha(N_d^+)^{1/3}$ as in Fig. 3.6 results in a hydrogenic donor-level depth of 0.014 eV. Here N_d^+ is the concentration of positively charged donors. Values of α are 3.8×10^{-8} cm-eV (Woodbury and Aven) and 2.5×10^{-8} cm-eV (Agrinskaya et al., 1975). The total concentration of shallow donors (N_d) and its ionization energy (E_d) along

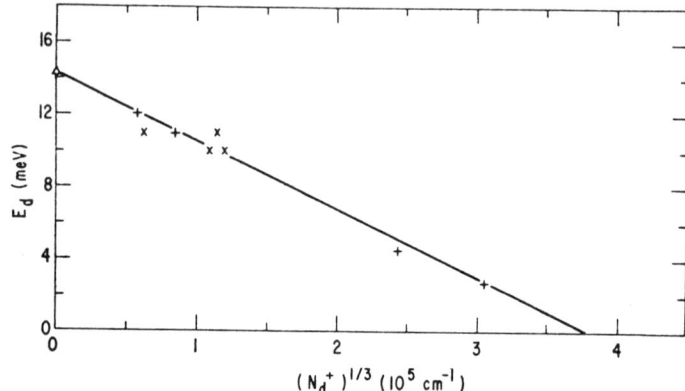

FIG. 3.6. Ionization energy (E_d) of hydrogenic donors in CdTe versus ionized donor concentration (N_d^+). The triangle refers to an optical measurement (Woodbury and Aven, 1974).

with the total concentration of compensating acceptors (N_a) are related by

$$n(N_a + n)/(N_d - N_a - n) = gN_c \exp(-E_d/kT) \qquad (3.10)$$

where g is the degeneracy factor and N_c is $2(2\pi m_e^* kT/h^2)$, the density of states in the conduction band. After an appropriate change of notation, Eq. (3.10) also applies when acceptors are the majority defect.

Woodbury and Aven (1974) show that n-type CdTe of even the highest mobility (140,000 cm²/V-sec at 20°K) and presumably of high purity is closely compensated. In the "best" material the shallow donor concentration is about 2×10^{14} cm⁻³ and the acceptor concentration is about 50% less. The addition of In also increases the concentration of electrons (de Nobel, 1959; Selim et al., 1975; Segall et al., 1963; Swaminthan et al., 1975). Yokazawa et al. (1965) find the electron density to be proportional to the concentration of In over the range 10^{17}–10^{18} cm⁻³. For intermediate doping there is a maximum in the Hall coefficient characteristic of impurity band conduction (Vodop'yanov and Abramov, 1968). This impurity band is responsible for a shift in the absorption edge (Brodin et al., 1970). For heavily doped samples the carrier concentration is independent of temperature.

The value of 0.014 eV for the activation energy is in excellent agreement with high resolution far-infrared absorption (Cohn et al., 1970, 1972; Wagner and McCombe, 1974) and photoconductivity (Simmonds et al., 1974; Wagner and McCombe, 1974; Bajaj et al., 1975) measurements. These measurements illustrate the 1s → 2p and 1s → 3p transitions of the shallow donor. Zeeman splitting confirms the assignment of these transitions to a simple hydrogen model and illustrates the interaction of bound electrons with LO phonons and the Stark effect in CdTe. Simmonds et al. associated shifts in the 1s → 2p transition energy to as many as six different chemical impurities.

The interaction of defects results in anomalous effects. At ionized impurity concentrations of 10^{14} cm^{-3}, long range interaction between compensating centers occurs resulting in a decrease in the concentration of effective scattering centers with decreasing temperature and a corresponding increase in the mobility (Woodbury, 1974). With photoconductivity measurements Bajaj *et al.* (1975) found a peak on the low energy side of the 1s → 2p transition that cannot be explained in terms of a simple hydrogenic model of isolated impurities. Instead it is associated with the interaction of donor impurities with one another. Pressure experiments suggest the transfer of donor atoms between nonequivalent lattice positions and their corresponding deionization (Baj *et al.*, 1976; Losee *et al.*, 1973). Although this is not proof of a long range interaction, it does suggest that atom migration occurs below room temperature when the lattice is not in thermal equilibrium.

Low temperature photoluminescence (PL) spectra in the exciton emission region (band gap down to about 1.56 eV) identify neutral and ionized donors. Figure 3.7 is a high resolution spectrum of a high purity *n*-type sample having a net electron concentration of 4.3×10^{13} cm^{-3} and a mobility of 2.5×10^4 cm^2/V-sec at 63°K. Distinct bands are present at 1.593, 1.592, 1.589, and 1.583 eV. Weak emission near 1.596 eV corresponds to the band gap energy at **k** = 0 less the exciton binding energy from Eq. (2.26). With the exception of free exciton emission, the lines are assigned to the recombination of excitons trapped at various impurities or defects. The binding energy of the exciton to the center is related to the ionization energy of the center when a single charge carrier is bound to it. For various II–VI

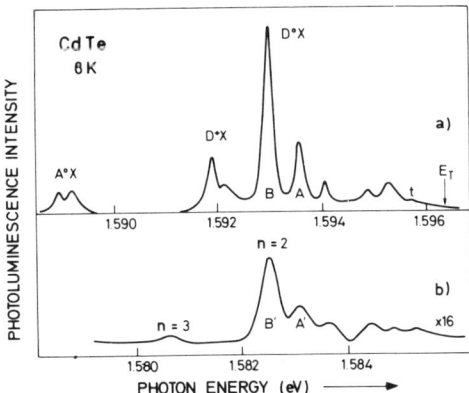

FIG. 3.7. Photoluminescence spectra of high purity CdTe. (a) Free and weakly bound excitons. (b) Two-electron transition replica of the exciton bound to a neutral donor. (Reprinted with permission from Suga *et al.*, 1974. © 1974, Pergamon Press, Ltd.)

compounds, Halsted and Aven (1965) have shown that the ratio of the exciton binding energy (E_b) to the ionization energy (E_i) is approximately 0.1 for excitons trapped at neutral acceptors and 0.2 for excitons trapped at neutral donors. Consequently, the peak energy of a bound-exciton band can be used to determine an approximate ionization energy for the center at which the exciton is trapped. The 1.593 eV exciton emission is commonly observed in CdTe and has been attributed to exciton trapping at a shallow chemical donor (Halsted and Aven, 1965; Noblanc et al., 1970). Using a binding energy of 0.003 eV ($E_b = 1.596 - 1.593$ eV) and a ratio of 0.2 gives an ionization energy of 0.015 eV which is characteristic of shallow hydrogenic donors in CdTe (Woodbury and Aven, 1974; Agrinskaya et al., 1975). Therefore, the 1.593 eV emission is associated with excitons trapped at impurity donors such as In or residual halogens and group III elements. Hiesinger et al. (1975), Cho et al. (1974a), and Suga et al. (1974) studied the Zeeman splitting and energy shifts of the exciton neutral donor complex in a magnetic field and identified lines at 1.59375 eV (line A) and 1.59320 eV (line B) in Fig. 3.7 as spin–exchange partners of the exciton bound to the neutral donor (D^0, X). A close replica of these lines occurs at about 1.583 eV and is associated with the radiative recombination of the exciton leaving the donor in the n = 2 excited state. Similarity of the Zeeman splitting patterns confirms this assignment. The line around 1.592 eV is associated with the recombination of excitons bound to an ionized donor (D^+, X).

The PL spectra of samples intentionally doped with In also show the 1.596, 1.593, 1.589, and 1.584 eV lines (Fig. 3.8). Free exciton emission is present only in the sample having the lowest concentration of In. The intensity of the remaining lines decreases with increasing In concentration, probably as a result of traffic through competing centers and a lower generation rate of excitons. The bound-exciton recombination spectrum is dominated by a band at 1.584 eV. Because the intensity is strong relative to the 1.593 eV line this line is ruled out as being the n = 2 replica of the exciton bound to the neutral donor.

Electron spin resonance (ESR) measurements indicate that the donors Cl, Br, I, Al, and Ga are present in the sulfur chalcogenides (Title, 1967). No hyperfine or superhyperfine structure is observed, presumably because of the high residual impurity content. Müller and Schneider (1963) calculated that the concentration at which delocalization occurs is about 10^{17} cm^{-3}. Alekseenko and Veinger (1974) observed resonance in n-type CdTe in the $5 \times 10^{15} - 5 \times 10^{16}$ cm^{-3} carrier range having an unknown residual in the $10^{16} - 10^{17}$ cm^{-3} concentration range. No fine structure was reported. Du Varney and Garrison (1975) did not find resonance due to B, Al, and In for material doped with these elements. However, in an analysis of Tl-doped crystals the same authors observed superhyperfine and hyperfine structure

II. DEFECTS AT LOW TEMPERATURE

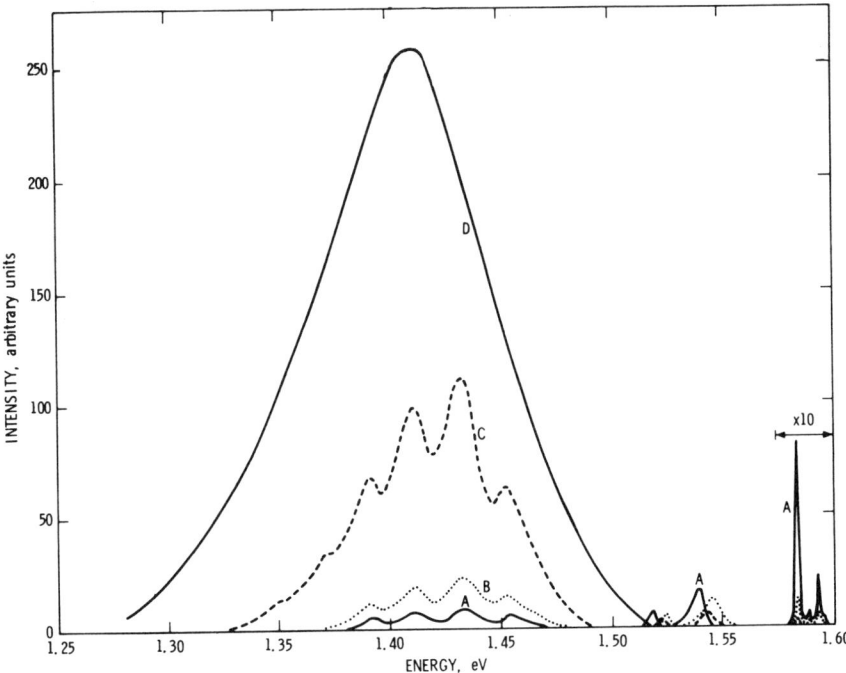

FIG. 3.8. Photoluminescence spectra of CdTe containing (A) 3×10^{15} cm^{-3}, (B) 10^{16} cm^{-3}, (C) 10^{17} cm^{-3}, and (D) 10^{18} cm^{-3} of In; $T = 4.2°$K (Barnes and Zanio, 1975).

and showed that Tl is in the Tl^{2+} state and that the ions are substitutional for Cd. A native defect–impurity complex, consisting of a sulfur vacancy V_S and a group I acceptor, Cu, Ag, or Au, is found in ZnS (Dieleman et al., 1964). No such defect is observed in CdTe by ESR.

Considering the available data it is likely that the hydrogenic donor level is an impurity from group III or group VII when the net residual impurity is a donor. However, when the net residual impurity is an acceptor, the hydrogenic donor level is likely to be a native defect.

There is some evidence indicating that the column III and VII elements are not simple hydrogenic donors, but are impurity states associated with higher-energy conduction band extrema (Foyt et al., 1966; Paul, 1968; Iseler et al., 1972; Strauss, 1971). When the resistivity of Cd-saturated samples containing Cl, Ga, In, and Br are examined versus hydrostatic pressure, as in Fig. 3.9, the electron concentration decreases by several orders of magnitude. Analysis of the linear portion of the curves result in energy levels for Cl (0.05 eV), Ga (0.05 eV), In (0.19 eV), and Br (0.26 eV) that are located just above the Γ minimum at atmospheric pressure. Application of pressure results in a displacement of the energy levels below the

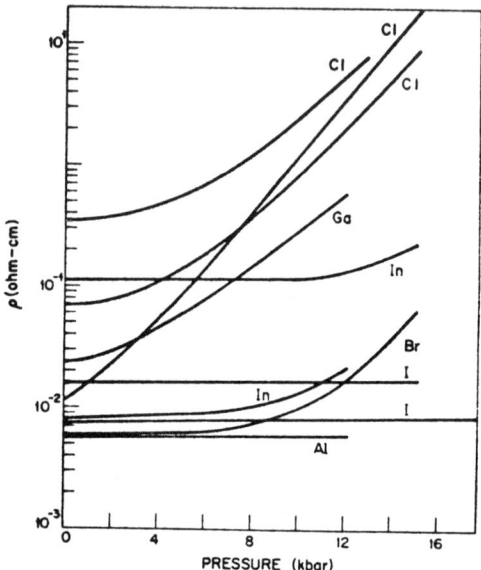

Fig. 3.9. Resistivity ρ at room temperature versus hydrostatic pressure for n-type CdTe. At atmospheric pressure, in order of decreasing ρ for each dopant, the carrier concentrations are, in units of 10^{17} cm^{-3}: Al, 15; Ga, 3.3; In, 1.1 and 11; Cl, 0.36, 0.94, and 8.5; Br, 16; I, 3.9 and 25. (Reprinted with permission from Iseler et al., 1972. © 1972, Pergamon Press, Ltd.)

Γ minimum and a corresponding transfer of electrons from the conduction band to these levels. Aluminum and iodine do not exhibit this behavior presumably because either these impurities are hydrogenic donors or because they introduce levels well above the Γ minimum and the extreme pressure applied is insufficient for appreciable electron transfer. At lower temperatures equilibrium is not readily established and there is a potential barrier to the transfer of electrons from the Γ conduction band to the non-Γ donor levels. The barrier height for Cl is 0.50 eV (Iseler et al., 1972; Losee et al., 1973). The 0.31 eV barrier height for Ga (Losee et al.) is not too different from the 0.27 eV barrier characteristic of the double acceptor lying 0.06 eV below the Γ minimum (Lorenz et al., 1964). Using Schottky barriers on Cl-doped and Ga-doped material, photocapacitance and transient capacitance measurements result in similar activation energies. Losee et al. argue that the associated decay rates are excessively long compared with that predicted by the mixing of the conduction band wave function with wave functions of substitutional impurities degenerate with the conduction band. Additional evidence that the non-Γ donor model is inoperative is based upon the field emission rates of electrons from filled centers. No field dependence is found for the Ga centers and the results could not be fit to a Poole–Frenkel model for the Cl-doped material.

TABLE 3.1

IONIZATION ENERGIES OF DONORS IN CdTe AS DETERMINED BY
HALL MEASUREMENTS

Assignment	ε_a (eV)	Reference
Residual impurities	0.014^a	Woodbury and Aven (1974)
Residual impurities	0.01	Segall et al. (1963)
Cl	0.014^a	Agrinskaya et al. (1975)
Column III residual	0.014^a	Agrinskaya et al. (1971a)
Residual impurities	0.017^a	Agrinskaya et al. (1971a)
Cd_i^+	0.02	de Nobel (1959) / Kröger (1974)
V_{Te}^+	0.023	Kröger and de Nobel (1955)
Ga non-Γ donors	-0.05	Iseler et al. (1972)
Cl non-Γ donors	-0.05	Iseler et al. (1972)
In non-Γ donors	-0.19	Iseler et al. (1972)
Br non-Γ donors	-0.26	Iseler et al. (1972)

[a] Extrapolated to infinite dilution.

Losee et al. imply that the slow kinetics of electron transfer is due to the formation of a donor-native defect complex. Further studies also provide evidence against the non-Γ donor as well as the double acceptor in explaining the relaxation behavior. Baj et al. (1976) propose the donor to be transferred by pressure between two nonequivalent lattice positions, each having a different ionization energy. Legros et al. (1977) propose that two Cl atoms are involved in each anomalous center with different configurations corresponding to different charge states and ionization energies. Anomalous phenomena also exist among other semiconductors, CdS:F, CdSe:F, GaSb:S and $GaAs_xP_{1-x}$:S. In the last case, slow kinetics occurs even when the band gap is indirect (Iseler, private communication). More research is necessary to fully understand the nature of the nonhydrogenic donor. Values of the energy levels for Γ as well as the proposed non-Γ donors as determined by electrical measurements are summarized in Table 3.1.

2. *Deeper Donors*

Deeper donor levels are not commonly seen in n-type material with electrical measurements. In high resistivity material ($\rho > 10^7$ ohm-cm) a shallow level at 0.025 and deeper levels at 0.05 and 0.6 are measured with the transient charge technique (Martini et al., 1972). These levels may not be apparent in the electrical measurements because their formation in significant concentrations may occur only in compensated high resistivity material. Although the transient charge technique is limited to high resistivity material a significant advantage of this approach is that the properties

TABLE 3.2

Trapping Parameters as Determined by the Transient
Charge Technique in High Resistivity CdTe

Dopant	Carrier	E_T (eV)	$N_T \times 10^{-15}$ (cm^{-3})	Reference
In	Electrons	0.05	100	a
Cl	Electrons	0.05	1–2	b
Cl	Electrons	0.027	24	b
Br	Electrons	0.026	190	b
In	Electrons	0.6	—	c
In	Holes	0.14	50	d
Cl	Holes	0.14	20	b
Br	Holes	0.14	160	b
Cl	Holes	0.36	6	b

[a] Canali et al. (1975b). [b] Canali et al. (1974).
[c] Zanio et al. (1968).   [d] Ottaviani et al. (1973).

of both electrons and holes can be studied in the same material. Table 3.2 summarizes the trapping parameters of CdTe for several traps in high resistivity material as determined by the transient charge technique. The trap concentration (N_T) and the activation energy (E_T) are obtained from various forms of

$$\mu_r = \mu_0[1 + (N_T/N_c)\exp([E_T - \beta\sqrt{E}]/kT)]^{-1}. \qquad (3.11)$$

Here μ_r is the measured drift mobility, μ_0 is the lattice mobility including scattering by defects, N_c is the density of states in the conduction band, E is the electric field, and β is the Poole–Frenkel constant.

Time of flight measurements with In-doped high resistivity CdTe show the electron drift velocity to decrease with increasing concentrations of In (Canali et al., 1975b). In the lightly doped material (≈ 1 ppm) the electron drift velocity is limited by Coulombic scattering by ionized centers and agrees with theory (Fig. 2.30). In the more heavily doped material electron transport is limited by trapping and detrapping effects by a level at 0.05 eV and is strongly dependent upon the electric field. Further evidence as to the presence of this center is given by fitting experimental data to Poole–Frenkel and tunnel theory. At low fields, less than 6 kV/cm, data are interpreted by a lowering of the Coulomb barrier around the trapping centers. A value of 2.75×10^{-4} V$^{-1/2}$ cm$^{1/2}$ eV calculated for the Poole–Frenkel constant agrees with that calculated for holes interacting with the 0.14 eV level (Ottaviani et al., 1972, 1973). When the electron drift mobility data are plotted versus the square root of the electric field a good fit is found at low fields to Eq. (3.11) thus verifying the Poole–Frenkel effect (Fig. 3.10). A plot

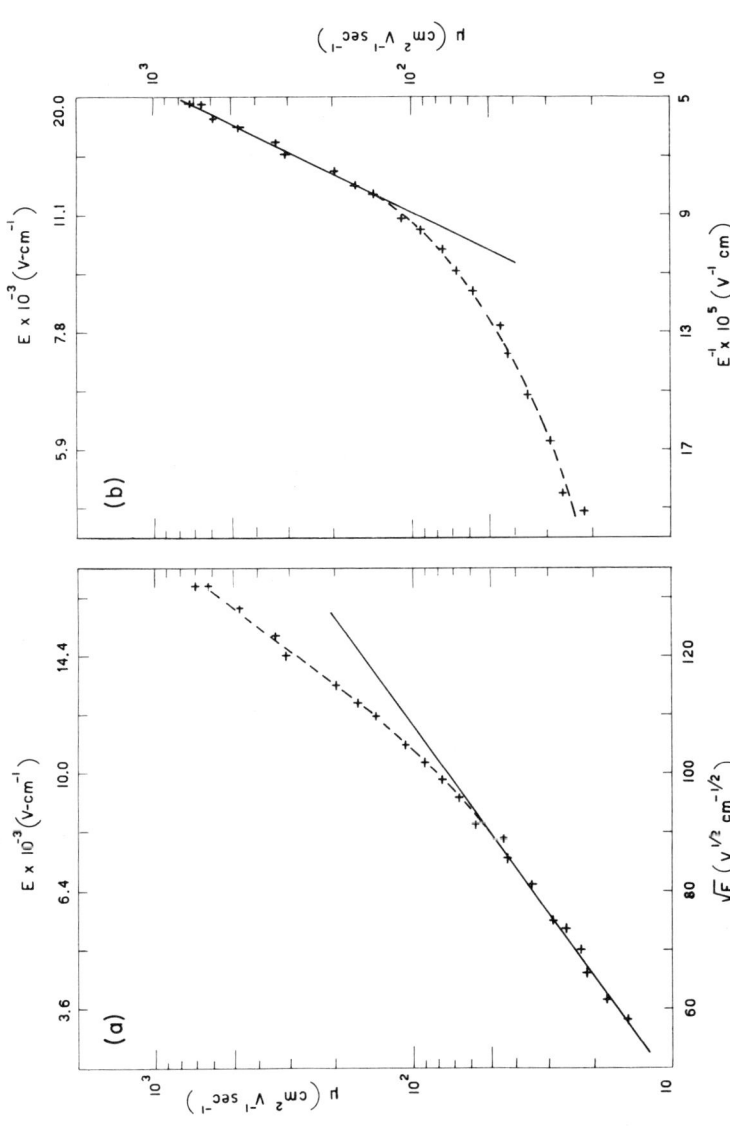

FIG. 3.10. Electron drift mobility at 77°K plotted as a function of (a) the square root of the electric field $\sqrt{E}$, and of (b) the reciprocal of the electric field E^{-1} in order to illustrate the Poole–Frenkel and tunnel effects, respectively. (Reprinted with permission from Canali *et al.*, 1973. © 1973, Pergamon Press, Ltd.

of the data versus the reciprocal of the electric field at high electric fields results in a good fit to tunnel theory. The value of 0.025 eV assigned to the shallower Cl and Br levels is larger than that of the hydrogenlike donor. Although not specifically reported by Canali *et al.* (1975b) shallower levels at about 0.02 eV in concentrations of about 10^{17} cm^{-3} in In-doped material act primarily as scattering centers especially at higher temperatures. The deeper 0.5 eV level may be associated with the 1.584 eV luminescence line assigned to an exciton bound to a neutral donor [(1.596 − 1.584 eV) × 5 = 0.06 eV] and/or the 1.545 eV edge luminescence line assigned to direct donor–valence band recombination. It is more generally accepted that the edge luminescence is associated with shallow acceptors and will be discussed later in Section II,B,3,b. The bound-exciton recombination spectrum in the In-doped samples of Fig. 3.8 is dominated by a band at 1.584 eV. Because its intensity is strong relative to the 1.593 eV band it is ruled out as being the n = 2 replica of the exciton bound to the neutral donor (Suga *et al.*, 1974).

In contrast with the frequent observations of the 1.589 and 1.593 eV bands, the 1.584 eV band has been reported in only a few studies (Triboulet *et al.*, 1974; Triboulet and Marfaing, 1973; Barnes and Kikuchi, 1975) and it does not dominate the exciton spectrum in any of these reported spectra. Since Triboulet and Marfaing observed band-to-band recombination at 1.605 eV, they attributed the 1.584 eV band to the first longitudinal optical (LO) phonon satellite of this emission. Band-to-band recombination is observed only rarely and only in very pure material as in the case of Triboulet and Marfaing. As expected, the 1.605 eV emission is not observed in compensated In-doped samples. Based upon additional line broadening and doping studies the 1.584 eV band is assigned to excitons trapped at a center associated with In. Except for differences in the line intensities, the general exciton structure is comparable between materials of different resistivity and purity. The exciton spectrum of Triboulet and Marfaing would be expected to be even more similar to that of Suga *et al.* since the electron concentration and mobilities are very similar. However, this is not so.

The physical nature of the 0.05 eV level is difficult to determine. However, it is most evident in high resistivity compensated material and is tentatively assigned in In-doped material to the donor $2\text{In}_{\text{Cd}}^+ \text{V}_{\text{Cd}}^{2-}$. Bell (1975) associates this level in Cl-doped material with the corresponding defect $\text{V}_{\text{Cd}}^{2-} 2\text{Cl}_{\text{Te}}^+$. Assignment of deeper levels is not straightforward. The level 0.6 eV from the conduction band has been associated with the second charged state of a double acceptor consisting of either a vacancy or a complex involving it (de Nobel, 1959; Lorenz and Segall, 1963; Agrinskaya *et al.*, 1969, 1970a). Lorenz *et al.* (1964) and Woodbury and Aven (1964) suggest that another native defect is involved and the 0.6 eV level could be the first level of the double acceptor observed electrically. The rather efficient trapping of electrons at this level precludes assignment of a charge state of a negative two. More than likely there are two different centers both having comparable activation energies.

Scharager et al. (1975) and Cumpelik et al. (1969) both measured levels at about 0.6 eV, using thermally stimulated current measurements. Using space-charge-limited currents Canali et al. (1975a) measure a center in In-doped semi-insulating material at the same depth in concentrations of 10^{12} cm^{-3} and having a trap cross section of 4×10^{-13} cm^2. Marfaing et al. (1974) using photocapacitance measurements in low resistivity material associated a 0.58 eV level with the doubly ionized native donor observed in high temperature electrical measurements. The large value of the electron trapping cross section ($\sigma \simeq 10^{-14}$ cm^2) is in agreement with this assignment. However, the depth of this level is much larger than the maximum value of about 0.3 eV as estimated from high temperature electrical measurements.

B. ACCEPTORS

1. *The 0.15 eV Complex*

An acceptor level at about 0.15 eV from the valence band is commonly seen in Hall measurements with *p*-type material (de Nobel, 1959; Yamada, 1960). In addition, this energy level is observed by luminescence, time of flight, and thermally stimulated current measurements and is induced either by radiation damage or by firing at an elevated temperature.

a. Luminescence Measurements. This level is presumably responsible for the 1.45 eV line commonly seen in the broad band impurity region ($hv <$ 1.48 eV) of photoluminescence (Halsted et al., 1961) and cathodoluminescence (Vavilov et al., 1964) spectra. In In-doped material this level is associated with the $V_{Cd}^{2-}In_{Cd}^{+}$ complex (Barnes and Zanio, 1975) and in halogen-doped material with the $V_{Cd}^{2-}Cl_{Te}^{+}$ complex (Agrinskaya et al., 1971b). The presence of this band in undoped material probably is due to a residual impurity such as Al. In order to define better the role of impurities in its formation, Barnes and Zanio analyzed the photoluminescence spectra of CdTe containing from 10^{16} to 10^{18} In atoms cm^{-3} (Fig. 3.8). Crystals were grown from excess Te and had *n*-type conductivity which increases with increasing In concentration. The most dramatic changes in the spectra are found in the broad band region. The area under this series of peaks increases directly with increases in the In concentration. Each series of peaks begins with the no-phonon peak at 1.454 eV and is separated by the LO phonon energy of 0.021 eV. For samples having donor concentrations greater than 10^{17} cm^{-3} Agrinskaya et al. (1970b) observed transverse optical (TO) phonon transitions to occur between the LO phonons. Violation of the selection

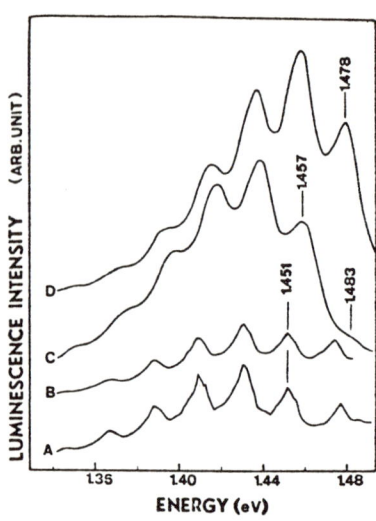

FIG. 3.11. Luminescence spectra at 4.2°K for CdTe grown by THM for (A) undoped crystals grown at 900°C and for crystals doped with (B) Al, (C) In, and (D) Cl. (Reprinted with permission from Furgolle et al., 1974. © 1974, Pergamon Press, Ltd.)

rules for electron–vibrational transitions in the presence of impurities explains this coupling. Such extra transitions are not apparent for the more heavily doped samples in Fig. 3.8. For example D, which contains the largest In concentration (10^{18} cm^{-3}), not even the phonon satellites are resolved, presumably because of poor crystal quality. Figure 3.11 shows the broad band luminescence spectra for crystals containing I (A), Al (B), In (C), and Cl (D) (Furgolle et al., 1974; Siffert et al., 1975). A fit of the distribution to $I_n = (\bar{N}^n/n!)I_0$ results in no-phonon lines ($n = 0$) at 1.478 eV and 1.457 eV for Cl and In, respectively. The value of 1.454 eV from Fig. 3.8 for In is in good agreement. The electron–phonon coupling strength or mean number of phonons created per transition ($\bar{N}$) is about 1.8 and 1.9 for Cl and In, respectively. Agrinskaya et al. (1971b) find that the zero phonon lines for both In and Al have energies of 1.45 eV. Unfortunately, the position of the no-phonon line for the Al-doped series in Fig. 3.11 is not sufficiently defined to make a comparison. Agrinskaya et al. (1971b) also observe a shift of about 0.03 eV between In and Cl. They base their results upon the difference in distances between the components of the complex. In the case of the Cl complex the distance between components of the complex is equal to the separation between nearest Cd and Te sites (2.7 Å) as Cl substitutes for Te. For the In complex the separation between Cd sites is 4.5 Å. The calculated values of the activation energy of a hole for the $(V_{Cd}Cl_{Te})^-$ and $(V_{Cd}In_{Cd})^-$ complexes are 0.14 and 0.17 eV and are in good agreement with these results. Comparable activation energies also have been obtained by thermal quenching of the luminescence (Agrinskaya et al., 1971b; Taguchi et al., 1973).

Halsted et al. (1961) also observed no-phonon lines of 1.453 eV and about 1.49 eV for p-type undoped and Ag-doped material. Lithium also shifts the activation energy of the 0.15 eV level (Desnica and Urli, 1972). A self-consistent picture of the nature of the 1.4 eV transitions is not available. Although it is accepted that these transitions terminate at deep acceptor levels, their origin is still controversial. The injection level and time-dependence measurements of Taguchi et al. (1973) are consistent with donor–acceptor (DA) pair transitions. However, this group, as well as Bryant et al. (1972) and Agrinskaya et al. (1971b) obtain about 0.13 eV activation energies from thermal quenching measurements that indicate not DA but conduction band to acceptor transitions. Agrinskaya et al. propose that these transitions occur within compact defect–dopant complexes with no Coulombic term involved. Norris and Barnes (1976) investigated this complex in more detail through the injection level dependence, frequency dependence, and thermal quenching of cathodoluminescence. The peak position of most samples was independent of injection level. The frequency response of the 1.4 eV band had a characteristic decay time of about 5 μsec, which is characteristic of DA transitions. However, such transitions were ruled out because of the absence of injection level shift. The fact that the dynamic response was independent of dopant concentration and thermal activation energies of 0.1 eV were measured are further evidence of conduction band to acceptor transitions. Except for the long decay times, the data does not support transitions from a donor level.

b. *Time of Flight Measurements.* The 0.15 eV level is seen in high resistivity In-doped (Ottaviani et al., 1972, 1973) and halogen-doped (Bell et al., 1974a; Canali et al., 1974) material with the time of flight technique. Evidence exists that the bonding for Cl complexes is stronger than for Br complexes. The concentration of the 0.15 eV traps in high resistivity Br-doped material is about an order of magnitude greater than the concentration of traps in high resistivity Cl-doped material. The stronger bonding of Cl with vacancies tends to increase the concentration of the more complex defects ($V_{Cd}^{2-} 2Cl_{Te}^{+}$) and reduce the concentration of isolated defects and more simple complexes ($V_{Cd}^{2-} Cl_{Te}^{+}$). The simpler defects are then associated with the trapping measurements. This analysis assumes that the total concentration of Cl and Br are comparable. Unfortunately, the concentration of these dopants is unknown and this model cannot be verified.

c. *Thermally Stimulated Current Measurements.* Scharager et al. (1975) report a 0.13 eV acceptor level from thermally stimulated current (TSC) measurements. A trap cross section of about 10^{-16} cm^2 for holes was measured. A level of about this energy is found in other TSC measurements. However, an analysis of other TSC measurements does not show a pattern to evolve (Toušek et al., 1967; Čumpelík et al., 1969; Agrinskaya et al., 1969; Gettings and Stephens, 1973; Zayachkivskii et al., 1974; Scharager et al., 1975; Barnes and Zanio, 1976). The measurements of Scharager et al. were singled out because the levels observed are somewhat consistent with those reported by other methods.

d. *Electron Spin Resonance Measurements.* Electron spin resonance has been associated with this complex in the Zn chalcogenides containing group III or VII atoms (Title, 1967). This complex is not observed in CdTe although surely present. Resonance has been associated with isolated cation vacancies in ZnSe but has not been reported for CdTe (Watkins, 1971).

e. *Thermal Introduction of the 0.15 eV Level.* The model for this complex should be consistent with the availability of Cd vacancies. For n-type undoped CdTe (Cd excess) Halsted et al. (1961) did not find this band, the PL band at 1.45 eV. However, for undoped p-type material presumably containing an excess of Te, the series with a no-phonon energy of 1.43 eV was found. The intensity of this band does not always behave consistently with firing conditions. Except for small energy shifts, Bryant and Totterdell (1972) find the 1.45 eV band in cathodoluminescence to be relatively inert to heat treatments. However, when the samples are irradiated with electrons capable of producing Cd displacements (i.e., energies greater than 260 keV) there is an increase in the intensity of emission. Irradiation by such energetic electrons at 70°K does not result in enhancement. Only when the Cd defects, presumably vacancies are allowed to migrate at room temperature, is the luminescence enhanced. The actual temperature at which radiation-induced Cd displacements are mobile is not well defined although in separate edge emission studies Bryant et al. (1972) assume Cd displacements to anneal out below 77°K. Taguchi et al. (1973) rule out isolated Cd vacancies or vacancy complexes in the formation of the 1.45 eV band since the intensity is increased and not decreased when material is fired in an overpressure of Cd. In impurity enriched material Rud' et al. (1971) also found an increase in the intensity with increasing p_{Cd} interstitials. When purer material was annealed, the emission intensity did not increase with increasing p_{Cd}.

Agrinskaya et al. (1971b) observed the 1.4 eV band to be weak after crystal growth of material containing In. However, after a subsequent 600°C anneal the 1.4 eV band became strong, corresponding to the diffusion of defects to form complexes. Rud' et al. (1971) also noted that for samples containing impurities the rate of cooling affected the intensity of the 1.45 eV band. Quenching from 850°C suppressed it whereas slow cooling enhanced it. Both results are consistent if the crystals of Agrinskaya *et al.* were quenched from well above 600°C after crystal growth. These results imply unusually slow diffusion constants for the native defects.

When samples contain high residual concentrations of donor impurities such as Al, large concentrations of vacancies may be generated regardless of the Cd vapor pressure. This could conceivably explain the results of Bryant and Totterdell (1972). However, the decrease in intensity with decreasing p_{Cd} is difficult to explain unless a comparable level is formed by acceptor impurity-native defect complexes. The presence of large concentrations of and recombination through a specific residual acceptor impurity could decrease the intensity of the complex emission at low p_{Cd}. Since the 1.45 eV band is observed in *n*-type, *p*-type, and high resistivity material, its formation appears to be more sensitive to the presence of impurities rather than the relative vapor pressures of Cd and Te.

f. Radiation-induced 0.15 eV Level. Irradiation with fast electrons changes the resistivity of both *n*- and *p*-type CdTe. When Matsuura *et al.* (1967) irradiated *p*-type CdTe containing 10^{18}–10^{19} Cu cm^{-3} at 210°K with monoenergetic electrons there was an increase in the net concentration of acceptors and the conductivity below a 400 keV electron energy and a decrease above 400 keV. The presence of large amounts of Cu make the results difficult to interpret but they are consistent with the threshold energy for atomic displacement. Barnes and Kikuchi (1968) analyzed *n*-type material and found little effect of ^{60}Co gamma irradiation. Chester (1967) also analyzed *n*-type material and found that irradiation by ^{137}Cs and ^{60}Co resulted in small changes which corresponded to about 10% increases in the net concentration of electrons and holes, respectively. These changes are associated with electron energies up to about 1.0 MeV for ^{60}Co and up to about 0.5 MeV for ^{137}Cs since most of the photon interactions are due to Compton events. The changes in conductivity are relatively small. However, it is worth noting that irradiation by the higher energy electrons generally compensates the material and drives it to higher resistivity. Although simultaneous electrical measurements were made in these studies, they were outside of the range of the 0.15 eV level.

Preferred displacement of Cd atoms occurs upon exposure to thermal neutrons. Electrical measurements show the 0.15 eV level to be clearly operative. The capture of a thermal neutron by ^{113}Cd results in the emission of about 9 MeV of gamma rays and recoil of the Cd nucleus. The degree of recoil and extent of damage, possibly involving Te displacements, depend upon the lifetime of the excited state. Based upon a displacement threshold energy of 15 eV the majority of the events result in only Cd displacements. There are only occasionally more energetic recoil events resulting in multiple displacements involving Te atoms. An analysis of the electrical properties before and after irradiation with thermal neutrons indeed indicates that the major changes are primarily associated with Cd rather than with Te defects. Barnes and Kikuchi (1968) irradiated n-type material having approximately 10^{15} cm^{-3} donors with fluences up to 10^{17} neutrons/cm^2 and obtained a threefold decrease in the carrier concentration. Abramov et al. (1970) and Chester (1967) obtained decreases over several decades and even attained type conversion. Abramov et al. did not obtain type conversion for samples having initial room temperature carrier concentrations greater than 5×10^{16} cm^{-3} even for doses up to 5×10^{18} neutrons/cm^2. In the work of Barnes and Kikuchi, type conversion was not observed, even though the initial carrier concentration was less than 5×10^{16} cm^{-3}. Although the neutron fluences were not as intense as those of Abramov et al., the carrier concentration appeared to saturate at the higher doses. These differences are presumably due to differences in the material. Nevertheless, this phenomenon is associated with Cd displacement since activation by the accompanying small fast neutron fluence resulted in insignificant changes. The p-type samples of Abramov et al. were much less sensitive to irradiation than n-type samples. Before irradiation, p-type samples had levels of 0.05 eV and 0.15 to 0.20 eV. After irradiation, both p-type and converted n-type samples had a single acceptor level of from 0.14 to 0.18 eV. To the contrary, Urli (1967) and Barnes and Kikuchi (1970) found dramatic increases in the resistivity of p-type material. Barnes and Kikuchi found the hole concentration to saturate after a decrease from 10^{16} holes/cm^3 to 5×10^9 holes/cm^3. In Barnes and Kikuchi's samples the 0.15 eV level dominates before irradiation and an 0.20 eV level after. In contrast, Urli observed type conversion of the originally p-type material, with a donor level at 0.02 eV dominating after irradiation. Levels at 0.3 and 0.4 eV are also revealed. Besides the 0.02 eV donor level, 0.09 eV and 0.35 eV levels were present in the p-type material of Urli before irradiation. Earlier work by Urli (1966) also resulted in p- to n-type reversal.

2. *Impurities*

The noble metals Au, Cu, and Ag primarily substitute for Cd and act as acceptors and secondarily fill interstitial positions and act as donors. Gold is probably the best studied of this group (de Nobel, 1959; Akutagawa et al., 1974, 1975). The covalent radius (1.50 Å) and the ionic radius (1.37 Å) of Au are close to that of Cd (1.48 Å), making substitution for Cd easy. The relatively small size also makes octahedral sites favorable interstitial positions. For CdTe containing 5×10^{16} cm^{-3} Au and fired at low p_{Cd}, de Nobel measured only 2×10^{15} holes/cm^3. For Cu, the concentration of holes was over two decades less than the amount added. Annealing Au doped samples at 100°C results in further decreases the hole concentration. De Nobel attributes this to the relocation of Au to interstitial sites where they act as donors. Akutagawa et al. (1974, 1975) found that for larger concentrations of Au, material is high resistivity. They attribute this to compensation by doubly ionized Cd interstitials. At large concentrations of Au it is likely that Au goes interstitial to act as a compensating center. De Nobel finds activation energies ranging from 0.27 to 0.35 eV for Au. A hole trap at 0.3 eV was also measured by Scharager et al. (1975) and Agrinskaya et al. (1969) using thermally stimulated currents. This level was not assigned to a specific defect or impurity.

Peaks corresponding to these energies are seen in luminescence. However, with increasing depth of the levels, the emission peaks become broader due to the stronger localization of the carrier at a center and consequently to an increase in the total number of phonons (i.e., increasing $\bar{N}$) that are allowed to interact with the center (Panossian et al., 1969; Gippius et al., 1974). For levels much deeper than 0.2 eV, resolution of the no-phonon line is difficult or impossible. De Nobel (1959) and Halsted et al. (1965) associate broad band peaks at 1.27 eV and 0.72 eV with Au. Garlick (1959) and Halsted et al. associate peaks at 0.92 eV and in the range 0.88 to 1.23 eV with Cu. Besides the no-phonon peak at 1.49 eV Halsted et al. associate a peak at 1.19 eV with Ag.

Several levels are identified specifically with Li. However, Desnica and Urli (1972) consider the presence of Li near defect sites to primarily alter the ionization energies of these levels. In analyzing both Li-doped and undoped samples irradiated with ^{60}Co gammas, they associate a 0.14 eV level with a complex requiring Li. Because of its small ionic radius (0.7 Å), Li diffuses rapidly and precipitates easily at room temperature. In large concentrations Li both substitutes for Cd to form acceptors and occupies interstitial sites to form compensating donors. Desnica and Urli saturated CdTe with up to 10^{21} cm^{-3} of Li at about 700°C and after cooling to room temperature associated changes in the electrical properties and the activation energies with different stages of precipitation of and the formation of complexes involving Li. The resistivity initially decreases corresponding to the precipitation of interstitial Li at room temperature. The resistivity changes up to an order of magnitude within a period of one day. Initial changes before measurement are probably several orders of magnitude. This rapid adjustment is followed by a slower process which reflects the subsequent adjustment of native defects and complexes thereof to compensate the residual acceptors. Different activation energies are associated with these stages. Usually low room temperature mobilities ($\mu_H \approx 0.1$ cm^2/V-sec) characterize severe disorder in Li-doped material. In higher mobility material, but material still dominated by impurity scattering, the mobility unexpectedly increases with decreasing temperature. When the temperature is lowered the association of complexes occurs thus reducing the concentration of scattering centers (Arkad'eva et al., 1969).

Both phosphorus and arsenic act as acceptors (Morehead and Mandel, 1964; Donnelly et al., 1968; Arkad'eva et al., 1966b; Kachurin et al., 1968; Agrinskaya et al., 1974) and would be expected to replace Te. Hall and Woodbury (1968) found P-doped CdTe to be *p*-type even when fired under saturated Cd pressure. The hole concentration was significantly less than the total amount of P added, suggesting a large degree of compensation. For samples fired at lower p_{Cd} the solubility is nearly independent of p_{Cd}. At higher p_{Cd} the solubility increases with p_{Cd} reaching a concentration of about 10^{20} cm^{-3} for samples fired at 900°C. At the higher pressures Hall and Woodbury suggest that P substitutes for Te and is highly compensated by a donor such as Cd_i. At lower p_{Cd} the difficulty in making stronger *p*-type material is associated with the formation of neutral substitutional complexes or the phosphorus occupying interstitial sites either as a donor or a neutral entity. Based on more detailed electrical measurements Selim and Kröger (1977) interpret the acceptor to be P interstitials and P on Te sites and the donor to be P on a Cd site. A range of activation energies are found for phosphorus. An upper value of about 0.05 eV is more characteristic because it represents a better estimate of infinite dilution. Gu et al. (1975) observed a maximum in the Hall coefficient for P-doped material which they attribute to impurity banding.

Electrical measurements indicate that the chromium acceptor level lies within 0.6 eV of the conduction band (Ludwig and Lorenz, 1963). The assignment of Cr as an acceptor is consistent with electron spin resonance measurements which identify chromium as $Cr(3d^5)$. Table 3.3 summarizes electron spin resonance results on the rare earth elements, donors, and other transition elements. Optical absorption measurements show that Fe^{2+} substitutes tetrahedrally for Cd (Slack et al., 1966). In fact for CdTe fired in Cd atmospheres, nearly all the Fe added in concentrations of about 10^{18} cm^{-3} is in this form (Slack et al., 1969). At concentrations of about 10^{19} cm^{-3} the Fe^{2+} concentration decreases with increasing Cd pressure.

TABLE 3.3

ELECTRON–SPIN RESONANCE STUDIES IN CdTe

Impurity	Investigator	Impurity	Investigator
Mn	Lambe and Kikuchi (1960)	Yb	Watts and Holton (1967)
Cr	Ludwig and Lorenz (1963)	Tm	Watts and Holton (1967)
Co	Ham et al. (1960)	Tl	Du Varney and Garrison (1975)
Er	Watts and Holton (1967)	Shallow donor	Alekseenko and Veinger (1974)
Nd	Watts and Holton (1967)		

Optical absorption studies show that hot electrons introduced by applying a high electric field decrease the populations of the lower Fe^{2+} ion levels and increases those of the higher levels (Fenner et al., 1968).

Levels introduced by other impurities are not so well understood. When Zayachkivskii et al. (1974) and Nikonyuk et al. (1975a,b) added Ge to CdTe semi-insulating material was obtained with an acceptor level at $E_v +$ (0.60 to 0.65 eV). Whether this level is directly due to a compensating effect of Ge is not known. These studies are important since there should be similarities between Ge and Si, the latter a contaminant present in the preparation of CdTe.

Oxygen, selenium, and sulfur replace Te. Relatively little work has been done on oxygen although it is common in CdTe and is expected to result in a local level which should act as an electron trap (Agrinskaya et al., 1974). Electrical measurements on n-type material doped with oxygen, however, do not show acceptor properties. Changes due to oxygen doping are apparent in photoluminescence.

Numerous levels are found with thermally stimulated current and thermally stimulated capacitor discharge methods. Some are consistent with ionization energies obtained from other techniques. However, the absence of an overall consistent pattern of levels prevents a good comparison with levels obtained from other techniques.

Ionization energies for p-type impurities determined by electrical measurements are listed in Table 3.4. Their interaction with native defects and other impurity atoms prevent precise determination of the activation energy and should be considered when referring to the table.

3. *The 0.06 eV Level*

a. *Electrical Measurements.* In undoped but highly compensated material ($N_d = 0.985 N_a = 1.7 \times 10^{18}$ cm^{-3}), Lorenz and Segall (1963) report a level 0.05 eV above the valence band edge, which they attribute to a native defect or complex thereof, although there is no evidence of its origin. Rud' et al. (1971) associate a hole density of about 5×10^{16} cm^{-3} at 300°C with this level that is not observed in high resistivity material, using time of flight measurements. The Cd vacancy would be a convenient assignment. However, considering the lack of evidence for isolated charged Cd vacancies in high temperature electrical measurements, it is unlikely that this level is associated with such a simple defect in the uncompensated state at such high concentrations. Watkins (1971) reports the presence of the cation vacancy in ZnSe, using electron spin resonance measurements (ESR), after irradiation with 1.5 MeV electrons. Although simple Cd vacancies have not been confirmed in either ESR or time of flight measurements in CdTe, they are surely present but in lower concentrations.

TABLE 3.4

Ionization Energy of Impurity Associated Acceptors in CdTe as Determined by Hall Measurements

Assignment	ε_A (eV)	Reference
Ag	0.3	Lorenz and Segall (1963)
Cu	0.35	Lorenz and Segall (1963)
Au	0.4	Lorenz and Segall (1963)
Cu, Ag, Au	0.33	de Nobel (1959)
Li	<0.05, 0.05, 0.14, 0.27	Desnica and Urli (1972)
Li	0.034	Crowder and Hammer (1966)
Li	<0.01, 0.22, 0.32	Vul and Chapnin (1966)
Li	0.028–0.17	Arkad'eva et al. (1969)
Li, Sb, P, Ag, Pb	0.27–0.51	Jenny and Bube (1954)
As	0.17^a	Ichimiya et al. (1960)
Na	0.028	Crowder and Hammer (1966)
P	0.03–0.04	Selim and Kroger (1977)
P	0.05	Arkad'eva et al. (1967b)
P	0.03	Arkad'eva et al. (1975)
P	0.036–0.052	Gu et al. (1975)
Cr	<0.6	Ludwig and Lorenz (1963)
V_{Cd}-complex	0.15	Abramov et al. (1970)
V_{Cd}	0.15	de Nobel (1959)
	0.20	Yamada (1960)
V_{Cd}, V_{Cd}-complex	0.05	Lorenz and Segall (1963)
Native defect	0.05	Rud' and Sanin (1971)
(halogen–V_{Te} complex)	$E_c - 0.06$	Woodbury and Aven (1965)
(Native defect–impurity complex)	$E_c - 0.06$	Lorenz et al. (1964)
DA^{2-b} (V_{Cd}–(complex))	$E_c - 0.06$	Abramov et al. (1970)
Native defect	$E_c - 0.06$	Lorenz and Woodbury (1963)
(2Cl–native defect complex)	$E_c - (0.09-0.12)$	Agrinskaya et al. (1975)
DA^{2-} complex	$E_c - 0.70$	Chapnin (1969)
DA^{-c} (Native defect complex)	$E_c - 0.6$	Lorenz et al. (1964)
DA^{2-} (Ge-induced)	0.62	Matlak et al. (1972)
Ge-induced	0.6–0.65	{Zayachkivskii et al. (1974) / Nikonyuk et al. (1975a,b)}
V_{Cd}^{2-}	$E_c - 0.6$	de Nobel (1959) and Kroger (1974)

[a] Authors note irreproducible results.
[b] DA^{2-}: Second charged state of a double acceptor.
[c] DA^-: Single charged state of a double acceptor.

II. DEFECTS AT LOW TEMPERATURE

b. Edge Emission. The edge emission region (1.48–1.56 eV) of the luminescence spectrum is generally rich in transitions that are often associated with free-to-bound recombination at a level near 0.06 eV. These transitions are associated with point defects of Cd. Bryant *et al.* (1968) irradiated samples at low temperature with monoenergetic electrons up to energies of 450 keV and analyzed the luminescence bands to determine the threshold energies for the displacement of Cd and Te atoms (Fig. 3.12). The onset of increased intensity occurs at about 235 keV for the 1.53 eV and 1.54 eV peaks and at about 340 keV for the 1.13 eV peak. This corresponds to threshold energies of 7.8 eV and 5.6 eV for Te and Cd, respectively. The defects are immobile at liquid helium temperatures. Annealing the samples at room temperature restores the spectra before irradiation and presumably results in the annihilation of the excess concentrations of simple isolated point defects of Cd and Te. These results do not imply that simple native defects are absent at room temperature but illustrate that such defects are quite mobile, and if uncompensated, will easily diffuse and may precipitate. Defects in CdTe or any other semiconductor are generally not in total equilibrium with the lattice at room temperature. However, electrically active isolated native defects are close to equilibrium with impurity associated defects in CdTe because of the high mobility of the native defect and the large driving force of self-compensation. Therefore, where the crystal is annealed above room

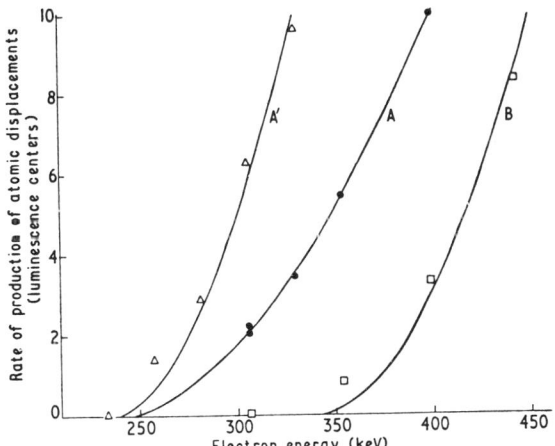

FIG. 3.12. Rate of production of atomic displacements as determined by the intensity of the (A′) 1.54, (A) 1.53, and (B) 1.13 eV luminescence lines versus electron energy. The 1.54 and 1.53 eV lines refer to the onset of Cd displacement at about 235 keV; the 1.13 eV line refers to the onset of Te displacement at about 340 keV. (From Bryant *et al.*, 1968. Copyright by The Institute of Physics.)

150   3. DEFECTS

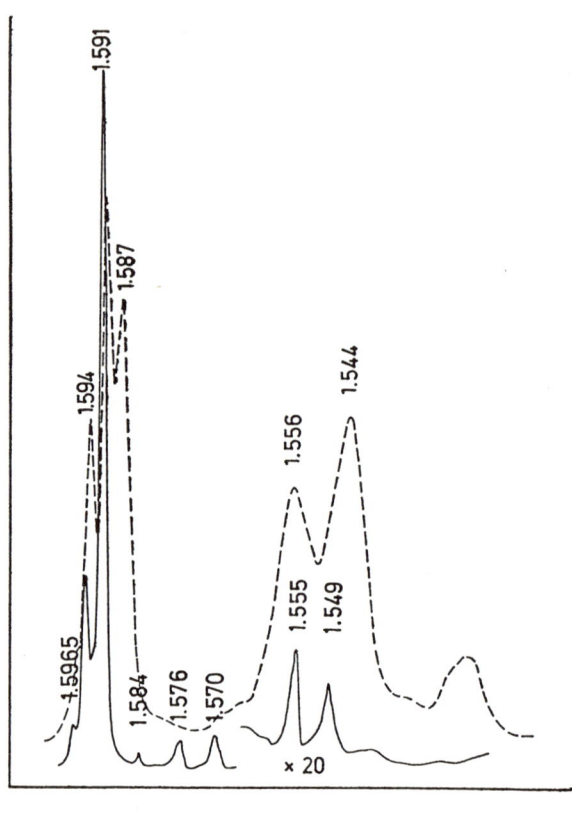

ENERGY (eV)

FIG. 3.13. Photoluminescence spectra of CdTe grown by the traveling heater method, using Cl-doped (---) and undoped (—) starting materials (Triboulet et al., 1974).

temperature uncompensated isolated native defects disappear. This annealing phenomenon suggests that electrically active isolated native defects are not the dominating defect species at room temperature. Therefore, although room temperature edge emission may be due to native defects it requires the presence of impurities.

In some cases the edge emission region (1.48 eV–1.54 eV) is void of transitions. This only suggests high purity material since impurities may be concentrated in precipitates. Figure 3.13 is a photoluminescence spectrum (PL) of undoped crystals grown by the traveling heater method (Triboulet et al., 1974; Siffert et al., 1975). The source material was zone refined CdTe. Except for the exciton region this PL spectrum is void of transitions. Material taken from the middle and end of the ingot show lines at 1.544 eV and 1.555 eV, the intensity of these lines increasing with impurity concentration. The intentional addition of Cl results in an increase in the intensity of the peak at 1.556 eV and an additional peak at 1.544 eV. The use of Bridgman feed material also results in the line at 1.544 eV. Earlier work by Marfaing and Triboulet (1971) on undoped and In-doped Bridgman grown material also shows transitions at 1.544, 1.550, and 1.556 eV. Barnes and Zanio (1975) found the intensity of the line at 1.544 eV to decrease with increasing In content. However, the intensity of the 1.45 eV peak simultaneously increased (Fig. 3.8). For this case, it is likely that the decrease in intensity with increasing In content is more related to an increase in the traffic through indium–vacancy complexes at 1.45 eV. Agrinskaya *et al.* (1971a) and Halsted *et al.* (1961) also found peaks in the edge emission region with doped and undoped material. Table 3.5 summarizes some of the edge luminescence results. The peak intensities are identified as weak (·), moderate (●), and intense (●). Phonon-assisted lines are not included. One energy does not dominate. It is more than likely that there is more than one defect which accounts for the range of transition energies from 1.52 V to 1.55 eV. Bryant *et al.* (1972) also assumed that Te defects, presumably vacancies, induce transitions in the region. They undertook additional studies at 77°K with monoenergetic electrons having sufficient energy to displace both Cd and Te. Since the irradiations were undertaken at 77°K the excess luminescence at 1.55 eV was attributed only to Te because at this temperature Cd defects are mobile and anneal out. If this is the case then the transitions in Table 3.5 may also be associated with Te vacancies.

Often transitions are characterized by doublets in this region. Agrinskaya *et al.* (1971a) classified transitions into two groups. The higher energy 1.54 eV group corresponds to free electron to acceptor transitions. The second group represents donor to acceptor transitions varying in energy from 1.52 to about 1.536 eV. An analysis of the lower energy lines provides information concerning the long range interaction between impurities and simple native defects. The peaks are shifted to shorter wavelengths when the excitation intensity and the doping level increase. The average distance between *all* the donor and acceptors, regardless of their state of charge, is fixed. However, the average distance between *neutral* donors and *neutral* acceptors which have trapped electrons and holes during the excitation process decreases with increasing excitation intensity. The energy of the emitted photon is given by the band gap less the donor and acceptor ionization energies plus a Coulombic term whose contribution increases with decreasing pair separation. Other explanations for transitions in this region have been proposed. Contrary to the results of Agrinskaya *et al.*, Taguchi *et al.* (1973) did not find any peak shift within this region, using time-resolved

TABLE 3.5

SELECTED EDGE LUMINESCENCE RESULTS IN CdTe

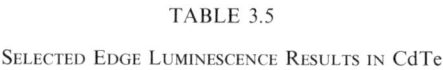

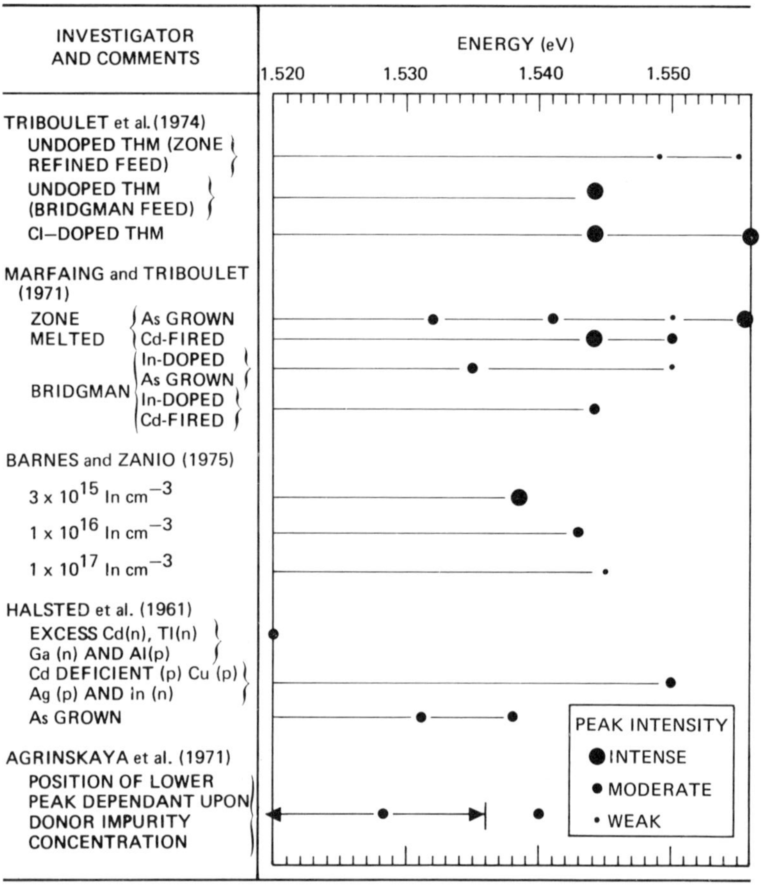

excitation measurements, and attributed both 1.54 and 1.528 eV peaks to free-to-bound transitions.

c. Double Acceptor. Most of the acceptors are located near the valence band. However, in the double acceptor model a higher energy transition in the doublet edge luminescence is associated with the transition of a free hole to a double acceptor located about 0.06 eV below the conduction band (Halsted and Segall, 1963). A lower energy transition is due to the recombination of an exciton bound to the singly charged state of the acceptor. The double acceptor is introduced in n-type material by either firing at elevated temperature in saturated cadmium atmospheres or by irradiating with 1.5 MeV electrons (Lorenz and Woodbury, 1963; Lorenz et al., 1963, 1964; Woodbury and Aven, 1964). The acceptor properties of this center after firing or irradiating are

(1) a decrease in the concentration of electrons,
(2) an increase in the concentration of electrons due to trapping of excess holes during photoexcitation,
(3) a simultaneous increase in the electron mobility during photoexcitation, presumably due to less effective scattering by a singly ionized center, and
(4) difficulty in attaining electronic equilibrium after cooling to lower temperatures.

Below 85°K several hours are required to attain thermal equilibrium with respect to the 0.06 eV level. Photoexcitation results in an increase in the concentration of electrons. When the light is turned off the excess concentration persists, because the conduction electrons are inhibited from returning to the original state by a 0.27 eV Coulombic repulsion barrier. Lorenz et al. (1964) associate this center with a native defect, most likely a vacancy associated with a residual impurity. The halogens are suggested as the residual impurity atom associated with the complex (Woodbury and Aven, 1964). Since the defects are formed under Cd-saturated conditions and are not observed after irradiation with thermal neutrons (Barnes and Kikuchi, 1968) it is not clear how acceptorlike Cd vacancies are energetically favorable. In CdSe, Matsuura et al. (1970) associate the double acceptor with the Se vacancy. The threshold energy for the production of Cd vacancies by electron irradiation is well above that for the production of Se vacancies. The generation of only Se defects is assured by irradiating at a lower energy. Also, the double acceptor in CdTe is annealed out above 300°C, whereas neutron induced and presumably cation associated defects are annealed out at about 130°C (Barnes and Kikuchi, 1970). Lorenz et al. consider a Te vacancy to act as an acceptor by modifying the covalent bonding about the vacancy. Hall measurements of CdTe:Cl annealed in Cd vapor also indicate the presence of an acceptor at $E_c - (0.09-0.12 \text{ eV})$ that has an electron repelling barrier of 0.5 to 0.6 eV. (Agrinskaya et al., 1975). The nature of the first ionized state is not specified. Phenomenon associated with the "double acceptor" may not be a result of a purely electron process. Losee et al. (1973), Baj et al. (1976) and Legros et al. (1976) associate relaxation processes, in particular the difficulty in attaining electronic equilibrium after cooling to lower temperatures, with atomic movements.

d. *Exciton Emission.* The 1.593 eV line dominates the PL exciton spectrum for the *n*-type CdTe shown in Fig. 3.7 (Hiesinger et al., 1975). However for

p-type material the same workers find the 1.589 eV line to dominate. The 1.589 eV line has been observed and reported by others in a variety of CdTe crystals (Halsted and Aven, 1965; Noblanc et al., 1970; Barnes and Kikuchi, 1975; Halsted et al., 1961; Bryant and Totterdell, 1971; Taguchi et al., 1973) and is associated with a neutral acceptor having an ionization energy of about 0.06 eV. Bryant and Totterdell (1971) have provided experimental evidence that the acceptor may be a neutral Cd vacancy by irradiating CdTe at low temperatures with electrons whose energy was sufficient to produce only Cd displacements. At an irradiation temperature of 4.2°K, where one would not expect the Cd defects to migrate, they observed large increases in 1.588 eV band intensity with increasing electron fluence. They concluded that the 1.588 eV band was due to an exciton trapped at a neutral Cd vacancy (V_{Cd}).

Analysis of high purity p-type material (Fig. 3.13) shows that the line at 1.591 eV dominates the exciton spectrum. Although significantly higher in energy than expected, this line is assigned to the neutral acceptor. Although the PL spectrum of the p-type THM crystal grown from a zone refined ingot is relatively clean as compared to the n-type material, band edge recombination is seen only in the latter. The p-type material was grown from Te solution whereas the n-type material was grown from a near congruent melt. Absence of band edge recombination may be due to the presence of inclusions in and the general poorer crystal quality of solution grown material versus material grown from the near-congruent melt.

C. Nonequilibrium Defect Structure

1. *Characteristic Temperature*

Brouwer plots (Fig. 3.2) are often used to describe the atomic disorder at room temperature. Their construction assumes that the defects are frozen in upon quenching from some elevated firing temperature and in the case of CdTe from temperatures as high as 900°C (de Nobel, 1959; Mandel, 1964; Kröger, 1965a,b, 1974, 1977; Chern and Kröger, 1975; Stuck et al., 1976, 1977; Marfaing, 1976). Evidence in this chapter supports the conclusion that the native defect structure cannot be frozen in by quenching from elevated temperatures. Instead defect migration and precipitation occurs. Because of the exponential and consequently strong dependence of the diffusion coefficient on temperature the disorder is characteristic of some lower temperature where the fastest moving defect becomes relatively immobile. Just what is the characteristic lower temperature is difficult to determine since resistivity changes occur even upon storage at room temperature. Any model based upon native defects being the majority species would be impossible to test because of the compensating effect of residual impurities. An alternative approach is to base a defect model on material containing known concentration of impurities. A familiar case to consider is In-doped material processed under Te-rich conditions. If V_{Cd}^{2-} and In_{Cd}^{+} are the majority defects, then the mobility of Cd vacancies determines the approach to the quasi-equilibrium state. In this case the diffusion coefficient of In (Kato and Takayanagi, 1963) is several orders of magnitude smaller than than that of Cd vacancies. When In_{Cd}^{+} and $V_{Cd}^{2-}In_{Cd}^{+}$ are the majority defects, the characteristic temperature is associated with the mobility of $V_{Cd}^{2-}In_{Cd}^{+}$ (Chern and Kröger, 1974). Complete equilibrium never occurs, but assigning the characteristic temperature to some temperature lower than the firing temperature is more realistic.

Evidence for characteristic temperatures in CdTe is available in the literature. Measurements of thermal neutron-induced damage in p-type CdTe revealed a major annealing stage between 400 and 500°K. (Barnes and Kikuchi, 1970). In addition, ion-implantation studies in CdTe by Gettings and Stevens (1974) indicated that raising the sample temperature to 473°K during implant resulted in a sharp increase in the fraction of implanted ions that were substitutional. The same authors (1973) also observed considerable annealing of ion-induced damage at 570°K in thermally stimulated current measurements of CdTe implanted with various ions. Meyer and Lang (1971) found that the average annealing temperature for the lattice damage in Bi- and Tl-implanted CdTe was 500°K. For Ar^+ implanted in In-doped CdTe at room temperature, the intensity of the 1.454 eV emission begins to increase sharply 480°K, reaching an intensity approximately ten times greater than the preimplant intensity (Barnes and Zanio, 1975). These results indicate that the number of 1.45 eV emission centers is growing above 480°K and that this growth is due to the onset of defect motion and the subsequent trapping of these defects at donor impurities.

2. Compensation Model

Because of the instability of electrically active native defects, they are assumed in this discussion to be present in substantial concentrations only in the presence of impurities. Schottky and Frenkel-type defect reactions are not considered here. Bell et al. (1974a) and Barnes and Zanio (1975) considered such a compensation model with halogen and In, respectively, as dopants in crystals grown at low p_{Cd}. Barnes and Zanio assume that the doubly charged cadmium vacancy, V_{Cd}^{2-}, forms two types of complexes: $V_{Cd}^{2-}In_{Cd}^{+}$, a singly ionizable acceptor, and $V_{Cd}^{2-}2In_{Cd}^{+}$, a fairly neutral entity, which can bind electrons or holes only weakly. Charge neutrality is maintained in the crystals primarily by the Cd vacancies (acceptors), In donors, and $V_{Cd}^{2+}In_{Cd}^{+}$ acceptors so that

$$2[V_{Cd}^{2-}] + [V_{Cd}^{2-}In_{Cd}^{+}] \approx [In_{Cd}^{+}] \quad cm^{-3}. \qquad (3.12)$$

The concentrations of free electrons and holes are negligible in these high-resistivity crystals. The two reaction equations of interest are

$$V_{Cd}^{2-} + In_{Cd}^{+} \rightleftharpoons V_{Cd}^{2-} In_{Cd}^{+} + \Delta H_1, \quad (3.13)$$

$$V_{Cd}^{2-} In_{Cd}^{+} + In_{Cd}^{+} \rightleftharpoons V_{Cd}^{2-} 2In_{Cd}^{+} + \Delta H_2, \quad (3.14)$$

and the corresponding equilibrium constants are

$$K_1 = \frac{[V_{Cd}^{2-} In_{Cd}^{+}]}{[V_{Cd}^{2-}][In_{Cd}^{+}]} = C_1 Z_1 \exp(-\Delta H_1/kT) \quad cm^3, \quad (3.15)$$

$$K_2 = \frac{[V_{Cd}^{2-} 2In_{Cd}^{+}]}{[V_{Cd}^{2-} In_{Cd}^{+}][In_{Cd}^{+}]} = C_2 Z_2 \exp(-\Delta H_2/kT) \quad cm^3, \quad (3.16)$$

where C_i is a constant and Z_i is proportional to both the configurational and vibrational portions of the entropy variation for each reaction (Kröger, 1974). Assuming that the reaction equation (3.13) usually occurs with the vacancy in the doubly charged state, then following Kröger, we take $\Delta H_1 \approx -1.0$ eV. The enthalpy for binding the second In atom to the complex is less than H_1 and is expected to be comparable to the value estimated by Kröger for the reaction equation (3.13) when the V_{Cd}^{-} is singly charged. Therefore, we use Kröger's value of -0.5 eV for ΔH_2. Equations (3.12), (3.15), and (3.16) contain four unknowns. The four unknowns can be determined after the following balance equation for the total indium concentration $[In_{Cd}]_T$ is included.

$$[In_{Cd}]_T = [In_{Cd}^{+}] + [V_{Cd}^{2-} In_{Cd}^{+}] + 2[V_{Cd}^{2-} 2In_{Cd}^{+}], \quad cm^{-3}. \quad (3.17)$$

Fig. 3.14 shows $[V_{Cd}^{2-} In_{Cd}^{+}]$ plotted versus $[In_{Cd}]_T$ with temperature as a parameter. Although not shown in the figure, the calculation indicates that $[V_{Cd}^{2-} In_{Cd}^{+}]$ predominates over $[V_{Cd}^{2-} 2In_{Cd}^{+}]$ at high temperatures and low $[In_{Cd}]_T$, while the opposite is true at low temperatures and high $[In_{Cd}]_T$.

Since the $V_{Cd}^{2-} In_{Cd}^{+}$ compensating acceptor center is responsible for the 1.454 eV transition, the concentration of luminescence centers can be calculated as a function of total In concentration using the above analysis. The recombination rate is proportional to the total area under the series of peaks beginning at 1.454 eV. Consequently, the comparison between the experimental results and the family of curves in Fig. 3.14 is made by normalizing the area under the spectrum for sample A to a value of $[V_{Cd}^{2-} In_{Cd}^{+}] = 1.25 \times 10^{15}$ cm^{-3} ($[In_{Cd}]_T = 3 \times 10^{15}$ cm^{-3} for sample A) and using this normalization factor to plot the areas for the remaining samples. The agreement between theory (the dashed line) and experiment is excellent if the characteristic temperature is near 500°K. A further analysis should consider In precipitation (Magee et al., 1975).

II. DEFECTS AT LOW TEMPERATURE

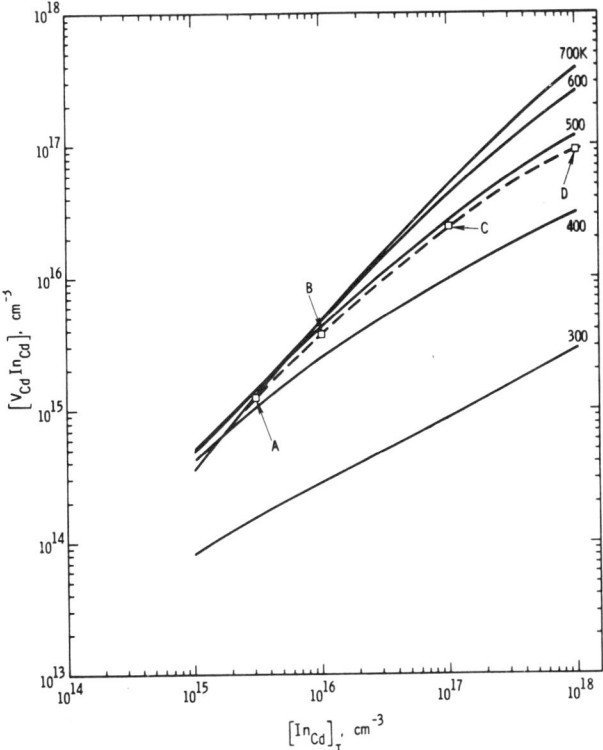

FIG. 3.14. Calculated concentration of $V_{Cd}^{2-}In_{Cd}^{+}$ acceptors $[V_{Cd}In_{Cd}]$ versus In concentration $[In_{Cd}]_T$ at different characteristic temperatures at which the atomic defects are "frozen in." Data points refer to normalized area under the 1.454 eV emission for (A) 3×10^{15} In cm^{-3}, (B) 10^{16} In cm^{-3}, (C) 10^{17} In cm^{-3}, and (D) 10^{18} In cm^{-3}. The data corresponds to a characteristic temperature of 500°K (Barnes and Zanio, 1975).

Only a qualitative description of the room temperature defect structure is possible for undoped material. For the case where the net concentration of electrically active residual foreign atoms are donors and material is processed at high p_{Cd}, Cd interstitials precipitate and the n-type conductivity is associated with foreign atoms. With decreasing p_{Cd}, a transition to the semi-insulating state occurs, where $(V_{Cd}^{2-} Al_{Cd})$, V_{Cd}^{2-}, and Al_{Cd}^{+} are the majority defects as in Eq. (3.17). Here Al_{Cd} is the residual impurity and V_{Cd}^{2-} is a deep level pinning the Fermi level near the center of the band gap.

When the net concentration of electrically active foreign impurities consists of acceptors, a definite $p-n$ transition occurs. At low p_{Cd} the residual acceptors dominate the conductivity. The n-type conductivity at high p_{Cd}

is associated with Cd interstitials, Cd_i^+ compensated by likely residual acceptor impurities such as Cu_{Cd}^-. The material is presumably n-type because the self-compensation mechanism prevents all of the excess Cd from precipitating. This results in a slight excess concentration of Cd_i^+ over Cu_{Cd}^-. Even if the concentrations were equal, material would be n-type since the 0.3 to 0.4 eV level for Cu is relatively deep. The difficulty in retaining large concentrations of deep native donors makes the semi-insulating state very difficult to attain by compensation with acceptor impurities. An additional difference between the group III and group VII donor impurities and the group I impurities is that the latter have amphoteric tendencies. It is likely that Cd_i-substitutional acceptor impurity complexes are present in high concentrations relative to the isolated specie because of the high mobility of both the group I impurities and native defects.

Because of precipitation and the presence of residual impurities, it is not possible to predict the pressure dependence of the electron concentration on p_{Cd} in undoped material. However, precipitation and compensation by residual impurities can explain the general behavior of the electron concentration. The anomalous decrease in the concentration of electrons with increasing firing temperature (de Nobel, 1959) may be due to increased concentrations of native defects with increased firing temperatures and their retention by compensating impurities after cooling to room temperature. For the case where a residual donor such as Al is present, firing at an elevated temperature, say 900°C, results in electrons, Cd_i^{2+}, Al_{Cd}^+, and V_{Cd}^{2-}, the concentration of these defects decreasing in this order. Quenching to room temperature results in the transformation of Cd_i^{2+} to Cd precipitates. The resulting concentration of electrons and compensating complexes $V_{Cd}^{2-} Al_{Cd}^+$ is less than the concentration of Al_{Cd}^+ as depicted in Fig. 3.15. Firing at a lower temperature, e.g., 700°C, results in a lower concentration of electrons, Cd_i^{2+}, and V_{Cd}^{2-} at the firing temperature. However, the final concentration of electrons is larger because of the lower concentration of V_{Cd}^{2-}, which forms $V_{Cd}^{2-} Al_{Cd}^+$.

A similar process also occurs in material containing residual p-type impurities such as Cu. The difference in this case is that the concentration of Cd interstitials is stabilized by foreign compensating acceptors. Just why the concentration of electrons is greater after the 700°C firing than after the 900°C firing is not clear. However, if the presence of Cd_i^+ partially stabilizes V_{Cd}^-, then a higher concentration of vacancies would be present after the 900°C firing than after the 700°C firing. This would result in a higher concentration of electrons in the latter case to maintain electroneutrality.

As a result of the precipitation of Cd and the stabilizing effect of the acceptor impurities, the room temperature p–n transition point moves to the right in Fig. 3.4 with increasing annealing temperature. For CdTe intentionally doped with Au as in Fig. 3.4 (middle) or other acceptors, the

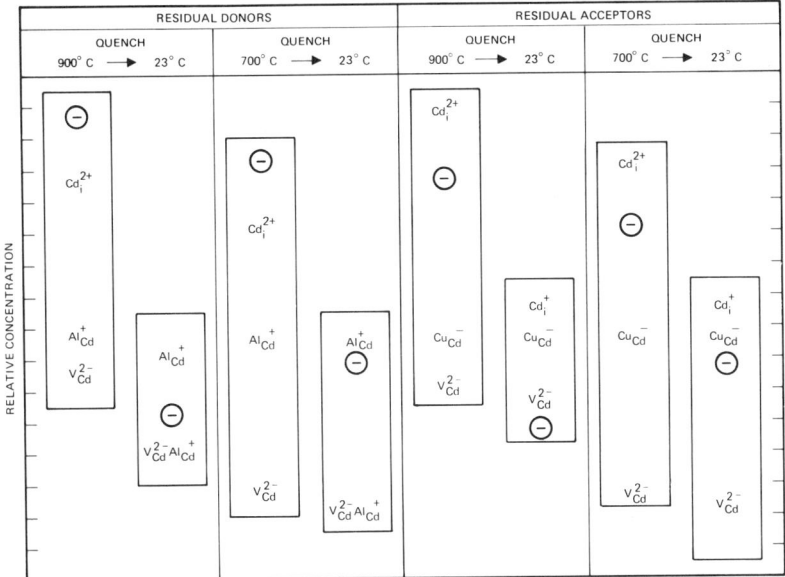

FIG. 3.15. Relative concentration (schematic) of electrons and atomic defects at two different firing temperatures and after rapid quenching to room temperature. Whether the residual impurities are donors (left) or acceptors (right), the room temperature electron concentrations are higher for the lower firing temperature.

transition is displaced even farther to the right. The transition point varies from sample to sample (de Nobel, 1959; Chern et al., 1975; Triboulet, 1968) and is therefore characteristic of the net residual acceptor impurity content and the degree of precipitation. Figure 3.4 (top) represents a sample containing about 10^{16} cm^{-3} residual Cu atoms. Although 5×10^{16} cm^{-3} Cl atoms were added to the sample in Fig. 3.4 (bottom), its transition point is to the right of the transition point for the sample containing residual Cu. Assuming 100% electrical activity for the impurities, the Cl-doped sample probably also contains a net residual concentration of acceptor impurities. Spectrographic analysis of the Cl-doped crystal showed the presence of Si (6×10^{17} cm^{-3}), Mg (2×10^{17} cm^{-3}), Cu (2×10^{16} cm^{-3}), and Fe, Pb, Ge, Ag, and Al in concentrations less than 10^{16} cm^{-3}.

Chern et al. (1975) pursued defect studies on undoped as well as In-doped crystals in further detail. In interpreting their results (Fig. 3.3) they assumed the concentration of excess Cd to be incorporated into the lattice by both V_{Te}^{2+} and Cd_i^{2+} with the ratio $[Cd_i^{2+}]/[V_{Te}^{2+}]$ increasing with increasing annealing temperature. Doubly ionized Cd interstitials are assumed to be the majority defect at high temperatures $\gtrsim 800°C$ and doubly ionized Te vacancies are assumed to be the majority defect at lower temperatures

$\lesssim 800°$C. Upon cooling to room temperature, interstitials precipitate, the concentration of vacancies remains unchanged, and the electrons are distributed over these vacancies and the remaining defects. The experimental results are in general agreement with the calculations. Since the concentration of nonprecipitating Te vacancies increases with increasing annealing temperature, the electron concentration should increase with increasing temperature, contrary to the results of de Nobel. A decrease occurs if the concentration of electrically active but compensating, nonprecipitating native acceptors increases to a greater extent with increasing firing temperature. The model of Chern *et al.* assumes that native defects are the majority electrically active defects. However, this chapter proposes that native defects, uncompensated by impurities, are unstable. More specific measurements indicate the Te vacancy to be unstable. Bryant and Webster (1967) observed a cathodoluminescence band near 1.1 eV following 77°K electron irradiation at energies above 340 eV, where both Te and Cd atoms are displaced. This band anneals out at about 125°K and did not appear when only Cd atoms were displaced by irradiating at lower energies. Taguchi *et al.* (1974) also find the 1.1 eV band from high energy electron irradiation to anneal out completely below 200°K. It is likely that this band is due to simple Te defects. The photoconductivity studies of Caillot (1972, 1975) indicate that simple V_{Te} gives rise to a level at $E_v + 0.46$ eV, which could be involved with the 1.1 eV band. This level anneals out rapidly above 77°K, giving rise to an $E_v + 0.33$ eV level that is stable at room temperature and is attributed to a complex involving V_{Te}. Other luminescence bands at 1.1 eV are stable at room temperature and also associated with Te complexes (Bryant *et al.*, 1968; Norris *et al.*, 1977). Radiation damage introduces both Te_i and V_{Te} and therefore the annihilation of Frenkel pairs is likely at higher temperatures where diffusion can proceed. Radiation damage studies only indicate that an above equilibrium concentration of Frenkel pairs is not stable at room temperature. This is not sufficient proof that V_{Te} alone is unstable but it does show that it is highly mobile and can diffuse readily to grain boundaries, dislocations, or to impurities to form complexes. If the large concentrations of Te_i proposed by Woodbury and Hall (1967) do exist at elevated temperatures, then the excess concentration at room temperature is likely to annihilate excess V_{Te}.

The tendency of radiation damage to drive CdTe to the higher resistivity state is also associated with the compensation processes that occur during the generation of native defects. Results vary among materials but the creation of Cd Frenkel pairs and the subsequent precipitation of defects and their association with residual impurities generally explain such phenomena. When Frenkel pairs are produced in material having a simple residual donor impurity such as Al and the temperature is high enough to permit diffusion, the following reaction equation proceeds to the right:

$$Al_{Cd}^+ + e + \overbrace{Cd_i^+ + V_{Cd}^-}^{\text{Frenkel pair}} \rightarrow Cd\text{ (precipitate)} + Al_{Cd}^+ V_{Cd}^{2-} + h. \tag{3.18}$$

Cadmium interstitials precipitate as a separate phase and Al associates with Cd vacancies to produce an acceptor level at about 0.17 eV. When p-type material presumably having a net residual impurity content is irradiated the following reaction equation also proceeds to the right:

$$Cu_{Cd}^- + h + \overbrace{Cd_i^+ + V_{Cd}^-}^{\text{Frenkel pair}} \rightarrow (Cu_{Cd}^- \, Cd_i^+) + V_{Cd}\text{ (precipitate)}. \tag{3.19}$$

Foreign acceptors associate with interstitials to form complexes. The degree of vacancy precipitation determines the degree to which the resistivity increases. Barnes and Kikuchi (1970) find that the defect responsible for increasing the resistivity anneals out at 130°C.

III. Summary

This chapter by no means presents a final discussion of defects in CdTe. However, it does highlight studies which provide insight into the structure. High temperature electrical measurements in excess Cd indicate that the Cd interstitial dominates the electrically active native defect structure above about 700°K. Consequently, material is n-type at elevated temperatures. High temperature diffusion studies and low temperature threshold measurements with high energy electrons show that native defects are extremely mobile at temperatures above approximately 150°K. The high mobility of the native defects prevents the defect structure from being frozen is at growth or postannealing temperature, and subsequent precipitation occurs during cooling to room temperature. Electrically active native defects are present near room temperature but only because of retention by compensating impurities. Material is n-type when prepared under cadmium overpressures. However, when prepared under tellurium overpressures, the material is p-type or high resistivity depending upon whether the excess residual impurities are, respectively, acceptors or donors. Electrical, luminescence, time of flight, and similar measurements shed some insight on the concentration, energy position, and nature of the native defects and impurities at room temperature. However, much is still unknown, especially with respect to the deeper levels.

Figure 3.16 summarizes more common levels observed in CdTe at room temperature. Unless otherwise noted, the levels are measured from the valence band. The level 0.014 eV from the conduction band refers to

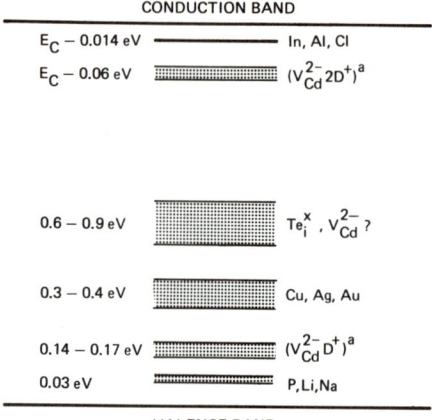

FIG. 3.16. Common defect levels seen within the band gap of CdTe at room temperature. (D^+ refers to a donor from group III or group VII.)

hydrogenlike In and Al donors at infinite dilution. Although not indicated, a value of 0.017 eV is associated with Cl and the other halides. There is no direct evidence that singly ionized Cd interstitials exist but it is likely that they are compensated majority defects when material is prepared at high p_{Cd} in the presence of an excess of residual acceptor impurities.

There is an electron trap at 0.06 eV observed in time of flight measurements in high resistivity material. It is surely present in low resistivity n-type material but in low concentrations. Although not apparent from electrical measurements this energy is observed in luminescence. The edge emission region is generally associated with transitions from the conduction band to an acceptor level 0.06 eV above the valence band. However, a transition from a level 0.06 eV below the conduction band to the valence band should be considered. Such an assignment is reinforced by the strong 1.584 eV luminescence found in donor-doped compensated material. Although there is no direct evidence for the nature of this level, it is assigned to a Cd vacancy–donor complex containing two donors. The second level of the double acceptor also occurs at this energy, although the nature of the defect is different. Levels near the center of the band gap are evident from electrical time of flight and TSC measurements. Although little or no evidence is available as to their identity, they have been previously assigned to V_{Cd}^- and V_{Cd}^{2-}. The latter is preferred here so the singly ionized state can be assigned close to the valence band. The assignment of the Cd vacancy close to the valence band is convenient to explain the strong edge emission luminescence and is more consistent with self-compensation mechanisms. To explain electron trapping at 0.6 eV, Te_i is assigned at this depth. The most prominent

III. SUMMARY

and best studied defect besides the hydrogenic donor is the vacancy–donor complex ($V_{Cd}^{2-}D^+$) located about 0.14 to 0.17 eV above the valence band. Many of the other levels, including the simple impurity acceptors, are not as accurately known due to the interaction among themselves and other defects. These uncertainties are reflected in the spread of the energy levels.

CHAPTER 4

Applications

I. Gamma-Ray and X-Ray Spectrometers

A. INTRODUCTION

Germanium and silicon are the common semiconductor materials used respectively for gamma-ray and x-ray spectrometers. However, because of their small bandgaps they must be cooled and in most cases operated near liquid nitrogen temperatures to avoid excessive thermal currents. At the other extreme, CdTe detectors operate at room temperature (Mayer, 1966; Akutagawa et al., 1967; Arkad'eva et al., 1966a), and in some cases as high as 100°C (Zanio, 1969a; Bell et al., 1970a,b).

Since the average atomic number (Z) of CdTe is 50 compared to 32 and 14 for that of Ge and Si, respectively, there is a more effective interaction of photons in CdTe than in either Si or Ge. Photons interact within the crystal by the photoelectric process, Compton scattering, and by electron–positron formation. In the photoelectric effect all of the photon energy is transferred to a tightly bound electron. This interaction dominates the absorption process at lower photon energies and varies as Z^5. From Fig. 4.1 the photoelectric absorption coefficients for CdTe, Ge, and Si at 100 keV are approximately 10, 2, and 0.06 cm^{-1} respectively. At intermediate photon energies (0.5–5.0 MeV), the Compton effect dominates the interaction mechanism. In the Compton interaction the photon transfers only a portion of its energy to an electron and unless multiple Compton or subsequent photoelectric interactions, occur, the energy of the initial photon is not entirely deposited within the crystal. At high enough photon energies (i.e., $hv \gtrsim 10$ MeV) the pair formation process dominates. However, these energies are well above those used in most applications for CdTe.

The high atomic number of CdTe often is of only incidental importance in many applications since a severe cooling requirement rules out the use of Ge or Si. Figure 4.2 shows the room temperature gamma-ray response of CdTe devices of various volumes to ^{137}Cs, which emits monoenergetic

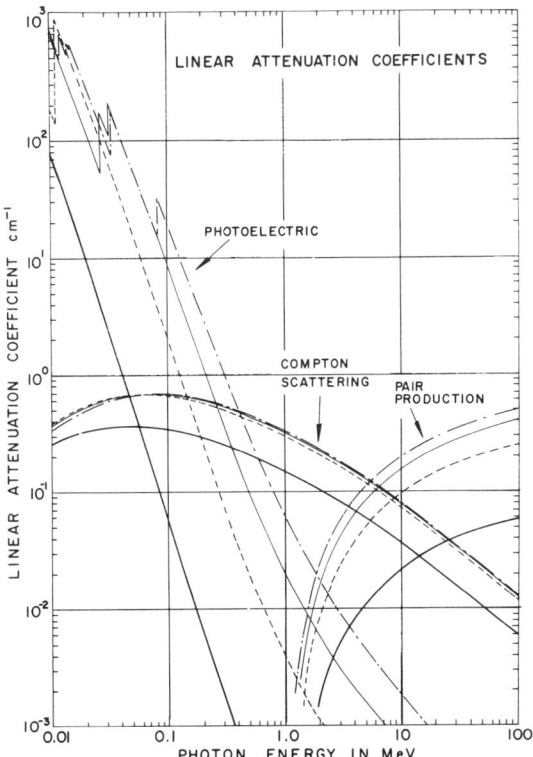

FIG. 4.1. Linear attenuation coefficients for photoelectric and Compton interactions and pair production in Si, —; Ge, ---; CdTe, —; and HgI$_2$, --- (Malm et al., 1973).

gammas at 662 keV. In all cases the photopeak is clearly resolved. The escape peak is also resolved for the smallest device (Jäger and Thiel, 1977). However, the majority of events occur by Compton interaction with an energy distribution having an upper energy at 478 keV. The peak at about 184 keV is due to the 180° backscattering of the photon from the surroundings into the detector. The resolution of the photopeak decreases with increasing volume. For devices approximately 5, 50, and 500 mm^3, the best resolutions are 2, 5, and 7%, respectively at full width half maximum (FWHM). (Full width at half maximum measured in percent is the ratio of the width of the photopeak in energy units to the position of the peak, also in energy units, times 100%.) Jones and Woollam (1975) obtained 1.3% FWHM resolution at 0°C by counting pulses with only a fast risetime. However, the effective volume was even smaller. Tranchart and Bach (1976) obtained about 1% resolution for active thicknesses of a few hundred microns. Larger volumes with comparable resolutions are possible with CdTe. However, it is unlikely

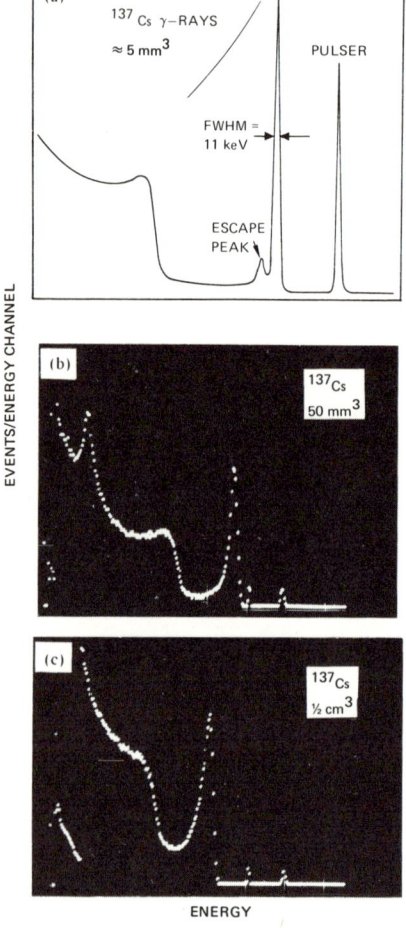

Fig. 4.2. Room temperature ^{137}Cs gamma ray spectra for single element CdTe gamma ray detectors having active volumes of approximately (a) 5 mm³ (Siffert et al., 1976), (b) 50 mm³ (Higinbotham et al., 1973), and (c) 500 mm³ (Zanio et al., 1974), corresponding to resolutions (FWHM) of 2, 5, and 7%, respectively.

that single element CdTe devices with volumes larger than 5 cm³ will ever have resolutions less than 7%. Regardless, there are numerous applications where only modest energy resolution is required. The NaI(Tl) scintillator system is commonly used where room temperature operation is necessary. However the energy resolution is only about 8% at 662 keV, and often it is necessary to restrict the volume of the device. Volume restrictions often rule out the use of a photomultiplier (PM) tube. Furthermore, the best CdTe detectors as well as detectors fabricated from Ge and Si have the important advantage that their pulse heights do not depend upon applied voltage and ambient temperature to the extent that the pulse height of the PM tube does.

B. X-Ray Spectrometer

Cadmium telluride is useful as an x-ray detector. The absorption coefficient is quite large at small photon energies and most of the electron–hole pairs are created at the surface of the detector. For example, with 59.6 keV photons, characteristic of ^{241}Am, about half the photons interact within the first 120 μm. Figure 4.3a shows the ^{129}I spectrum for a CdTe detector fabricated from high resistivity material having an active area of 10 mm^2 and operated at 0°C. The resolution of the 28.6 keV K_α peak is 1.9 keV FWHM. The 3.9 keV L_α peak is also apparent. Resolution degrades with increasing detector area. The resolution for the ^{241}Am 59.4 keV peak with

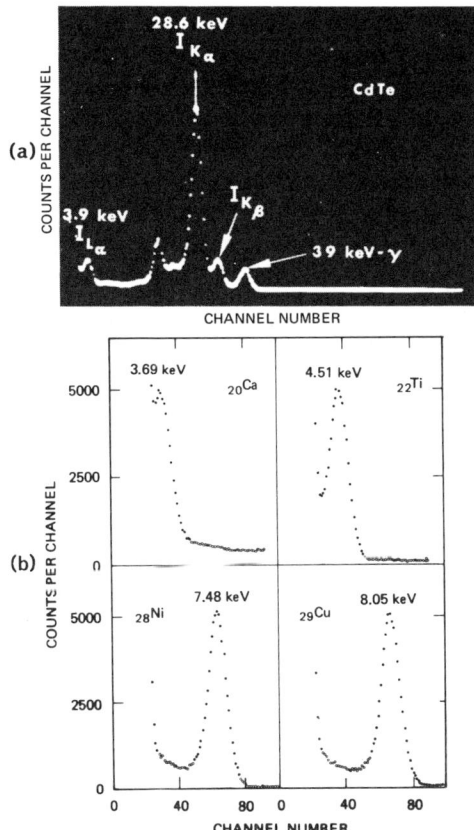

FIG. 4.3. X-ray spectra for (a) ^{129}I with a CdTe device fabricated from semi-insulating material having an irradiated area of 10 mm^2 and operated at 0°C (Barnes and Zanio, 1976) and (b) $_{20}$Ca, $_{22}$Ti, $_{28}$Ni, and $_{29}$Cu obtained with a ^{238}Pu excitation source using a Schottky barrier on n-type material (Dąbrowski et al., 1976, 1977).

an 80 mm² active area device is 3 keV. With a 3 cm² active area device operated at room temperature, 14 keV resolution is obtained at 59.6 keV (Zanio 1977).

Gold surface barriers on low resistivity n-type material is an alternative approach in fabricating x-ray detectors. Dąbrowski et al, (1976, 1977) obtained resolutions of 1.7 keV and 1.1 keV at room temperature for 59.6 keV gamma rays from ^{241}Am and 5.9 keV K x-rays from ^{55}Fe. The detector contribution to the peak width at 5.9 keV was 240 eV. Figure 4.3b shows x-ray spectra of various elements obtained with a ^{238}Pu excitation source. The active area of these detectors, which are operated at about 30 V, is 3 mm².

For low energy x-rays, small active areas and operation at room temperature, the low resistivity surface barrier approach to the fabrication of x-ray detectors is more attractive. However, for cm² areas and higher energy x-rays the high resistivity approach with modest cooling is more attractive and may even be necessary due to capacitance effects. Both approaches have practical use. Large area devices would also be useful in the determination of environmental contamination of ^{239}Pu, which emits photons at 13.5, 17.0, and 20.2 keV (Beck et al., 1976). Zanio et al. (1975a) fabricated a well with five CdTe detectors to count ^{125}I in radioimmunoassay testing. Economou and Turkevich (1976) suggest smaller detector to be useful for chemical analysis by x-ray fluorescence.

Comparable resolutions have been attained with a HgI_2 detector at room temperature. For a 14 mm² device, Slapa et al. (1976) measured a detector contribution of 318 eV to the 5.9 keV ^{55}Fe peak.

C. GAMMA RAY SPECTROMETER

1. *Safeguards*

Gamma ray spectrometry is widely used in the nondestructive identification and assy of nuclear fuel and scrap materials (Augustson and Reilly, 1974). Since most of the heavy isotopes of interest emit characteristic gamma rays it is easy to distinguish U, Pu, and Th from each other and make accurate isotopic analyses with Ge detectors. However, the use of Ge for a portable analytical tool in this application is impractical due to the necessity of cooling the detector for operation. Higinbotham et al. (1973) showed that the resolution of small (20 mm³) CdTe detectors is superior to that of NaI for Pu and U analysis. Of particular interest in the analysis of Pu are the three peaks at 333, 375, and 414 keV which smear together with NaI spectrum but are effectively resolved by CdTe. Figure 4.4a shows the CdTe gamma ray spectrum for a 91% ^{239}Pu plate. The 333 keV peak contains contributions from ^{237}U, ^{241}Am, and ^{239}Pu, the 375 keV peak is mostly due to ^{239}Pu and the 414 keV peak results only from ^{239}Pu.

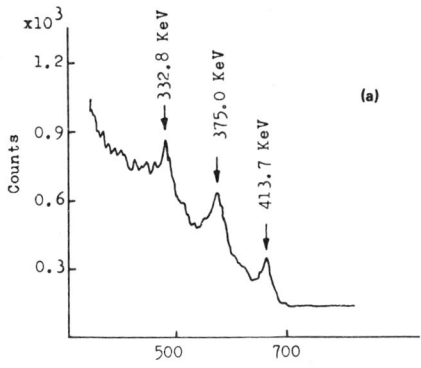

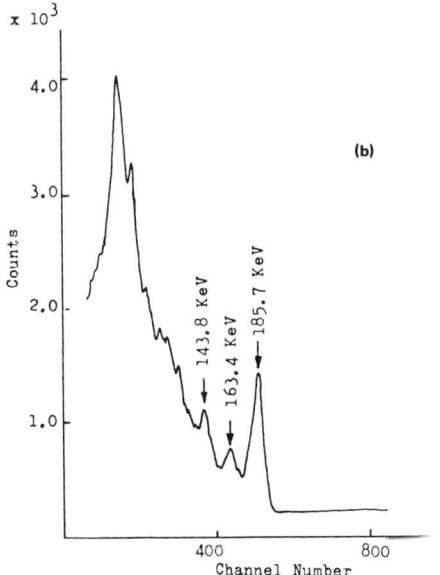

FIG. 4.4. Room temperature CdTe gamma ray spectra for (a) a 91% ^{239}Pu plate and (b) a 90% enriched U powder (de Carolis et al., 1976).

There are at least two analyses for U that are important. The first is for the measurement of the enrichment (^{235}U/total U) in bulk samples. Uranium 235 gives off a dominant photon at 185 keV, whereas ^{238}U emits no gamma rays in this energy region. Figure 4.4b shows the gamma response of a CdTe detector for a 90% enriched U powder. These results were taken for detectors of about 50 mm^3. The usefulness of CdTe detectors in these two applications is limited by their small size. However, the small size of the CdTe detectors is particularly suitable for determination of the ^{235}U enrichment of inner rods of a complete fuel assembly. The International Atomic Energy Agency has also developed small preamplifiers that are compatible with CdTe detectors for insertion between the rods (de Carolis et al., 1976).

There are other applications for small probe-type detectors in the monitoring of nuclear fuels. The Canadian deuterium–uranium pressurized heavy water reactors utilize mechanical scanners with small Geiger tubes and LN$_2$ germanium detectors for locating failed-fuel bundles (Lipsett and Stewart, 1976). Considering such applications Jones and Woollam (1975) and Jones (1977) used pulse-shape discrimination to select pulses with risetimes less than 100 nsec and obtained 8 keV resolution for ^{137}Cs at 0°C. Although the effective volume of this detector was about 3 mm^3 it is adequate in situations where extremely high photon fluxes are encountered.

2. *Gauging*

Gauging by the attenuation of nuclear radiation in matter has been used in industrial applications and quality control processes for a number of years. In conventional gauging the incident and transmitted flux of a monochromatic gamma-ray beam is measured. Knowing the corresponding attenuation coefficient in the medium, the thickness of the medium can be calculated to a high degree of accuracy. The use of large propellant tanks in the manned space program has created the requirement for gauging systems capable of operating with high precision and reliability under zero g conditions. At the present time, accurate spacecraft propellant gauging systems are suited to operate only when an acceleration is applied, thereby settling the liquid propellant in a predetermined geometric configuration; the propellant level is then measured with point sensors. This method of gauging, however, is generally unsuited to operation under extended space flight conditions due to the uncertainty in the spatial distribution of the propellant. Bupp et al. (1973) assessed the use of nuclear techniques to gauge large quantities of propellant under zero g conditions with several source–detector pairs. The CdTe detectors were positioned opposite collimated ^{137}Cs sources on tanks containing up to 19,000 lb of salt water as a fuel simulant. Propellant loading was determined by counting photoevents above a fixed energy threshold. This arrangement discriminates against photons from other sources that are Compton scattered in the propellant and subsequently interact within the detector. Modest energy resolution in this application is helpful in improving the accuracy. Using six 200 mm^3 detectors, an accuracy of approximately 0.6% of full tank loading was obtained under 1 g conditions. A comparable system using 0.5 cm^3 detectors was successfully flight tested at zero g in a KC-135 Zero-G aircraft and was accurate to about 2% (Bupp, 1976).

Cadmium telluride detectors monitor the thickness of heat shields and nosetip shape changes on rocket launched test vehicles during reentry by backscatter techniques (Droms et al., 1976). Airborne measurements are required since no ground testing facility can simulate the extreme and complicated reentry conditions. Gauging by backscattering is ideal because the radioactive material is not incorporated in the ablative material and consequently does not perturb the measurements. Unfortunately, this method is limited to gauging materials 2 in. or less. In order to gauge thicker materials and in addition obtain superior spatial resolution, it is necessary to implant gamma emitting isotopes in the ablative material (Fig. 4.5) and measure the decrease in count rate as the isotopes, heat shield and nose tip materials ablade away. With multiple source detector pairs and efficient detectors capable of some energy resolution this technique offers precise spatial resolution. Bar-shaped detectors provide a small frontal area for good spatial resolution and the necessary length for good stopping power. By shaping the electric field, and thus taking advantage of the superior transport properties of electrons as compared to holes as discussed later, modest energy resolution is also possible. Fewer detectors are required if modest energy resolution and appropriate collimation are available (Zanio 1977). The gamma spectrum of a 2 cm long single element bar-shaped detector operating at 250 V is capable of resolving the ^{137}Cs photopeak from the ^{60}Co photopeaks. These spectra represent the response to point sources with no Compton scattering within the ablative material. As a result of Compton scattering as described in Section I, C, 3, better resolution may be required. The severe vibration environment during reentry also requires special packaging to prevent microphonics and possibly a piezoelectric contribution from the CdTe to the signal. Baxter (1976) packaged the detector along with all of the associated analog electronics and some acoustic isolation in a single hybrid package.

3. *Nuclear Medicine*

Applications in nuclear medicine utilize the unique properties of small probes. Probes are used for either dosimetric or tracer uptake measurements. For dosimetric studies devices should have air or tissue equivalent response. Since semiconductor detectors, especially CdTe, show greater than tissue equivalent response, the use of the probes in this application is restricted. Cadmium telluride probes with volumes ranging from 0.5 mm^3 to 100 mm^3 have been constructed (Martini, 1973; Martini et al. 1972b; Zanio et al.; 1972; Walford and Parker, 1973). In some cases the CdTe probe is ideal for the local uptake of tracers. Meyer et al. (1972) demonstrated the possibility of detecting corneal lesions in animals by measuring the disappearance rate of a gamma emitting ^{75}Se–sodium selenite solution with a CdTe probe.

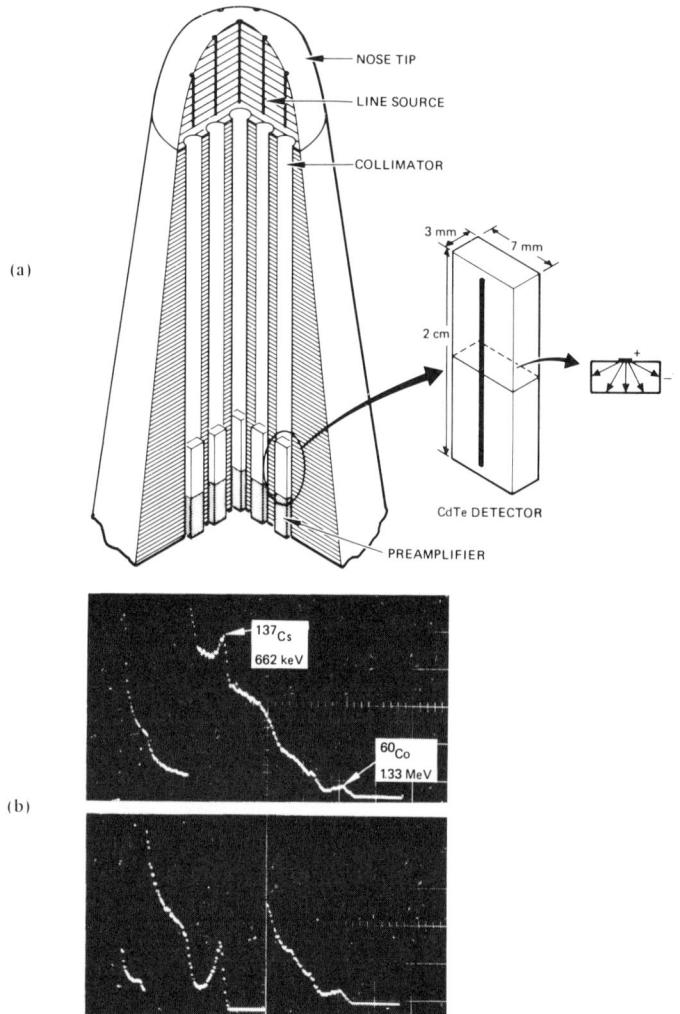

FIG. 4.5. (a) Heat shield (schematic) for reentry vehicles showing collimated source–detector pairs with the sources implanted in the ablative material. (b) Simultaneous ^{137}Cs and ^{60}Cs spectra taken with a bar-shaped detector utilizing a shaped electric field (Zanio, 1977).

Comparative studies with a Si(Li) probe showed that meaningful results could not be obtained. Garcia *et al.* (1974) used a larger probe to detect acutely infectious teeth in dogs following the administration of ^{99m}Tc-polyphosphate, a 144 keV gamma ray emitter. The performance of the probe was compared with that of a NaI(Tl) rectilinear scanner. The spatial resolution of the CdTe probe was superior since it could easily distinguish between normal and infected root tips less than 1 cm apart, whereas the NaI detector system was unable to distinguish the normal root tips of an uninfected tooth located between those of two adjacent infected teeth. Garcia *et al.* (1976) also used a hand-held CdTe probe to test for venous thrombosis of the legs by measurement of the local uptake of ^{125}I fibrinogen. The CdTe probe was equal in diagnostic capability to a standard NaI probe and because of its smaller size was more versatile.

Increases in the liquid accumulation in the lung result in an increase in the local density and an increase in the scattered fraction of an incident collimated beam. By examining the transmitted and scattered fractions of a 100 keV collimated beam with a pair of CdTe detectors Kaufman *et al.* (1976) constructed a dosimeter which could conceptually be of importance in the evaluation of patients with pulmonary edema. Vogel *et al.* (1977) also propose that a simple ^{125}I source–CdTe detector pair could be useful in a scanning system to monitor changes in bone mineral loss.

Nuclear medicine uses biotelemetry techniques to monitor physiological parameters. This approach minimizes interference from measurement systems. Bojsen *et al.* (1977) used small (2–3 mm^3) CdTe detectors to replace implantable GM tubes in the detection of radiotracers in small animals.

An important objective of nuclear medicine is to image vital organs and other regions of the body with finer spatial resolution and at faster rates. Several approaches to imaging are used, all of which complement one another and all of which potentially involve CdTe detectors. Remarkable results are possible using a single thin large area NaI scintillator and an array of photomultipliers and collimators to monitor the distribution of radioactive isotopes in the tissue, such as with the Anger camera. The usual isotope is ^{99m}Tc which emits monoenergetic photons at 140 keV. McCready *et al.* (1971) demonstrated that a Ge semiconductor camera has superior spatial resolution compared to that of the Anger camera. Although uneconomical, the superior performance of this system is due to its fine energy resolution. The ultimate energy resolution of the Ge detector is not required. Therefore, the requirement of only modest energy resolution plus the convenience and economy of room temperature operation makes CdTe a promising material for this application.

Compton scattering is detrimental to imaging and limits the accuracy of gauging. Good energy resolution is helpful. In the imagining system (Fig. 4.6a) the detector should ideally only respond to primary events (P) emanating from the geometrical field of view defined by the collimator. These events are recorded in the higher energy peak of Fig. 4.6b. However, gamma rays emitted from outside of the field of view (S and S′) may undergo Compton scattering within the field of view, be scattered into the detector, and be recorded in the spectrum at a lower energy. Including scattered radiation

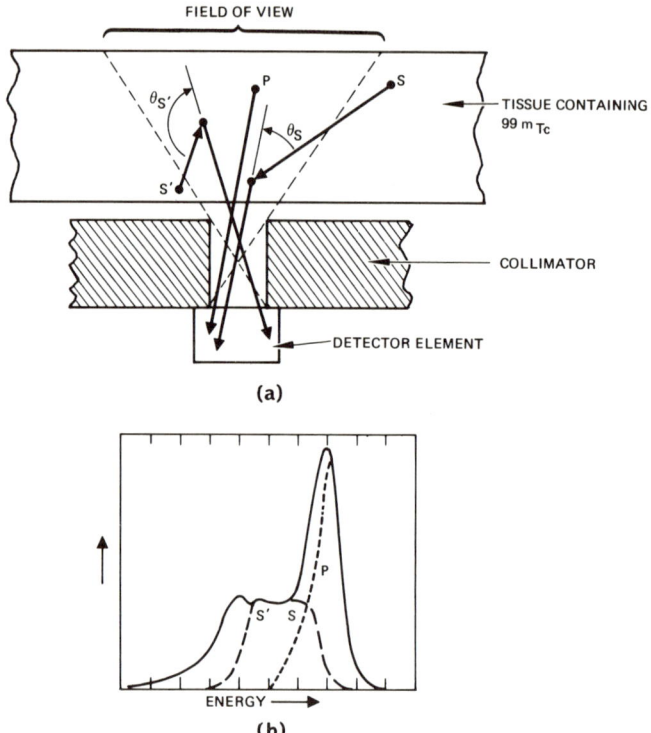

FIG. 4.6. (a) Compton scattering in tissue. The detector directly views photon emission from radioactive decay occurring within the field of view (P) defined by the collimator and views lower energy photons scattered within but emanating from outside of the field of view (S and S'). (b) Gamma ray spectra due to direct and scattered photons.

in the analysis degrades the spatial resolution. By only counting events above a fixed energy threshold, the scattered fraction can be reduced but only at the expense of eliminating a portion of the useful fraction. However, the superior energy resolution of the Ge detector permits the energy threshold to be increased significantly without a serious loss in the primary fraction (Hoffer and Beck, 1971). Resolution (FWHM) for NaI and Ge are 25 keV and 8 keV, respectively, in the rectilinear scanner. Although this latter resolution is typical of poor Ge detectors, it is adequate for imaging application. This resolution has also been obtained in small CdTe 10 mm^3 detectors.

Preliminary work has been undertaken in fabricating arrays with small CdTe detectors (Zanio, 1977). Figure 4.7 shows results from a 4 × 4 matrix of 3mm × 3 mm × 2 mm CdTe detectors. Line scans across rows with a collimated ^{57}Co beam shows the spatial resolution of the array to be charac-

I. GAMMA-RAY AND X-RAY SPECTROMETERS 175

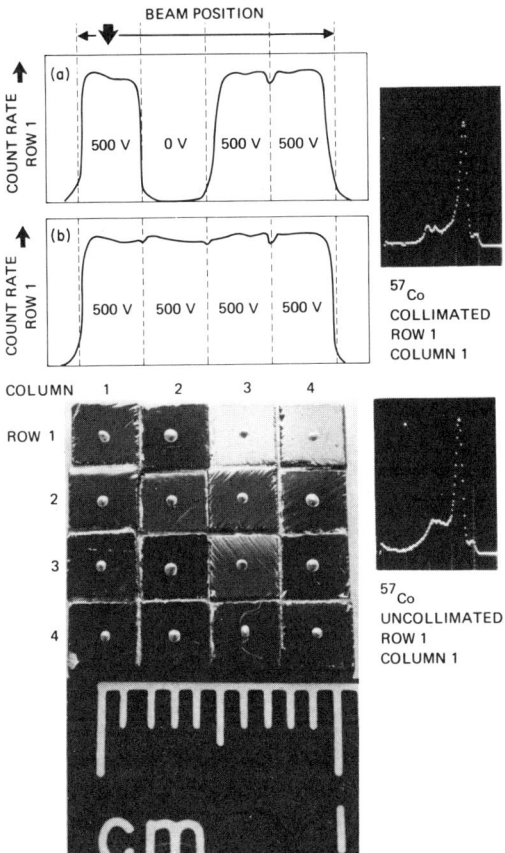

FIG. 4.7. (Bottom) A 4 × 4 matrix of (3 mm × 3 mm × 2 mm) CdTe detectors. (Top) Line scan with a $\frac{1}{2}$ mm diameter ^{57}Co beam across a row (a) with four elements operated at 500 V and (b) with only elements 1, 3, and 4 operated. Included on the right hand side are ^{57}Co spectra from a detector element using (bottom) the collimated beam and (top) from an uncollimated point source (Zanio, 1977).

teristic of the dimensions of the detector element. Typical energy resolution of a detector element in the array is 10 keV (FWHM) for ^{57}Co. The low yield of detector material and the necessity of arrays in a monolithic configuration to accommodate information processing presently prevents the economic construction of large arrays for practical use.

Positron annihilation cameras (Llacer and Cho, 1973; Phelps et al., 1976) offer another approach to imaging. In this application, pairs of detectors are located diametrically opposite positron emitting tissue. Coincidence counting specifies the location of the event and the degree of uptake of radioactive isotopes by the tissue. The dimensions of present NaI detectors for this application are of the order of inches. The smaller CdTe detectors are not practical for this application unless patient dose can be increased several orders of magnitude.

An important development in the field of medicine is the three-dimensional imaging of an object by computerized axial tomography. The object is located between an x-ray generator and an array of detectors. After applying appropriate algorithms to various scanning schemes, a three-dimensional picture can be constructed. Bar-shaped devices, as in Fig. 4.5, are typical of the element size. In a discussion of materials for tomography applications, Cho et al. (1974b) considered CdTe and operated smaller devices fabricated from In-doped material at 2 MHz. Since count rates several orders of magnitude higher than this are required to obtain good signal-to-noise ratios, detectors must be operated in the integrated current mode. Preliminary results with Cl-doped material grown by THM (Allemand et al., 1977) show pulses generated by high photon fluxes to have tails with decay times of the order of milliseconds. Excessive pulse decay times could interfere with the use of CdTe for this application.

Some production results on Cl-doped material grown by THM (Brelant et al., 1977) indicate that although the yield of material for counters is improving, the yield of spectroscopy grade material is poor. Detectors in the 100–200 mm^3 volume range with little or no energy resolution are priced at about $1000 per unit. This is to be expected since compared to Si, Ge, and GaAs, relatively little materials work has been undertaken on CdTe Also, the quality of CdTe detectors is not presently limited by intrinsic properties and future work on other growth methods should result in material of such reproducibility and quality as to make room temperature imaging with CdTe possible.

D. LIMITATIONS

Trapping restricts the size and energy resolution of CdTe gamma-ray and x-ray detectors. In the case of Cl-doped material, polarization is an additional trap-related problem. Leakage currents are also excessive, especially for x-ray detectors fabricated from high resistivity material. However, lower trap concentrations should eventually result in lower leakage currents and permit higher bias voltages and better collection efficiencies.

1. *Trapping*

A brief description of charge transport in a plane–parallel device (Fig. 4.8) illustrates the effects of trapping on the resolution. Here the material is fully depleted and the resistivity is high enough ($\rho > 10^7$ ohm-cm) so that the electric field is uniform between the electrodes, which typically have a

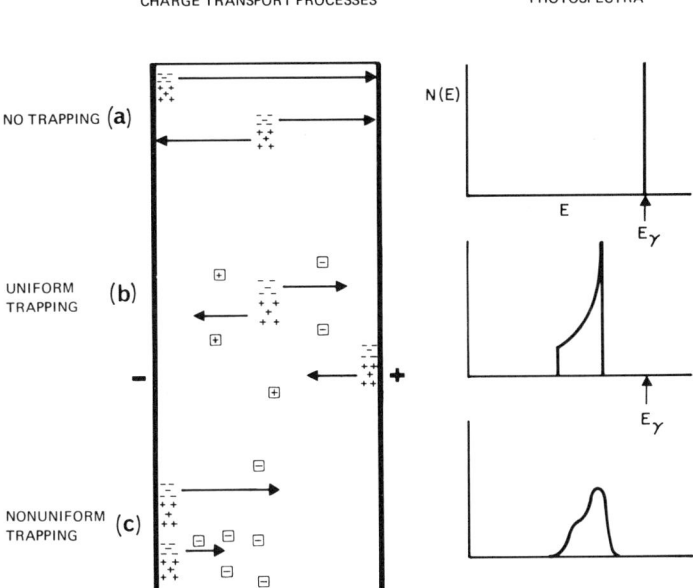

FIG. 4.8. Charge transport of electrons (−) and holes (+) in a plane-parallel detector and the resulting shape of the photopeak for the case of (a) an ideal detector with no trapping, (b) uniform trapping at electron traps (⊟) and hole traps (⊞), and (c) nonuniform trapping of electrons and holes.

separation (L) of a few millimeters. The photon energies of interest range from a few keV to a few MeV. Since it requires 4.43 eV to create an electron–hole pair at room temperature in CdTe (Quaranta et al., 1970; Cornet et al., 1970a; Dąbrowski et al., 1974), anywhere up to 10^6 electron–hole pairs are created per event.

Electron–hole pairs are generated primarily near the surface in x-ray detectors (Fig. 4.8a). Irradiation at the negative electrode of a device fabricated from semi-insulating material results in charge transport primarily due to electrons. With no trapping, charge is completely collected and the photopeak is a sharp line at full collection E_γ. Since the drift length for electrons ($\lambda_e = \mu_e \tau_e^+ V/L$) is quite good in CdTe, respectable charge transport occurs and the actual peak position is just below E_γ. Here μ_e and τ_e^+ are the drift mobility and trapping time. In low resistivity material, good charge collection also occurs for both electrons and holes (Dąbrowski et al., 1975; Iwańczyk and Dąbrowski, 1976) because the drift lengths of electrons and holes are much greater than the depletion width.

For more penetrating photons interacting in a gamma ray spectrometer, electron–hole pairs are created throughout the crystal. For the extreme case illustrated in Fig. 4.8, the distribution is uniform. To more clearly illustrate the effects of trapping on the resolution, only the photoelectric process is considered. With no trapping, electrons and holes are completely collected regardless of the position at which they are created in the detector. In this case (Fig. 4.8a) the gamma spectrum is represented by a sharp line or peak at full collection (E_γ) typical of cooled Ge detectors. When trapping is severe (Fig. 4.8b), the entire electron (hole) cloud does not reach the positive (negative) electrode. Consequently, the photopeak or pulse height is shifted downward. A spread in pulse height occurs even though the concentrations of electron and hole traps are uniform. If the drift length of the electrons and holes are equal, then the maximum pulse height will occur when electron–hole pairs are created midway between the electrodes; the minimum pulse height will occur when the interaction occurs at either of the two electrodes. For the case of nonuniform trapping (Fig. 4.8c), the pulse height distribution will broaden even further.

Several groups have calculated the effect of trapping of electrons and holes on the shape of the photopeak (Bell, 1971; Zanio et al., 1972; Siffert et al., 1974; Swierkowski, 1976). In some cases Compton interactions were also considered. A precise calculation is not possible because of nonuniform material (Zanio, 1970b), detrapping (Martini et al., 1972a) and the dependence of the mobility and trapping time on the electric field, especially at high electric fields (Canali et al., 1971b; Alekseenko et al., 1971). The shapes agree reasonably well with actual spectra. Good resolution in a plane parallel device requires the trapping time to transit time ratio (τ^+/T_R) for both electrons and holes to be significantly greater than one. Here T_R equals $L^2/\mu V$. For comparable electron and hole trapping times the resolution is limited by the trapping of holes or τ_h^+ because of the smaller mobility and consequently the longer transit times of holes with respect to electrons. Consider $\tau^+ \approx 10^{-6}$ sec, an upper value reported by several groups (Zanio et al., 1974; Bell et al., 1974a, 1975a; Alekseenko et al., 1974). Assuming 200 V across a 3 mm thick device, the transit time for electrons $T_R(e)$ is 0.45×10^{-6} sec and that of holes $T_R(h)$ is 4.5×10^{-6} sec. Typical pulse shaping time constants are about 1 μsec. Even if τ_h^+ were larger, the charge collection of holes would be limited by the pulse shaping network. Either decreasing the electrode separation, and consequently the volume, or increasing the applied voltage may not be desirable. An alternative approach to decrease the effective charge collection time and improve resolution is to take advantage of the superior mobility of electrons by forcing the charge collection process to be primarily due to electrons. This has been undertaken by properly shaping the electrodes (Zanio et al., 1972; Zanio, 1977; Malm et al., 1975, 1976). Even for severe trapping of the holes (i.e., $\tau_h^+/T_R(h) \ll 1$) good resolution is possible. The effect of the $\mu\tau^+$ product and of the electrode configuration on the shape of the spectrum for only one carrier collection

I. GAMMA-RAY AND X-RAY SPECTROMETERS 179

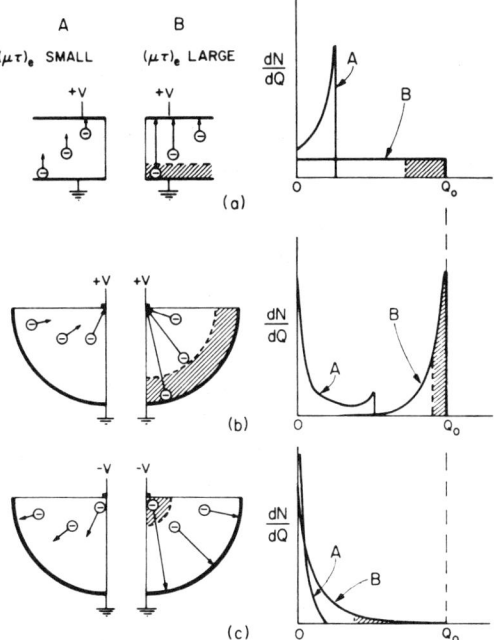

FIG. 4.9. Charge displacement and photopeak formation for electron collection only and small (A) or large (B) values of the $\mu\tau^+$ product. The pulse height spectra are shown in (a) for a plane-parallel configuration, (b) for a spherical configuration with the inner contact positive, and (c) with the inner contact negative. In the spectra, Q_0 represents full charge collection. Gamma-ray interactions within the shaded region of the detectors correspond to pulses in the shaded region of the spectra (Malm et al., 1975).

is shown schematically in Fig. 4.9. Both small (A) and large (B) values of the electron $\mu\tau^+$ product are considered. For the plane-parallel configuration with a uniform electric field E (Fig. 4.9a), the drift length is uniform through the device. When λ_e is much less than the electrode separation L, most of the electrons drift an equal distance and the spectrum shows a peak at a small fraction of the full collection amplitude (Akutagawa and Zanio, 1969). When $\lambda_e \gg L$, most of the electrons reach the positive contact. Since the charge induced in a uniform electric field is proportional to the electron path length, a rectangular spectrum extending to full collection amplitude occurs.

With a spherical (or hemispherical) configuration, the drift length and velocity of electrons is a function of radial position in the detector because of the nonuniform electric field (Canali and Malm, 1976). In such a nonuniform field most of the induced charge is produced by carrier traversal in the high field region. With the small contact positive (Fig. 4.9b), in case A (small $\mu\tau^+$) where partial charge collection occurs, only a small photopeak is observed. However, for case B, where all the electrons are collected, a clear photopeak appears in the spectrum. This contrasts with the plane-parallel case, where only a steplike spectrum occurs for complete electron collection.

With the small contact negative (Fig. 4.9c), the number of pulses of large amplitude is small because of the reduced volume near the negative contact. For both positive and negative polarities gamma-ray interactions in the shaded regions of Fig. 4.9 produce the largest pulses.

Regardless of the electrode configuration, reducing the trap concentration is important. Traps are a natural consequence of attaining high resistivity material through compensation. High-resistivity ($\rho > 10^7$ ohm-cm) compensated material in itself is relatively easy to obtain by preparing donor-doped material at or near Te-saturated conditions. However, simultaneously attaining the high resistivity state and a low uniform concentration of traps is a formidable problem. Only the deeper traps, greater than about 0.2 eV from the conduction (valence) band for electrons (holes), limit the performance of CdTe at room temperature. Electrons (holes) trapped at shallower levels are thermally reemitted to the conduction (valence) band in times much shorter than the transit time and therefore these levels do not hinder charge collection. Although there is evidence as to the nature of the deep traps, positive identification has not been made. In some cases their energy levels and concentrations are known. Chapter 3 discusses the more common energy levels. Levels discussed here pertain to high resistivity material and refer to those observed in the course of device operation.

Detrapping studies show that a level approximately 0.6 eV from the conduction band limits electron transport in In-doped material (Zanio et al., 1968; Canali et al., 1971a). Canali et al. (1975a) measure a trap concentration of about 3×10^{11} cm^{-3} at the same energy with space charge limited current measurements in In-doped material having a τ_e^+ of about 100 nsec. Although the depth of the electron trap at 0.6 eV is well defined, its origin is not certain. Strong trapping by a negatively charged species such as the Cd vacancy is ruled out. Neutral interstitials or a complex of the vacancy with donors are more likely candidates.

A level about 0.35 eV above the valence band in the 10^{15}–10^{16} cm^{-3} concentration range limits hole transport in Cl-doped material (Bell et al., 1974a; Canali et al., 1974). A center deeper than 0.14 eV, and likely to be present in larger concentrations, limits hole transport in In-doped material (Ottaviani et al., 1973). Residual impurities such as Cu or a complex thereof are likely candidates.

Relating τ^+ to the trap concentration (N_T) is straightforward through

$$\tau^+ = 1/N_T \sigma \langle v \rangle, \tag{4.1}$$

where σ is the trapping cross section and $\langle v \rangle$ is the thermal velocity of the carrier. Proceeding one step further, Ottoviani *et al.* (1975) and Mayer (1968) estimated the maximum uniform trap concentration required to attain 1 keV resolution. However, in practice, relating the resolution to trapping and therefore predicting the ultimate resolution is difficult because fluctuations in the trap concentrations are always present (Zanio, 1970b). In fact if fluctuations are neglected, calculations show that with microsecond trapping times for electrons and holes, resolutions of less than a few keV are possible with 10 cm^3 devices. Resolutions of only a few percent have been measured with mm^3 devices.

One method of estimating the transport properties required of larger devices, which would have the same resolution as that of smaller devices, is by scaling; i.e., assuming that if uniformity can be maintained, comparable resolutions require comparable τ^+/T_R ratios. Siffert *et al.* (1976) obtained approximately 2% resolution with ^{137}Cs for a 0.5 mm thick and approximately 5 mm^3 volume device. These results are characteristic of Cl-doped material that has a $\mu\tau^+$ product for holes in the 10^{-5}–10^{-4} cm^2/V range and a $\mu\tau^+$ product for electrons of about 10^{-3} cm^2/V (Bell *et al.*, 1974a; Alekseenko *et al.*, 1974). The τ^+/T_R ratios for electrons and holes are approximately 100 and 10, respectively. Therefore, the resolution is limited by hole trapping. (In the simplest approximation, values of 10 and 100 for τ^+/T_R correspond to collection efficiencies of about 95% and 99%, respectively.) Assuming that the volume scales as the cube of the electrode separation, an electrode spacing of about 2.5 mm is necessary to achieve the $\frac{1}{2}$ cm^3 volume. For a 1000 V bias this predicts that a $\mu\tau^+$ product of about 10^{-3} cm^2/sec or an order of magnitude decrease in the concentration of hole traps is required to obtain the same collection efficiency, and presumably the same resolution as that of the smaller plane parallel device. This approach is optimistic since in the $\frac{1}{2}$ cm^3 hemispherical device only 7% resolution for ^{137}Cs has been attained although the $\mu\tau^+$ product for the important carrier, electrons, is 10^{-3} cm^2/V. (The transport properties for holes in the hemispherical devices are not an important consideration as long as their $\mu\tau^+$ products are at least an order of magnitude less than that of the electrons.) A salient feature of the full energy peak resolution of a hemispherical detector collecting only electrons is that the full energy peak resolution is better than that in planar devices having comparable device parameters and $\mu\tau^+$ products for both electrons and holes (Malm *et al.*, 1977). A more realistic approach is to consider the uniformity problem to be more severe in larger volume devices and to require a $\mu\tau^+$ product of 10^{-2} cm^2/V for both electrons and holes in the plane-parallel device and 10^{-2} cm^2/V for electrons in the hemispherical device to attain a 2% resolution.

A reduction in the concentration of hole traps is important in both configurations. It is of direct concern in the plane-parallel device where hole trapping degrades resolution directly. It is of a more indirect concern in the hemispherical device. The general disorder and concentration of electron traps decrease with decreasing residual impurity content in compensated high resistivity material and reducing the impurity content should improve the electron as well as the hole trapping time. Compared to Si, Ge, and the III–V compounds, very little work has been undertaken in purifying CdTe. Further purification should result in considerable improvement in the performance of this material as a gamma-ray and x-ray spectrometer. Regardless of the degree of purity that is attained, unless the THM technique can be scaled up to larger diameters, the Bridgman approach or other solution growth techniques offers the best chance of making large single element devices.

2. *Polarization*

Wald and Bell (1972) reported 4% resolution for detectors fabricated from Cl-doped CdTe grown by THM. Malm and Martini (1973, 1974) showed these devices to degrade due to a deep center (N_P) which changes its charge state. The pulse height and counting rate both decreased with time after the bias was applied. These polarization effects in Cl-doped material are also confirmed by Siffert *et al.* (1976), Fabre *et al.* (1976), and Triboulet *et al.* (1974). Just after the bias is applied ($t = 0$) the electrode separation (W) defines the depletion width and the polarization centers (N_P) are neutral. Although the device depletes from the positive electrode, the electric field, as in Fig. 4.10, is nearly constant because the concentration of ionized acceptors N_a^- is only slightly larger than the concentration of compensating ionized donors N_d^+. After some time ($t > 0$), either holes are thermally released from the polarizing centers and swept out or electrons are injected from the opposite contact and trapped at the same centers. The width of the space charge region is now given by

$$x = [2\varepsilon V/q(N_a^- + N_P^- - N_d^+)]^{1/2} \qquad (x \lesssim W). \qquad (4.2)$$

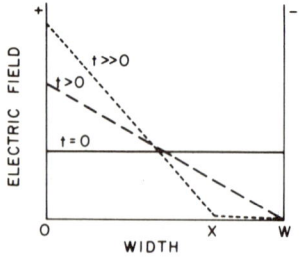

FIG. 4.10. Electric field distribution as a function of time in a polarizing CdTe device using blocking contacts for Cl-doped semi-insulating material grown by the traveling heater method (Malm and Martini, 1973).

Malm and Martini (1974) calculated approximately 6×10^{11} cm^{-3} for N_P. Siffert et al. (1976) obtained values in the 10^{11}–10^{12} cm^{-3} range. These concentrations correspond to trapping times of a few milliseconds. Since the total trapping time in an operational device is in the 10^{-7}–10^{-6} sec range, it is unlikely that the polarization centers in Cl-doped material seriously affect the trapping times. It is possible that the deep centers are related to the compensation mechanism and therefore completely eliminating them is unlikely. A partial solution to this materials problem is to allow limited injection of holes through the contact. Good gamma spectra without polarization are attained using this approach, which is discussed later in more detail under the topic of contacts (Bell et al., 1974b; Serreze et al., 1974; Siffert et al., 1976). Eichinger et al. (1974) avoided polarization by inserting thin insulating layers of mylar between the crystal surfaces and the metallic electrodes and operating the device periodically with the bias on and off. When the bias is off, the electrodes are shorted and a light turned on to assist depolarization. A resolution of 12 keV for 662 keV with 0.15 cm^3 crystals at $-20°$C was obtained. Although these latter two approaches result in good resolution, the most promising long range approach is to eliminate polarization so the detector can be operated in a conventional manner.

3. *Radiation-Induced Degradation*

Polarization also occurs at high radiation levels where the rate of formation of electron–hole pairs is greater than the rate at which they can either be swept out from the depletion region or recombine. Radiation-induced polarization is observed in halogen-doped material (Siffert et al., 1976) at room temperature and in In-doped material below $-75°$C (Zanio et al., 1970). Radiation-induced polarization becomes insignificant with heating and/or reducing the radiation flux.

When detectors fabricated from halogen doped material are irradiated with 33 MeV protons, changes in the collection efficiency, resolution, and leakage current occur (Nakano et al., 1976). Radiation effects are more pronounced in Cl-doped material than in In-doped material. Cathodoluminescence studies of ion-implanted material show that Cl-doped material is more susceptible to damage than In-doped material (Norris et al., 1977).

E. BLOCKING CONTACTS

Room temperature carrier concentrations as high as 3×10^{17} cm^{-3} for holes (Gu et al., 1975a) and 10^{18} cm^{-3} for electrons (Yokazawa et al., 1965) have been achieved in CdTe. It is not unreasonable therefore to consider both p–n and p–i–n structures with CdTe (Arkad'eva et al., 1967a). When Gu et al. (1975b) prepared p–n^+ diodes by the vapor diffusion of Al into a

p-type region containing about 10^{17} cm^{-3} holes, the forward and reverse currents at 77°K were due to tunneling via states within the band gap. Hunsperger (unpublished) prepared p–i–n structures using indium-doped material for the i-regions. However, the performance of these detectors was limited by trapping and generation-recombination currents via deep traps. Good blocking contacts on Cl-doped material grown by THM results in polarization.

Schottky barriers are an alternative approach to forming blocking contacts on CdTe devices. For ionic materials the barrier height depends strongly on the work function of the metal. [The election affinity in CdTe is 4.28eV Milnes and Feucht, 1972).] The presence of surface states in covalent semiconductors such as Si, Ge, and GaAs result in barrier heights for Schottky diodes that are nearly independent of the work function of the metal. From photovoltage measurements at a Te-CdTe interface, Zitter and Chavda (1975) associate a level at 0.59 eV with surface states. The barrier heights of various metals on low resistivity CdTe are shown in Table 4.1. A strong dependence on work function is not evident. A study of the barrier heights of various metals on various semiconductors (Kurtin *et al.*, 1969; and McGill, 1974) indicates CdTe to be more covalent than indicated by the Phillips scale of ionicity. However Phillips (1973a, 1974) considers the valence–electron charge distribution and shows that the effect of polarization is to reduce the useful barrier height. The relative independence of diode characteristics on metals of different work functions is more pronounced for CdTe because of the high polarizability.

TABLE 4.1

BARRIER ENERGIES FOR VARIOUS METALS ON LOW RESISTIVITY CdTe

Metal	CdTe type	Barrier energy (eV)
Al	n	0.35[a]
Mg	n	0.40[a]
In	n	0.50[a]
Ni	n	0.53[a]
Cu	n	0.57[a]
Ag	n	0.65[a]
Pd	n	0.65[a]
Au	n	0.69[a], 0.63[b], 0.6[c]
Au, Ni	p	0.3–0.5[d]

[a] Parker and Mead (1969).
[b] Swank (1967).
[c] Toušková and Kůzel (1972).
[d] Bube *et al.* (1976).

Regardless, there is some flexibility in choosing a more appropriate metal to alter either the blocking or Ohmic behavior of the contact. Dąbrowski et al. (1974, 1975, 1976, 1977) used Au evaporations on low resistivity n-type material for room temperature x-ray detectors. Previous to the evaporations, the wafers were etched in a bromine–methanol solution. Forward to reverse current ratios at room temperature are about $10^6/1$. Marfaing et al. (1974) utilized photocapacitance measurements with this contact to study deep and shallow levels. Parker and Mead (1969) also prepared such barriers by cleaving crystals in vacuum and evaporating contacts without exposure to air. Only slight differences in the barrier heights are found between devices prepared in vacuum and those cleaved in air. Parker and Mead (1969) also show that for carrier concentration of approximately 10^{17} cm^{-3} at room temperature, tunneling occurs when such diodes are operated at lower temperature in both the forward and reverse directions. States in the forbidden gap sometimes increase the tunneling probability and increase the forward current.

The precipitation of Au from $AuCl_3$ solution (Toušková and Kužel, 1972; de Nobel, 1959) results in diodes whose characteristics also can be explained at room temperature by Schottky's theory of a metal–semiconductor junction with recombination in the space charge region. At lower temperatures, the reverse current is explained by tunneling through a high resistivity layer.

The deposition of Au from a $AuCl_3$ solution onto mechanically polished but not etched n-type CdTe results in diode properties at room temperature that are characteristic of heterojunctions (Toušková and Kužel, 1973; Toušková and Toušek, 1971). At lower temperatures and higher voltages, carriers tunnel from the p-type Au_xTe_x region into the n-type CdTe in the reverse direction, either directly or by means of interfacial states within the gap. Toušková and Kužel (1972) show that the room temperature I–V characteristics of Au plated from $AuCl_3$ onto etched surfaces are in good agreement with heterojunction theory at room temperature only.

Akobiroba et al. (1974) obtain a blocking contact for Au on both n- and p-type CdTe. Rectification occurs independent of the deposition method and the nature of the surface preparation. They attribute this phenomenon to the formation of an n-type CdO layer sandwiched between the Au and a p-type layer formed in either the n- or p-type base material by the out diffusion of Cd.

Evaporated Al contacts on undoped high resistivity ($\rho > 10^7$ ohm-cm) p-type material (Triboulet et al., 1974) and aguadag on high resistivity In-doped material (Zanio et al., 1975b), both on polished surfaces, results in leakage currents lower than expected from those based upon the bulk resistivity. Similar contacts on high resistivity but p-type Cl-doped material grown by the traveling heater method not only results in blocking action but also in the ionization of a deep acceptor and time-dependent depletion regions (Malm and Martini, 1973, 1974; Siffert *et al.*, 1976; Fabre *et al.*, 1976; Triboulet *et al.*, 1974). In Cl-doped material a neutral acceptor lies just above the Fermi level. For Al blocking contacts on high resistivity *p*-type material the bands bend downward at the metal–semiconductor interface. When the bias is initially applied holes are swept out and the field is nearly uniform across the entire device as in Fig. 4.10. Subsequently, either the center thermally ionizes holes or captures injected electrons, resulting in polarization and a more intense space charge region at the positive contact. The total space charge is due to the contribution from the normal band bending at zero bias and the contribution from the ionization of deep traps. Bell *et al.* (1974b) and Serreze *et al.* (1974) avoided polarization by using a more *p*-type or Ohmic contact at the anode so that the Fermi level does not cross the acceptor level. The electroless deposition of Pt, Au, or Ir from Cl solutions on Br–CH_3OH etched surfaces replaces evaporated Al and eliminates polarization. However, because the resulting barrier height is small, the leakage currents are high. Bodakov *et al.* (1960) observed changes in the surface conductivity of high resistivity material upon exposure to air. Siffert *et al.* (1976) intentionally oxidized the surfaces of high resistivity material and attained limited control of the degree of band bending and injection with a metal–oxide–semiconductor structure. Enough band bending is allowed to provide some blocking action but not to such an excess that the Fermi level crosses the trap level.

II. Modulators

A. Electrooptic Modulators

Modulated laser beams are potentially useful in deep space and terrestrial communications systems (Nussemeier *et al.*, 1974; Forster *et al.*, 1972; Bonek *et al.*, 1974; Popa, 1974), as well as for high-resolution IR spectroscopy (Corcoran *et al.*, 1973) and the generation of intense nanosecond pulses (Figueira, 1974; Smith and Davis, 1974). Although CdTe has been considered for modulation at 1.06 μm (Popa, 1973), its lower loss at 10.6 μm makes it more attractive at the latter wavelength. Cadmium telluride is optically isotropic. However an electric field induces a change in the optical indices of refraction giving rise to a birefringence that is proportional to the electric field. When oriented properly, the crystal therefore becomes an optical wave plate with a voltage-controlled retardation and can electrically modulate the intensity, phase, or polarization of a light beam. Bell *et al.* (1975a) utilized the electrooptic effect to observe the electric field strength in a Cl-doped radiation detector. Figure 4.11 is a schematic of an amplitude modu-

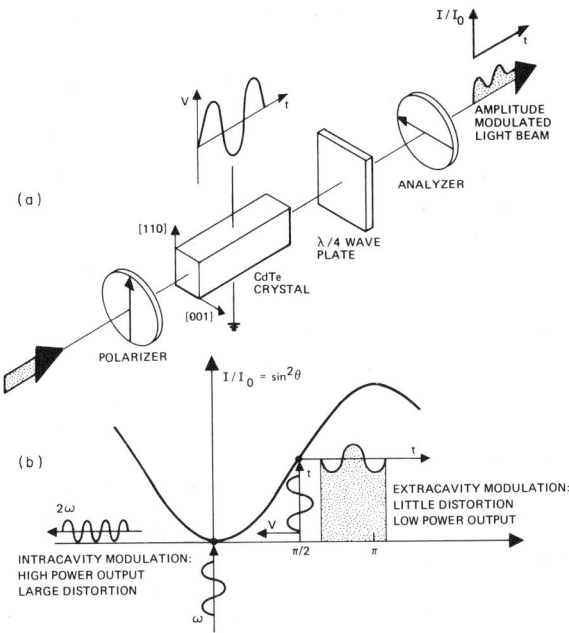

FIG. 4.11. (a) Extracavity amplitude modulation of a laser beam with a $\bar{4}3$ m electrooptic crystal. (b) Transmitted intensity (I/I_0) versus retardation (θ). Operating with a quarter wave plate ($\theta \approx \pi/2$) in the extracavity mode results in good linearity of the transmitted beam. Operating in the intracavity mode without a wave plate ($\theta \approx 0$) results in distortion but a potentially higher power output.

lation system. Light passes through the polarizer parallel to the [110] edge of the CdTe crystal. After passing through the crystal in the [110] (long) direction, the relative phase retardation induced by the birefringence for an applied voltage V is (Pankove, 1971)

$$\phi = (2\pi l/\lambda)n_0^3 r_{41}(V/d), \qquad (4.3)$$

where λ is the free space wavelength of the incident radiation, n_0 is the unperturbed refractive index, r_{41} is the electrooptic coefficient, l is the device length, and d is the electrode separation. After passing through a quarter wave plate and a crossed polarizer, the intensity is

$$I/I_0 = \sin^2 \theta = \sin^2(\phi/2 + \pi/4), \qquad (4.4)$$

where $\pi/4$ refers to the retardation of a quarter-wave plate. Consequently, by varying the applied voltage the incident light beam is intensity modulated. A half-wave voltage of 750 V is required to obtain 100% amplitude modulation about the half-power point at 10.6 μm for a 1.5 mm by 50 mm rod. Spears and Strauss (1977) report low-voltage electrooptic phase modulation at 10.6 μm in waveguides formed by proton bombardment of n^+ material. A device, composed of a 22-μm thick guide layer covered with a 14 mm long electrode produced 110° phase modulation with a 100 V bias. The modulation factor of over 1°/V is more than twice that reported for any other CO_2 laser modulator.

The $n_0^3 r_{41}$ product determines the effectiveness of an electrooptic material as a modulator. The electrooptic coefficients measured at different wavelengths are given in Table 4.2. Kiefer and Yariv (1969) found the electrooptic coefficient to be independent of frequency from 0 to 100 kHz at both 3.39 and 10.6 μm. Gallium arsenide is a commonly used electrooptic material for modulation with infrared. However, CdTe is superior to GaAs, due not only to a higher $n_0^3 r_{41}$, but also to a lower value of the absorption coefficient. Since $n_0^3 r_{41}$ is a factor of two larger for CdTe than for GaAs, a fourfold decrease in modulation power is allowed with CdTe to attain the same intensity output. Also the absorption coefficient for CdTe (Kuhl, 1973) is about an order of magnitude less than that for GaAs (Klein and Rudko, 1968). The extinction ratios (ER) in both materials are comparable with the best CdTe and GaAs, approaching $10^4:1$ (Figueira, 1974). However, as the effective areas of the modulators are increased, the measured ER decreases, suggesting that most of the residual birefringence in the crystal occurs at the edges, possibly due to unrelieved strain introduced in processing. Using gating voltages, single nanosecond input pulses are selected from pulse trains for large CO_2 amplifier systems. Crystals and their coatings have withstood energies of 0.25 J in a 70-nsec pulse and peak powers of 40×10^6 W/cm^2.

The low absorption coefficient of CdTe ($\alpha \approx 0.001$ cm^{-1}) makes intracavity modulation attractive. In the intracavity approach the modulator is located inside the laser cavity where the light beam is circulated or multiply reflected. In this approach the optical modulator transfers a fraction of the

TABLE 4.2

ELECTROOPTIC COEFFICIENTS OF CdTe (r^{41}) VERSUS WAVELENGTH

Wavelength (μm)	Coefficient r_{41} (m/V 10^{-12})	Reference
1.0	2.2	Stafsudd et al. (1967)
1.0	5.3	Bagaev et al. (1970)
3.39	6.8	Kiefer and Yariv (1969)
10.6	6.8	Kiefer and Yariv (1969)
10.6	6.2	Nikolaev and Koblova (1971)
23.35	5.5	Johnson (1968)
27.95	5.0	Johnson (1968)

circulating power completely out of the cavity by means of a polarizing element such as a Brewster-angle window. This technique requires significantly less driving power to attain the same output intensity than that required when the modulator is located outside of the cavity. There are some disadvantages, however. In the extracavity device the system operates with optimum linearity near $\theta = \pi/4$ as in Fig. 4.11. (The retardation is small compared to $\pi/4$.) The coupling intracavity system on the other hand operates at or near $\theta = 0$ with poor linearity. Moreover due to a fundamental resonance in which the stored energy oscillates between the inverted atoms and the electromagnetic field (Yariv et al., 1973), the usefulness of the system is limited to frequencies above 1 MHz.

Kiefer et al. (1972) nonetheless consider coupling modulation to be the most useful form of modulation for CO_2 lasers for wideband space-communication systems. A useful application for this wideband capability is that of compensation for Doppler shift using a tunable optical carrier. Tunable carriers are achieved by employing the coupling modulation approach and filtering out one of the side bands. Huang et al. (1974), for example, have demonstrated efficient coupling modulation at 1 GHz with CdTe, in which an optical beat method was used to test the effectiveness of the coupling.

B. ACOUSTOOPTIC MODULATORS

The interaction of light with acoustic waves offers an alternative approach to modulation (Dixon, 1967; Pinnow, 1970). Acoustic waves of wavelength λ_s in Fig. 4.12 form an optical grating and scatter light in a manner similar to the first order Bragg scattering of x-rays. A shift in the frequency of the incident photon beam occurs by an amount equal to the acoustic frequency. For the relative direction of the phonon and photon momentum vectors in Fig. 4.12, the frequency of the diffracted beam decreases. The fraction of power transmitted from the incident beam (I_1) to the transmitted beam

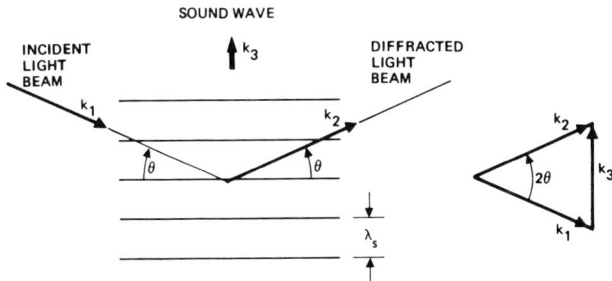

FIG. 4.12. Bragg scattering of an incident light beam (k_1) through 2θ by sound wave (k_3) of wavelength λ_s.

(I_2) is given by

$$\frac{I_2}{I_1} = \sin^2\left[\frac{\pi}{2}\left(\frac{2}{\lambda^2}\frac{L}{H}M_2 P_{ac}\right)^{1/2}\right], \quad (4.5)$$

where L is the dimension in the direction of optical beam propagation, H is the orthogonal dimension of the acoustic column, P_{ac} is the acoustic power, and λ is the wavelength of light. For small efficiencies in Eq. (4.5), the intensity of the diffracted beam is proportional to the intensity of the acoustic wave. The figure of merit of the effectiveness of the material in coupling the incident and diffracted beams is given by

$$M_2 \equiv n^6 p^2/\rho v_s^3, \quad (4.6)$$

where ρ is the density, n is the index of refraction, v_s is the acoustic velocity, and p is the appropriate strain–optic coefficient of the material. Joiner et al. (1976) determined the strain–optic coefficients through Eq. (4.6). Rather than measuring M_2 directly from Eq. (4.5), a comparative measurement relative to fused quartz was made. Weil and Sun (1971) measured the piezo-optic coefficients (q_{ik}) (See Chapter 2, Section V,F) and calculated the strain–optic coefficients through

$$p_{ij} = \sum_k q_{ik} c_{kj}, \quad (4.7)$$

where c_{kj} are the elastic constants. The strain–optic coefficients for CdTe are listed in Table 4.3. The figure of merit (M_2) for CdTe is 32×10^{-18} sec^3/gm for CdTe. This value corresponds to longitudinal acoustical waves traveling in the [110] direction, with the optical waves propagating in the [010] direction. The figure of merit is about a factor of three less than that of GaAs and a factor of about thirty less than Ge (Abrams and Pinnow, 1970).

In view of the relatively low absorption coefficient of CdTe at 10.6 μm, it is an interesting acoustooptical material. Spears et al. (1976) fabricated

TABLE 4.3

STRAIN-OPTIC COEFFICIENTS IN CdTe

Reference	Wavelength (μm)	p_{11}	p_{12}	p_{44}
Weil and Sun (1971)	10.6	−0.152	−0.017	−0.057
Joiner et al. (1976)	10.6	−0.07	−0.01	−0.06
Bendow et al. (1974)[a]	10.6	−0.05	−0.104	−0.104
Bendow et al. (1974)[a]	3.9	−0.001	−0.002	0.002

[a] Theoretical values. Electronic contribution to p_{ij} not included.

an acoustooptic modulator in an n/n^+ CdTe waveguide which was formed by proton bombardment. Using Rayleigh surface waves, a modulation efficiency of 10% at 10.6 μm and 27 MHz was obtained with 0.5 W of acoustic power. This acoustooptic modulator is similar in efficiency to bulk Ge devices ten times its length (Spears and Strauss, 1977).

Another scheme of modulating light is through the Franz–Keldysh effect. Berozashvili and Dundua (1975), however, show this approach to have several disadvantages. One disadvantage is the nonlinear quadratic dependence of the amplitude of the modulated signal on the electric field. An additional disadvantage is that modulation occurs at a frequency twice that of the applied field.

III. Optical Elements

A. Uses

High resistivity CdTe ($\rho > 10^6$ ohm-cm) is commercially available in the form of lenses, Brewster windows, and partial reflectors. Partial reflective mirrors are used as laser output couplers, as interferometer plates, or as beam splitters. Besides offering the lowest transmission loss among the nonhydroscopic materials, CdTe has a relatively flat transmittance from about 1 to 30 μm. Hot pressed polycrystalline material commonly known as Irtran 6 (Harvey and Wolfe, 1975) is available with a 5000 psi fracture strength and a hardness of about 60 on the Knopp scale, which is about as hard as hard coal. However, the absorption characteristics of CdTe in this form are relatively poor, being about 0.01 cm^{-1} at 10.6 μm. Single or large grain crystals are commercially available with absorption coefficients of about 0.002 cm^{-1}.

New system concepts using high power lasers require infrared windows having lower absorption coefficients, higher strength, greater uniformity, and lower thermal and stress-induced optical distortion. Material having yield strengths of 10,000 psi and absorption coefficients of 10^{-4} cm^{-1} are desired. Kuhl (1973) reported a few CdTe laser windows to have absorption coefficients in the 0.0006–0.001 cm^{-1} range with fracture strengths in the 2000–4000 psi range. There does not appear to be a significant difference in the absorption coefficient in low absorption material grown from the melt by the Bridgman technique (Gentile *et al.*, 1973a,b) and material grown from the vapor by physical vapor transport (Shiozawa *et al.*, 1973). Zinc selenide and KCl are superior to CdTe for high power window applications. Although the absorption coefficient of CdTe is about a factor of two better than ZnSe, its mechanical properties are about a factor of three poorer. The mechanical properties of KCl and KBr are better than those of low loss CdTe and their absorption coefficients are at least an order of magnitude lower. The comparison with respect to strength is not altogether fair regarding CdTe since the work on ZnSe and KCl refers to fine grained material. High power laser windows may fail either by thermal fracture or by optical distortion. Deutch (1975) established figures of merit for selected laser window materials for failure by both modes. The figure of merit of CdTe for failure by thermal fracture is larger than that of GaAs, ZnSe, and KCl. However it rates poorly with KCl and ZnSe for failure by optical distortion. Although contingent upon adequate coatings for protection against moisture, the halides are the most promising materials for high power laser windows. However, CdTe is still intriguing because it is the material that is most likely to be susceptible to an order of magnitude decrease in the absorption coefficient.

B. Absorption Mechanisms

Deutsch (1973) compiled data on CdTe, ZnSe, GaAs, and the halides and showed that the absorption coefficient follows

$$\beta = A \exp[-\gamma\omega/\omega_{LO}] \qquad (4.8)$$

when the frequency ω is well above the Reststrahl frequency (ω_{LO}). Here A and γ are constants. Figure 4.13 shows β versus reduced frequency (ω/ω_{LO}) for GaAs, ZnSe, and CdTe. Since the shape of β is exponential, it is consistent with a multiphonon process. A straight line on the high side of the spectrophotometer data is useful for a conservative estimate of β at 10.6 μm. Laser calorimeter data at 10.6 μm shows that while there is reasonable agreement for ZnSe and GaAs the experimental values for CdTe are much higher than projected. In ionic insulators, McGill et al. (1973) consider the effect of an anharmonic potential on the multiphonon process for frequencies higher than several times the Reststrahl. They substantiate the exponential dependence as in Fig. 4.13 and argue that a similar dependence is to be expected in CdTe. Cadmium telluride is amenable to such a comparison because of its highly ionic nature. It is therefore unlikely that multiphonon interactions limit 10.6 μm absorption in CdTe. Multiphonon cooperation, resulting in electron transitions from the valence band to the conduction band as in Fig. 2.24, are even more unlikely. If β is proportional to the electron concentration for FCA and intraband absorption as in Fig. 2.21, then these mechanisms for resistivities greater than 10^4 ohm-cm should also be negligible.

Crystals containing small concentrations of impurities give rise to vibrational modes. Infrared absorption measurements show a local mode at 391 cm^{-1} (25 μm) for Be-doped material (Hayes and Spray, 1969; Sennett et al., 1969). Doping with 300 ppm by weight of Be was sufficient to observe

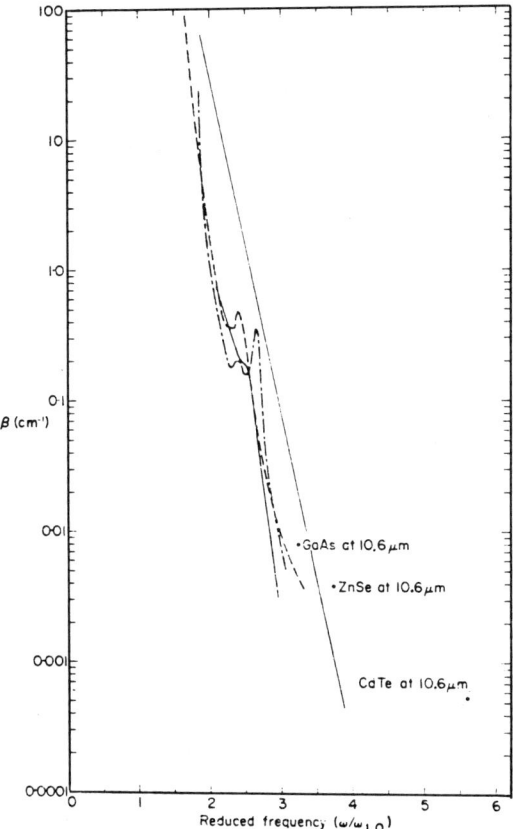

FIG. 4.13. Absorption coefficient versus reduced frequency $\beta = A \exp \gamma(\omega/\omega_{LO})$ for CdTe, —; ZnSe, ---; and GaAs, ---. (Reprinted with permission from Deutsch, 1973. © 1973, Pergamon Press, Ltd.)

the second harmonic clearly. Considering the degree of doping necessary to observe absorption, it is unlikely that point defects will limit absorption in relatively clean window material at 10.6 μm.

Lattice-dynamical calculations are in reasonable agreement not only for Be but also for other isoelectronic impurities such as Mg, S, and Se (Talwar and Agrawal, 1974; Balkanski and Beserman, 1968; Gaur et al., 1971). Talwar and Agrawal (1975) changed the force constants for their calculations to fit the data for Al in II–VI compounds. These calculations may not be realistic due to compensation by native defects. Dutt et al. (1976) measured local mode frequencies at 282, 287, and 326 cm^{-1} for $Al_{Cd}V_{Cd}$ and at 399 cm^{-1} for Al. Other frequencies in this range are proposed for $Al_{Cd}Cu_{Cd}$, $Al_{Cd}Au_{Cd}$, $Al_{Cd}Au_i$ (or $Al_{Cd}Au_{Te}$), $Al_{Cd}Ag_i$ (or $Al_{Cd}Ag_{Te}$), and

$Al_{Cd}Sb_{Te}$. From 10^{18} to 10^{19} Al cm^{-3} was added in these studies. However, the fraction of Al on substitutional sites is now known. Further studies are necessary to measure cross sections and determine the influence of point defects and complexes on absorption at 10.6 μm.

The absorption coefficient is most likely limited by the scattering of photons by particles whose dimensions are small compared to the wavelength of the incident radiation. Sparks and Duthler (1973) show that volume fractions in the 10^{-7}–10^{-8} range can result in an absorption coefficient of 10^{-4} cm^{-1}. Local heating at inclusions may cause thermal defocusing through stress-induced birefringence (Joiner *et al.*, 1976; Bendow *et al.*, 1974). The local heating effect can eventually lead to failure of the material at high power levels. The entrapment of inclusions during crystal growth and the formation of precipitates due to the retrograde solubility are discussed in Chapter 3, Section I. Hall *et al.* (1975) apply the theory of light scattering by small particles in a homogeneous matrix to CdTe and calculate absorption coefficients that are larger than that measured. For material prepared by physical vapor deposition having a precipitate volume fraction in the 0.003–0.01 range, they measured absorption coefficients in the 0.003–0.4 cm^{-1} range. The density, size, and distribution of small particles, presumably Te, and voids were determined by TEM and are described in more detail in Chapter 3, Section I. Magee *et al.* (1975) identified Te precipitates as well as dislocation tangles, $CdCl_2$ and In_2Te_3 platelets in melt grown material, and associated increased absorption with general disorder in the crystal. Evidence of micron-sized voids, observed by Hall *et al.* (1975) and hypothesized by others (Kuhl, 1973; Gentile *et al.*; 1973a,b; Shiozawa *et al.*, 1973), was not detected. With windows grown by physical vapor transport, Shiozawa *et al.* made calorimetric absorption measurements on uncoated samples and samples painted with ordinary fingerpolish. The absorption coefficient for the painted sample increased because the paint directly absorbed a large fraction of the photons scattered by precipitates and voids and also increased the internal angle for total internal reflection and subsequent absorption. Precipitates (>1 μm) were observed with an infrared microscope. Small absorption coefficients are usually determined by calorimetric absorption measurements. However, Nurmikko *et al.* (1975) also utilized the temperature-dependent shift in the optical absorption edge [Eq. (2.31)] to indicate slight temperature increases induced by laser heating to measure small absorption coefficients.

After the heat treating of both vapor growth and melt growth material, fine precipitates were removed (Zanio, 1971) and larger precipitates became smaller, resulting in a reduced absorption coefficient. Gentile *et al.* (1973a,b) also found that annealing reduced the absorption coefficient. Magee *et al.* (1975) found that Te precipitates (≈ 60 Å) decomposed in melt grown material after anneals above 500°C. In principle, lower absorption coefficients are to be expected by further reduction of the concentration and size of small particles and voids.

C. MECHANICAL PROPERTIES

The fracture strength of low absorption CdTe is in the 2000–4000 psi range (Kuhl, 1973). A factor of three improvement in the strength with lower absorption would make CdTe useful for high power laser window applications. More basic studies on the fracture and slip mechanisms are necessary to determine if these strengths can be attained. Some mechanical properties studies have been made. In cleavage studies on single crystals, Wolff and Broder (1959) associated cleavage and resulting surface structure with the type of bonding; whereas the covalent diamond structure fractures along (111), the more ionic sphalerite structures such as CdTe fracture along (110). Although CdTe shows distinct (110) microcleavage by the "razor blade technique," there is also microcleavage in the planes (hhl) $(h \geqslant l)$ and in all planes of the zone [001]. From a ratio of (011) to (111) type cleavage Wolff and Broder assign a 60% ionicity to CdTe. Of the sphalerites CdTe, ZnSe, and GaAs, CdTe has the lowest yield strength and is also the most ionic on the Phillips' scale. Cleavage in planes of zone [001] suggest that the surface is deformed so that the atoms of the first, and probably to a small extent the second, surface layer shift to new positions characteristic of the NaCl structure. In low resistivity material, cleavage more readily occurs on the polar (111) planes. When the Cd and Te atoms separate from one another, there is an adjustment of the electronic charge. Presumably, in low resistivity material the differences in charge between the two (111) surfaces are effectively neutralized through the last point of contact during separation.

Carlsson and Ahlquist (1972) and Buch and Ahlquist (1974a) showed that the electron concentration affects the yield stress. The yield stress in Fig. 4.14 for undoped CdTe both in the dark and with light is least for high resistivity material and increases with increasing electron or hole concentration. Swaminathan et al. (1975) find that for In-doped material the hardness also increases with increasing concentration of electrons. The results are explained in terms of the pinning of dislocation by point defects. In CdTe the 60° dislocation shown in Fig. 4.15 is assumed to determine plastic deformation. The angle between the dislocation line and the Burgers vector is 60°, giving the dislocation both edge and screw properties. From etch pit studies, Inoue et al. (1963) and Teramoto and Inoue (1963) postulate two types of 60° edge dislocations. One type is a Cd dislocation whose extra half-plane terminates in an extra row of Cd atoms at the slip plane. The other type is the Te dislocation whose extra half also terminates in an

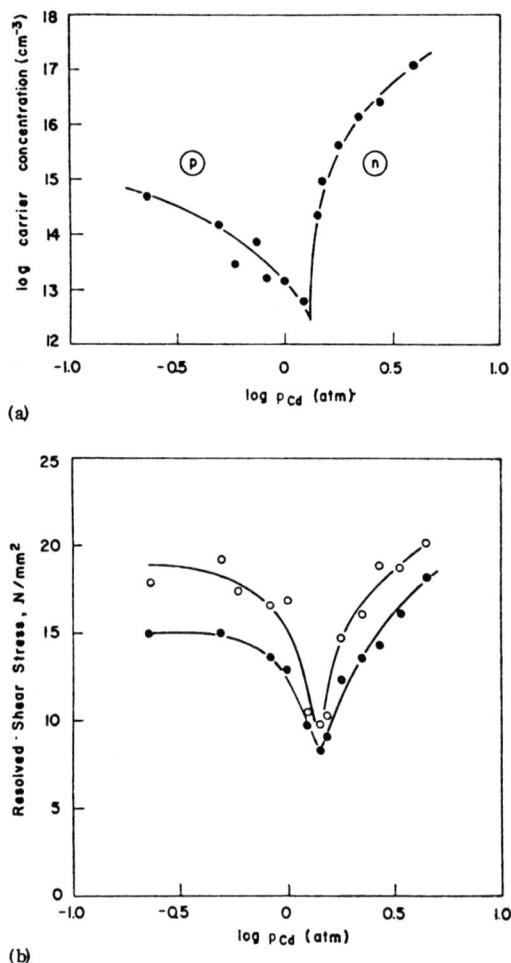

FIG. 4.14. (a) Room temperature concentrations of electrons and holes in CdTe single crystals as a function of p_{Cd}. (b) Corresponding yield stress in compression in darkness (●) and during illumination (○) (Buch and Ahlquist, 1974a).

extra row of Te atoms at the slip plane. Buch and Ahlquist assume that the Te (Cd) dislocation acts as a donor (acceptor). The interaction of the ionized donor- (acceptor-) type dislocations with ionized acceptor (donor) defects results in an electrostatic attraction leading to pinning of the dislocation and an increase in the yield stress. A change in mechanical properties with illumination is proof of the electrostatic interaction between the dislocations and point defects.

The most effective approach in increasing the yield strength is to decrease the crystallite size. Buch and Ahlquist (1974b) attained strengths up to 9000

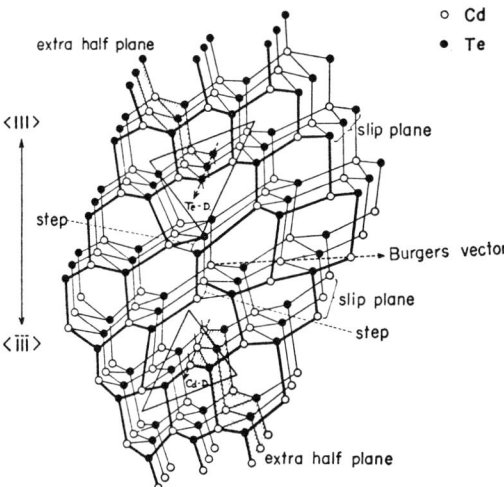

FIG. 4.15. Atomic model showing Cd (○) and Te (●) 60° edge dislocations intersecting the (111) surface. Steps associated with the 60° edge dislocations are indicated. On the (111) surface, two triangles are drawn schematically whose sides are parallel to the $\langle 110 \rangle$ directions and apexes are oriented along the [211] and [$\bar{2}1\bar{1}$] directions (Teramoto and Inoue, 1963).

psi with a 31 μm grain size in hot pressed material. Holt (1962, 1966, 1969) reviews grain boundaries in the sphalerite structures. Whether the low absorption coefficients and higher yield strength can be simultaneously attained in the same material remains to be seen.

IV. Solar Cells

Silicon single crystal solar cells are critical components in the operation of satellites. It is likely that the more efficient $GaAs/Ga_xAl_{(1-x)}As$ cells will also play a key role in this application and may even eventually replace the Si cell. With rising energy costs, terrestrial photovoltaics are becoming more attractive as a supplementary energy source. However, the present cost of Si and $GaAs/Al_{(1-x)}As$ cells prevents widespread usage. Consequently, the Energy Research and Development Administration is undertaking research and development on terrestrial photovoltaics so as to provide several percent of the domestic energy needs of the United States of America by the year 2000. Economic considerations dictate the use of thin film cells. This makes the compound semiconductors CdS, InP, CdTe, Cu_2S, and $CuInSe_2$ attractive for this application. The efficiency in converting available solar energy to useful means is an important criterion in choosing the best material. A 10% efficient cell with a cost of about \$200/peak kW in megawatt production quantities is a desired goal.

Efficiencies in excess of 14% have been attained under terrestrial illumination with single crystalline Si and GaAs/Ga$_x$Al$_{(1-x)}$As cells. The best all thin film results are about 8% for both the Cu$_{2-x}$S/CdS (Mattox, 1976; Meakin et al., 1976) and CdTe/CdS (Nakayama et al., 1976) heterojunction systems. Considering that Yamaguchi et al. (1976) prepared 12% efficient single crystal heterojunction cells with CdTe makes this material even more important for terrestrial applications.

A. HOMOJUNCTION CELLS

The maximum power that a cell can theoretically deliver is a product of its open circuit voltage times its short circuit current, which increase and decrease respectively with increasing bandgap. [Anomalous photovoltages, up to several hundred volts in some cases, are generated along the surface of CdTe thin films (Goldstein and Pensak, 1959). Such voltages maybe discounted, however, as they generate negligible power.] The maximum theoretical efficiency is about 27%, which corresponds to a bandgap of 1.5 eV (Loferski, 1963; Hovel, 1975). This is the bandgap of CdTe.

Vodakov et al. (1960) and Naumov and Nikolaeva (1961) report efficiencies of 4% and 6% for homojunctions. The details of their cell fabrication are not apparent. Bell et al. (1975b) discuss the (n) on (p) CdTe homojunction solar cell and conclude that a minority carrier lifetime of at least 10^{-7} sec is required to obtain a 10% efficient cell. Since minority carrier lifetimes in (n)CdTe are less than 10^{-8} sec and lifetimes are poorer in (p)CdTe than in (n)CdTe (Artobolevskaya et al., 1967; Cusano, 1967; Cusano and Lorenz, 1964), an economical CdTe homojunction terrestrial cell is unlikely. Radiative lifetimes are about $1-5 \times 10^{-9}$ sec (Picus et al., 1968b; Morehead, 1967). Trapping times of better than 10^{-7} sec for both electrons and holes have been measured in semi-insulating CdTe (Zanio et al., 1974; Bell et al., 1974a). However, these values are not minority carrier lifetimes and refer only to deep traps. When electrons (holes) are trapped at shallow levels in semi-insulating material, they are reemitted back to the conduction (valence) band rather than recombining with holes (electrons). The total trapping time, which includes trapping at shallow traps, is undoubtedly much smaller.

Because most of the photons are absorbed in the first micron in CdTe and the lifetimes are low, the junction must be located close to the surface. The diffused layer must also be of low resistivity to avoid high series resistance. Excessive doping or deviations from stoichiometry, however, further reduces the lifetime. Even if the series resistance and low lifetime problems could be overcome, it is likely that surface recombination would be excessive (Yamaguchi et al., 1975). Artobolevskaya et al. (1967) measured values of 10^4 and 10^5 cm/sec for the recombination velocity using etched and optically polished surfaces.

B. HETEROJUNCTION CELLS

1. n-Type CdTe

Improved efficiencies are possible by using heterojunctions. Cusano (1963, 1966, 1967) prepared CdTe thin film heterojunctions by sequentially vacuum-depositing CdS onto a molybdenum or glass backing to form the n-type contact, vapor-depositing n-type CdTe with an apparatus such as in Fig. 1.21, treating the resulting composite film in a warm aqueous cuprous solution to form copper telluride and evaporating a Au grid. Bernard et al. (1966) also formed copper telluride by flash evaporation. Figure 4.16 (top) shows the construction of such a cell. Radiation is incident on the copper telluride side and nearly all absorbed by the CdTe. The bandgap of copper

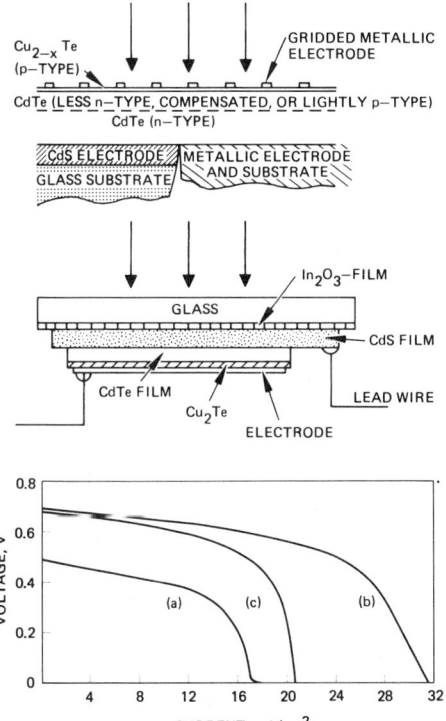

FIG. 4.16. (Top) The $Cu_{2-x}Te(p)/CdTe(n)$ solar cell. (Middle) The $CdS(n)/CdTe(p)$ solar cell. (Bottom) Load characteristics for the (a) thin film $Cu_{2-x}Te/CdTe$ cell which has an efficiency of 5.1% for 87 mW/cm^2 of sunlight (Cusano, 1963), (b) $CdS(n)/CdTe(n)/CdTe(p)$ thin film cell which has an efficiency of 8.1% for 140 mW/cm^2 using a solar simulator (Nakayama et al., 1976), and (c) $CdS(n)/CdTe(n)/CdTe(p)$ single crystal cell which has 12.1% efficiency for 68 mW/cm^2 of sunlight (Yamaguchi et al., 1976).

telluride is about 1.04 eV (Sorokin et al., 1965) and is believed to be indirect (Cusano, 1963). Indium solder makes good n^+ contact to the single crystal cells. The p-type contact to the single crystal cell is prepared in a way similar to that of the thin film devices. Figure 4.16 (bottom (a)) shows the load characteristics for an approximately $\frac{1}{2}$ cm^2 thin film cell in sunlight. Correcting for surface reflectively, Cusano estimates 9 to 10% efficiency as possible for the $7\frac{1}{2}$% single crystal cells. The best results are obtained by increasing the resistivity of the CdTe near the heterojunction so that the electric field in the extended depletion regions assists in the charge collection.

$Cu_{2-x}Te$ films change with time (Cusano, 1966; Justi et al., 1973; Lebrun, 1971; Guillien et al., 1971). This is associated with the presence of moisture, absorbed oxygen, and the diffusion of Cu. Transient problems have been solved in $Cu_{2-x}S/CdS$ (Stanley, 1975) and solutions are likely to be applicable to CdTe. Care to avoid pinholes during any deposition process is important. Although not reported in other CdTe work, Justi et al. found that it was necessary to increase the film thickness to about 20 μm to avoid shorting. In this system the short circuit current are smaller than theoretical values and, as in the case of $Cu_{2-x}S/CdS$ cells, may be due to recombination through interface states (Hovel, 1975).

2. *p-Type CdTe*

In light of these problems, more effort is being devoted to various II–VI heterojunction combinations with p-type CdTe being the smaller bandgap component. The (p)CdTe/(n)CdS system is attractive since it has a theoretical efficiency of 15% and no energy spike (Adirovich et al., 1969). Besides providing a low resistivity n-type contact to CdTe, the CdS serves as a window and eliminates surface recombination. Although the minority carrier lifetime in p-type material (Bube et al., 1976) is probably less than that in n-type material (Cusano and Lorenz, 1964), the higher electron mobility is sufficient to make the diffusion lengths comparable. Specific approaches to this pair are listed in Table 4.4. The reader should refer to the conditions of measurement and specifics of the cell before comparing the efficiencies in this table. Yamaguchi et al. (1976) obtained 12% efficiency by the vacuum evaporation of CdS on (111) P-doped CdTe grown by the Bridgman technique. An In–Ga alloy and Ni were used to make contact to the CdS layer and CdTe substrate, respectively. Although these results refer to an active area of only 0.5 cm^2, they are impressive. An analysis of this cell shows the structure to be (n)CdS/(n)CdTe/(p)CdTe and therefore contains some elements of the homojunction. The load characteristics are shown in Fig. 4.16. Bube et al. (1976) attained efficiencies up to 7.9% with the same approach. However, Bube et al. (1975) used a three point measurement technique to measure the efficiency and therefore this number does

TABLE 4.4

THE CdTe/CdS HETEROJUNCTION PAIR

Approach	Investigators	Solar cell efficiencies (%)
Screen printing method	Nakayama et al. (1976)	8.1[a]
Vacuum deposition of CdS on CdTe	Yamaguchi et al. (1975, 1976)	10.5[b] (12.0)[c]
	Adirovich et al. (1969, 1971)	
	Bonnet and Rabenhorst (1972)	5–6
	Fahrenbruch et al. (1975)	
	Bube et al. (1976)	7.9
Spraying CdS on CdTe	Bube et al. (1976)	5.6
Close space vapor transport CdTe on CdS	Fahrenbruch et al. (1974, 1975)	4
	Mitchell et al. (1975)	
Co-evaporation of Cd and S on CdTe	Fahrenbruch et al. (1974)	
Evaporation of Te onto CdS with subsequent heat treat	Dutton and Muller (1968)	

[a] All thin film.
[b] Power conversion efficiency based upon total area.
[c] Power conversion efficiency based upon active area.

not represent the effective solar cell power. An estimate of the actual solar power using a two-point rather than a three-point measurement with existing contact technology shows that the 7.9% efficiency would be reduced to approximately 7.8%. However, the contacting procedure is not without hazard (Fahrenbruch, 1976, private communication). An indium–tin oxide film on the CdS serves as a conducting layer to reduce the series resistance. Glycerol serves as an antireflection coating. Spraying CdS on CdTe to form the heterojunction results in an efficiency of 5.6%. Spraying CdS is simple. However, the economics of the cell depend upon how easily the CdTe can be deposited in thin film form. The close space vapor transport of CdTe onto CdS has resulted in a cell with a 4% solar efficiency, an open circuit voltage of 0.61 V, and a quantum efficiency of 0.85.

Using a screen printing method Nakayama et al. (1976) obtained an 8.1% power conversion efficiency for an all-thin-film cell. The structure is a back-wall cell and has the elements of both the CdS/CdTe and Cu_2Te/CdTe heteropairs (Fig. 4.16 middle). Transparent In_2O_3 is first deposited on glass. A 20-μm CdS film with a resistivity of about 0.2 Ω-cm is next deposited by the screen printing method. A 10-μm thick CdTe paste of about the same resistivity is prepared by mixing In-doped n-CdTe and $CdCl_2$ with a binder in a mixer and is applied by the same method. The structure is dipped into a hot aqueous cuprous ion solution for 10 min at 200°C to form Cu_2Te. Copper diffusion along the CdTe grains promotes the formation of p–n junctions near the CdS/CdTe interface. Microprobe analysis of single crystal cells prepared by the same cuprous ion solution method substantiates the formation of homojunctions with a (n)CdS/(n)CdTe/(p)CdTe/$(p^+)Cu_2Te$ heteroface structure in the thin films. The load characteristics of this cell are also shown in Fig. 4.16.

Preliminary work on an (n)indium–tin–oxide/(p)CdTe heterojunction is interesting because of the large open circuit voltage, the large bandgap window, and the simplicity of construction (Bube et al., 1976). Additional CdTe cells utilizing other chalcogenide windows have been prepared. The ZnSe/CdTe cell is potentially the most efficient II–VI combination. By using ZnSe (E_g = 2.67 eV) instead of CdS (E_g = 2.43 eV), about 18% more photons can potentially be collected, resulting in higher short circuit currents. The 0.41 eV higher diffusion potential also should result in higher open circuit voltages. Only 1 to 2% quantum efficiencies have been reported (Bube et al., 1975). A conduction band spike which impedes charge transport is present in this systems. Also it is likely that ZnTe or CdTe or CdSe interlayers form near the junction due to the higher processing temperatures. The solubility of Te in CdS is appreciable (Vitrikhovskii and Mizetskaya, 1959; Bonnet, 1970). Bonnet and Rabenhorst (1971) obtained a continuous series of CdS_xTe_{1-x} thin films. Although this system is processed at a lower temperature, it is likely that the interface in the CdS/CdTe system is also not abrupt. However, the system is composed of 3-elements rather than 4-elements and therefore presents fewer unforeseen technological problems.

The presence of these windows may eliminate surface recombination. However, because of the large lattice mismatches, 9.7% and 17.5%, respectively, for the CdS and ZnSe windows on CdTe, large recombination currents are likely to limit cell performance severely. Tunneling has proved to be the dominant component to the forward dark current in CdS/CdTe (Adirovich et al., 1971) and other heterojunctions except those with a very small lattice mismatch (Hovel, 1975). For the vapor deposition of CdS on {110} CdTe the [0001] and [111] axes are nearly parallel. Igarashi (1971) interprets the small angle to accommodate dislocations in the small angle boundary and to eliminate the interfacial mismatch. More ideal cells are GaAs/Ga$_x$Al$_{1-x}$As and InP/CdS which have lattice mismatches of less than 0.4% and solar efficiencies of better than 14%.

C. CONTACT TO p-TYPE CdTe

In the Cu$_{2-x}$Te/CdTe cell, Ohmic contact is easily made to CdTe with In or CdS. The contact between CdS and Mo is also Ohmic at all temperatures (Guillien et al., 1971). However, in the heterojunction cell utilizing p-type CdTe single crystals, Ohmic contact is not as easily made to the CdTe and the series resistance is significant (Bonnet and Rabenhorst, 1972). Consequently, Bube et al. (1975), in their evaluation of such cells using p-type CdTe, do not report useful power. Instead, they use a three-point measurement technique to calculate the efficiency.

Approaches to attain high-conductivity surfaces on p-type CdTe include special etching treatments and ion implantation. Many etching treatments leave an excess of Te (Ichimiya et al., 1960). Fahrenbruch et al. (1975) report that with Ni contacts and 2–4 ohm-cm material, they achieved contact resistance of 0.15 ohm-cm^2. Prior to depositing Ni, the surface is etched with K$_2$Cr$_2$O$_7$:H$_2$SO$_4$:H$_2$O to provide a thin Te layer (Zitter, 1971; Zitter and Chavda, 1975). The excess Te is then diffused at 200°C to form a p^+ layer. Conduction consequently occurs by tunneling from the Ni Schottky barrier. Gu et al. (1975a) report the formation of Ohmic contacts to undoped and P-doped p-type single crystals in the 3×10^{14}–3×10^{17} cm^{-3} range with a K$_2$Cr$_2$O$_7$:HNO$_3$:H$_2$O etch and subsequent heat treatment in H$_2$ from 200 to 250°C for 3 min. The introduction of Cu, Ag, and Au and the alkali metals into CdTe also results in p-type material. However, the diffusion of the group V elements in the II–VI compounds is slower (Woodbury, 1967; Hall and Woodbury, 1968) and therefore they are preferred for more stable solar cells.

Although Ohmic contact can be made to single crystals the preparation of Ohmic contacts on low resistivity p-type films is a more difficult problem. Ion implantation is a possible approach for further decreasing the contact resistance and increasing the conductivity of the surface (Kachurin et al., 1968). Donnelly et al. (1968) achieved type conversion by implanting 400 keV As$^+$ into n-type CdTe. Type conversion did not occur when Ar instead of As was the implanted ion. Bean (1976) implanted As, Kr, Cd, Te, and Cr into either high resistivity or n-type CdTe and obtained a p-type surface with mobilities of about 40 cm^2/V-sec. Only Cs implantation produced true chemical doping. Changes with the other ions were related to defect doping. The lowest sheet resistivity attained was still larger than 10^2 ohm/□. When Arkadeva et al. (1975) implanted Cd$^+$ along with As and P, the electrical activities were several orders of magnitude lower than when implanting without Cd^+. Implantation has been undertaken with In (Kachurin et al., 1967; Gettings and Stephens, 1973); however, since Ohmic contacts can be easily made to n-type CdTe, implantation with group III and VII ions is not likely to be pursued in great detail. Interpretation of implantation data is not always straightforward. When samples are implanted at temperature or annealed afterwards without encapsulation, changes in the surface composition may occur. In addition, native defects are generated several thousands of angstroms beyond the extrapolated range of the implanted ion (Norris et al., 1977).

V. CdS/CdTe Heterojunction in Liquid Crystal Imaging

The CdS/CdTe heterojunction (Fraas et al., 1976a,b) is critical to the operation of the liquid crystal light valve, a potentially useful device for large screen multicolor projection displays (Grinberg et al., 1975). Figure 4.17a is a schematic of such a display system that consists of a cathode ray writing light source, an image intensifying light valve, and the display optics. The writing light source is projected onto the light valve (Fig. 4.17b), whose primary components are the CdS/CdTe heterojunction, a dielectric mirror, and a 4 to 6 μm-thick liquid crystal. The writing light, incident on the 14 to 16 μm-thick CdS film, capacitively divides the voltage between the liquid crystal and the heterojunction. In one form of the light valve, polarized white light from the projection lamp passes through the liquid crystal, is reflected from the dielectric mirror, and passes through the second polarizer (crossed with respect to the first). With the writing light on, the voltage is transferred from the heterojunction to the liquid crystal, which has a negative dielectric anisotropy. The applied voltage rotates the molecules from their initial state parallel to the field and introduces phase retardation that changes the polarization of the projected light and due to dispersion allows selected colors to pass the crossed polarizers. The effectiveness with which the voltage is divided between the heterojunction and the liquid crystal depends upon the ease with which the light-injected charge modifies the depletion width and capacitive impedance of the heterojunction. The charge on the Ohmic indium–tin oxide contact next to the CdS film is proportional to the back bias of the diode which in the dark is fully depleted. However, when light falls on the diode, electron–hole pairs are created in the CdS near the CdTe/CdS interface. The holes are trapped at recombination centers in the CdS and the electrons are swept back and forth at the frequency of the applied voltage waveform. An image is formed because the local width of the depletion region and the local photocapacitance decrease and increase respectively with increasing light intensity. The liquid crystal voltage

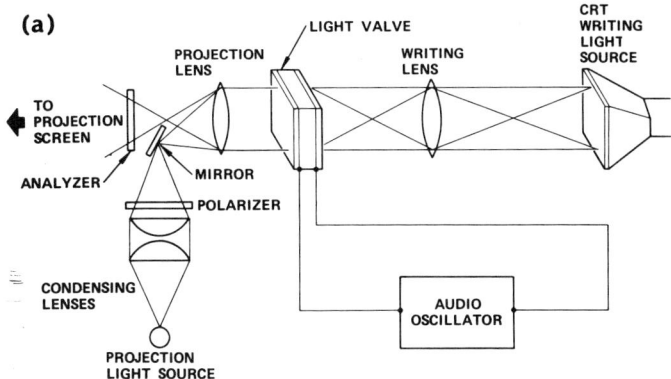

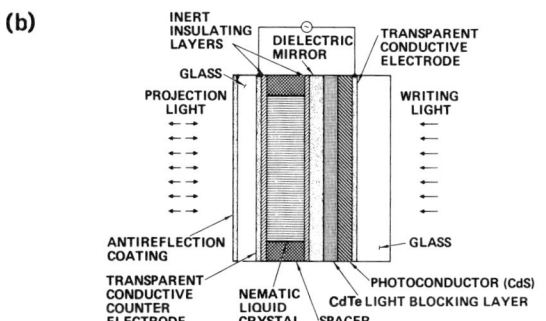

FIG. 4.17. (a) Schematic of the ac light valve display system and (b) Various layers comprising the light valve (Grinberg et al., 1975).

correspondingly increases with the intensity of the writing beam. Thus local variations in the writing light image are transformed to local variations in the liquid crystal voltage with a resultant projected replica image. The CdTe film (≈ 2 μm) besides acting as a blocking contact, prevents the writing light beam from passing through the entire structure. These functions are not unique to CdTe. The unique feature of the CdTe is that the high resistivity at the grain boundaries prevents the lateral spread of light-induced charge. By incorporating suitable dopants at the heterojunction, Fraas (1976) was able to operate the light valve at TV rates.

VI. Miscellaneous Applications

Some of the more interesting applications have considered CdTe for use as a photoconductor, as microwave and light emitting devices, and for use in integrated optics, optical coatings, and nonlinear optics. In some cases basic materials limitations prevent practical use. In other cases either the material is not sufficiently developed or alternate materials offer advantages. Regardless, related measurements have provided important contributions to the understanding of solid state and device physics in compound semiconductors.

A. Photoconductors

Since the early work of Frerichs (1947) numerous photoconductivity studies have been undertaken (Nikonyuk *et al.*, 1975a,b; Sokolova *et al.*, 1969; Kireev *et al.*, 1969; de Nobel, 1959; Svechnikov and Aleksandrov, 1957; Zaitsev, 1971; Potykevich *et al.*, 1971; Konorov and Shevchenko, 1960). A wide range of energy levels, cross sections, etc. are found in these and other works referred to in these studies. Although the photoresponse at the band edge is well defined, further work is necessary to establish a constant pattern with respect to defects. Farrell *et al.* (1974) took advantage of the photoresponse at the band edge and proposed that a photoconducting sensor be mounted on the firewall inside an aircraft turbine engine and operate as an aircraft fire detector. For a detector operating at 400°C the sharp photoconductive response at 1 µm would be able to discriminate white gas flame at 1920°K from a 1000°F background as in Fig. 4.18.

B. Nonlinear Optics

Cadmium telluride and other $\bar{4}3m$ compounds such as ZnSe and GaAs possess properties that are favorable for nonlinear optical applications. They have very high resistance to damage by high-power laser sources, low absorption coefficients, wide regions of optical transparency, and very high nonlinear coefficients (Patel, 1966; Tang and Flytzanis, 1971; Weiss, 1975). Unfortunately, cubic crystals do not exhibit natural birefringence and for this reason the effects of dispersion in limiting the length of coherent nonlinear optical effects cannot be compensated to achieve phase-matched generation over the entire length of the crystal sample. Second harmonic generation (SHG) has been observed at 10.6 µm (Patel, 1966) and in the 26 to 33 µm range (McArthur and McFarlane, 1970). Using a continuous wave CO_2 laser Stafsudd and Alexander (1971) were able to obtain only about 40 µW of output power for 28 W of input power. This experiment illustrated the durability of the material but also illustrated the inefficiency of SHG with this approach.

The difficulty of obtaining high-quality birefringent nonlinear materials in the infrared regions of the spectrum has renewed interest in $\bar{4}3m$ materials and methods of increasing the apparent coherence length (200 µm for CdTe). In 1962, Bloemberger and his co-workers (see Armstrong *et al.*, 1962) suggested that a series of plates with alternating crystallographic orientation could be used to obtain periodic phase matching and substantial nonlinear

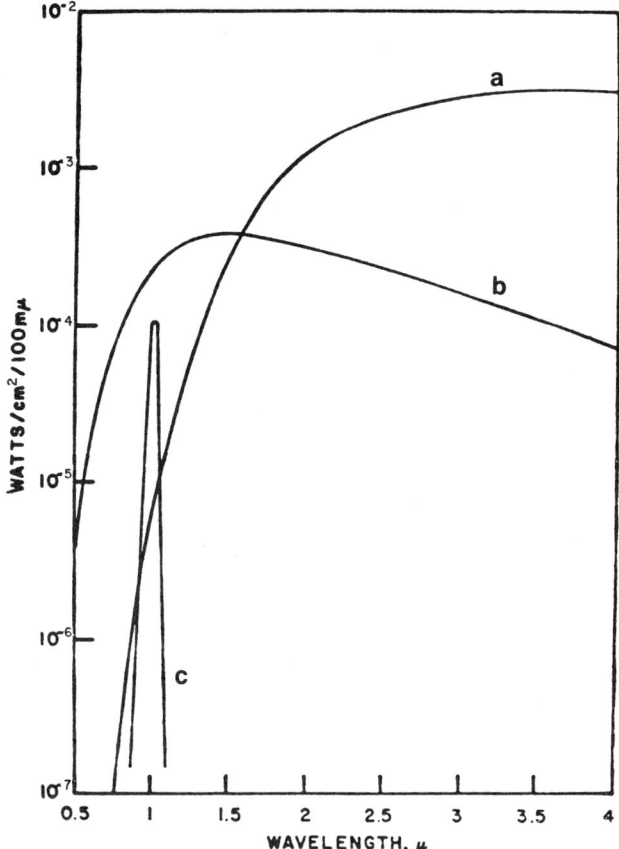

FIG. 4.18. Spectral distribution of the radiation intensity incident on a 1 cm² CdTe photodetector from (a) a black body 10 ft away with 60° field of view at 1000° F. and (b) a white gasoline–air flame 6 in. in diameter. (c) The photoresponse of the CdTe detector. (Farrell et al., 1974.)

effects. An early experiment by Boyd and Patel (1966) and more recent experiments by Piltch et al. (1976) have demonstrated that this technique is practical for several applications, notably doubling 10.6 μm radiation and mixing midinfrared wavelengths to obtain sum- and difference-frequency radiation.

A related method of achieving effects equivalent to multiples of oriented plates was recently demonstrated by Dewey and Hocker (1975). They found that rotational twins that are frequently found in $\bar{4}3m$ materials such as ZnSe and CdTe produce the same optical effects as would be achieved in a stack of optically contacted and oriented plates of thicknesses equal to the twin spacing. The experiment of Dewey and Hocker was to mix a ruby laser and a ruby-pumped dye laser in ZnSe, yielding continuously tunable infrared radiation from 4 μm to 21 μm. This wide tuning range has not been achieved by any other simple system and demonstrates a potentially important use of CdTe, ZnSe, and related $\bar{4}3m$ materials in nonlinear optics. Frequency doubling of near-infrared radiation in rotationally twinned ZnSe has also been demonstrated (Hocker and Dewey, 1976), as has doubling of 10.6 μm radiation in rotationally twinned CdTe (Dewey, 1977). Further doubling experiments by the same group produced powers comparable to those achieved with five CdTe plates orientated to produce periodic phase matching. These experiments suggest that doubling efficiencies of several percent with rotationally twinned crystals are possible. Success in attaining large nonlinear optical conversion efficiencies is possible if twinning can be controlled to some extent.

C. Gunn Effect

Several groups have studied the Gunn effect in CdTe (Ludwig, 1967; Oliver *et al.*, 1967; Oliver and Foyt, 1967; Picus *et al.*, 1968a). Independent measurements (Fig. 2.30) on high resistivity material show the maximum in the drift velocity characteristics to agree with the threshold values in low resistivity material. Unfortunately, after a few seconds of observing the Gunn effect, current runaway occurs, presumably due to carrier injection induced by moving high field domains.

D. Luminescent Devices

In heavily compensated samples, the Gunn effect is suppressed and recombination radiation occurs as a result of impact ionization. Picus *et al.* (1968b) and Van Atta *et al.* (1968) associate simultaneous line narrowing with the onset of superradiance. Excitation with fast electron irradiation, results in narrowing of the emission line to a value less than kT, a sharp increase in its intensity, and directional emission. Vavilov and Nolle (1965, 1966, 1968) associated this with stimulated emission. Quantum efficiencies of up to 12% have been observed in p–n electroluminescence devices operated at 77°K (Mandel and Morehead, 1964; Naumov, 1967). However, contact resistance and heating may prevent laser action (Moorhead, 1967). Gu *et al.* (1973 and 1975b) have prepared n^+–p diodes by the Al vapor-diffusion method resulting in abrupt and smooth junctions. However the resistivity of the bulk p-type region was still too high and stimulated emission was not achieved. The success of the III–V compounds for use as light emitting and microwave devices and the limitations of the II–VI compounds have deemphasized further work with the chalcogenides of Cd and Zn in this direction.

E. $Hg_{1-x}Cd_xTe$ Substrates

The alloy system $Hg_{1-x}Cd_xTe$ is an important infrared material in the 0.9–12 μm wavelength range and is a good lattice match to CdTe. Spears et al. (1976) consider this combination for special integrated optics applications. Solid state diffusion (Almasi and Smith, 1968), ion bombardment (Foss, 1968), vapor transport (Kireev et al., 1971), and epitaxy are all possible approaches to achieve this coupling.

F. Antireflection Coatings

Since it is relatively easy to evaporate, CdTe is useful as an antireflection coating. Calculations indicate that at 10.6 μm CdTe films on KCl with either ZnS or CdS should result in a reflectivity loss of less than 0.01%. Knox et al. (1975) sputtered CdTe onto KCl. Cadmium telluride is preferred to Ge because the latter (1) penetrates several hundred angstroms beyond the KCl–Ge film interface, (2) oxidizes upon exposure to normal atmospheres, and (3) has thermal runaway problems. Lotspeich (1974) also coated single crystals of GaAs with high resistivity CdTe to provide a lower index buffer region in an optical waveguide modulator.

VII. Summary

Cadmium telluride will never make as significant an impact on the semiconductor market as Si has. However, certain properties make it important and unique for several applications. The combination of its high atomic number, wide bandgap, and good transport properties make it useful for room temperature gamma-ray and x-ray detectors.

The high electrooptic coefficient combined with its low absorption coefficient make it useful for electrooptic modulators, especially at 10.6 μm. Acoustooptic modulators have also been fabricated, even though the figure of merit of CdTe as an acoustooptic device is less than that of Ge or GaAs. Low absorption from 1 to 30 μm make it ideal for lenses, Brewster windows, partial reflectors, fiber optics, etc. Unless the mechanical properties can be improved, presently available material is not suitable for high power laser windows.

Solid solutions of $Hg_xCd_{1-x}Te$ are important as infrared detectors. Cadmium telluride is used either as a substrate for epitaxy, diffusion, or ion implantation in the formation of micron-thick layers of this ternary.

Energy conservation has become a serious issue and CdTe is an ideal candidate for this application because of the optimum match of the band gap to the solar spectrum. Respectable efficiencies (8%) for all thin film cells have been obtained. Unless reproducible contact can be made to thin film *p*-type material for the CdS/CdTe heterojunction, an economical cell is unlikely. The CdS/CdTe heterojunction has been invaluable to the development of the liquid crystal light valve. Limitations imposed by the *p*-type contact has prevented development of luminescent devices.

The development of CdTe as a Gunn device, and transistor laser, is discouraging. However, breakthrough in materials development can either revitalize old or open up new fields. Control of twinning during crystal growth would permit efficient mixing of coherent light in nonlinear optic applications. Improved trapping times would result in the use of solid state detectors in the field of whole body scanning using computerized axial tomography. Cadmium telluride is therefore an important material to pursue, not only from the viewpoint of basic materials research but also for its technological importance.

References

Abdalla, M. I., and Holt, D. B. (1973). *Phys. Status Solidi A* **17**, 267.
Abdullaev, G. B., Shilkin, A. I., Shakhlakhtinskii, M. G., and Kuliev, A. A. (1965). *Selen. Tellur. Ikh Primen.* **42**.
Abramov, A. A., Vavilov, V. S., and Vodop'yanov, L. K. (1970). *Fiz. Tekh. Poluprovodn.* **4**, 270 [*Engl. Transl.: Sov. Phys. Semicond.* **4**, 219].
Abrams, R. L., and Pinnow, D. A. (1970). *J. Appl. Phys.* **41**, 2765.
Adirovich, E. I., Yuabov, Yu. M., and Yagudaev, G. R. (1969). *Fiz. Tekh. Poluprovodn.* **3**, 81 [*Engl. Transl.: Sov. Phys. Semicond.* **3**, 61].
Adirovich, E. I., Yuabov, Yu. M., and Yugudaev, G. R. (1971). *Proc. Int. Conf. Phys. Chem. Semicond. Heterojunc. Layer Struct. 1970* **2**, 151.
Agrinskaya, N. V., Arkad'eva, E. N., Matveev, O. A., and Rud', Yu. V. (1969). *Fiz. Tekh. Poluprovodn.* **2**, 932 [*Engl. Transl.: Sov. Phys. Semicond.* **2**, 776].
Agrinskaya, N. V., Arkad'eva, E. N., and Matveev, O. A. (1970a). *Fiz. Tekh. Poluprovodn.* **4**, 370 [*Engl. Transl.: Sov. Phys. Semicond.* **4**, 306].
Agrinskaya, N. V., Arkad'eva, E. N., and Matveev, O. A. (1970b). *Fiz. Tekh. Poluprovodn.* **4**, 412 [*Engl. Transl.: Sov. Phys. Semicond.* **4**, 347].
Agrinskaya, N. V., Arkad'eva, E. N., and Matveev, O. A. (1971a). *Fiz. Tekh. Poluprovodn.* **5**, 863 [*Engl. Transl.: Sov. Phys. Semicond.* **5**, 762].
Agrinskaya, N. V., Arkad'eva, E. N., and Matveev, O. A. (1971b). *Fiz. Tekh. Poluprovodn.* **5**, 869 [*Engl. Transl.:* (1971). *Sov. Phys. Semicond.* **5**, 767].
Agrinskaya, N. V., Alcksandrova, G. I., Arkad'eva, E. N., Atabekov, B. A., Matveev, O. A., Perepelova, G. B., Prokof'ev, S. V., and Shmanenkova, G. I. (1974). *Fiz. Tekh. Poluprovodn.* **8**, 317 [*Engl. Transl.: Sov. Phys. Semicond.* **8**, 202].
Agrinskaya, N. V., Alekseenko, M. V., Arkad'eva, E. N., Matveev, O. A., and Prokof'ev, S. V. (1975). *Fiz. Tekh. Poluprovodn.* **9**, 320 [*Engl. Transl.: Sov. Phys. Semicond.* **9**, 208].
Aitkhozhin, S. A., and Temirov, Yu. Sh. (1971). *Kristallografiya* **15**, 1057 [*Engl. Transl.: Sov. Phys. Crystallogr.* **15**, 916].
Akobirova, A. T., Maslova, L. V., Matveev, O. A., and Khusainov, A. Kh. (1974). *Fiz. Tekh. Poluprovodn.* **8**, 1701 [*Engl. Transl.:* (1975), *Sov. Phys. Semicond.* **8**, 1103].
Akutagawa, W., and Zanio, K. (1969). *J. Appl. Phys.* **40**, 3838.
Akutagawa, W., and Zanio, K. (1971). *J. Cryst. Growth* **11**, 191.
Akutagawa, W., Zanio, K., and Mayer, J. W. (1967). *Nucl. Instrum. Methods* **55**, 383.
Akutagawa, W., Turnbull, D., Chu, W. K., and Mayer, J. W. (1974). *Solid State Commun.* **15**, 1919.
Akutagawa, W., Turnbull, D., Chu, W. K., and Mayer, J. W. (1975). *J. Phys. Chem. Solids* **36**, 521.

Aleksandrov, B. N. (1961). *Fiz. Met. Metalloved.* **11**, 588 [*Engl. Transl.: Phys. Metals Metallogr.* **11**, 99].
Aleksandrov, B. N., and D'yakov, I. G. (1962). *Fiz. Met. Metalloved.* **14**, 569 [*Engl. Transl.: Phys. Met. Metallogr.* **14**, 81].
Aleksandrov, B. V., and Udovikov, V. I. (1973). *Izv. Akad. Nauk SSSR Met.* **2**, 17.
Alekseenko, M. V., and Veinger, A. I. (1971). *Fiz. Tekh. Poluprovodn.* **5**, 2233. [*Engl. Transl.:* (1972). *Sov. Phys. Semicond.* **5**, 1952].
Alekseenko, M. V., and Veinger, A. I. (1974). *Fiz. Tekh. Poluprovodn.* **8**, 215 [*Engl. Transl.: Sov. Phys. Semicond.* **8**, 143].
Alekseenko, M. V., Arkad'eva, E. N., and Matveev, O. A. (1970). *Fiz. Tekh. Poluprovodn.* **4**, 414 [*Engl. Transl.: Sov. Phys. Semicond.* **4**, 349].
Alekseenko, M. V., Arkad'eva, E. N., Veinger, A. I., and Matveev, O. A. (1971). *Fiz. Tekh. Poluprovodn.* **5**, 1310 [*Engl. Transl.:* (1972), *Sov. Phys. Semicond.* **5**, 1156].
Alekseenko, M. V., Arkad'eva, E. N., Kisilenko, V. S., Maslova, L. V., Matveev, O. A., Prokof'ev, S. V., Ryvkin, S. M., and Khusainov, A. Kh. (1974). *Fiz. Tekh. Poluprovodn.* **8**, 550 [*Engl. Transl.: Sov. Phys. Semicond.* **8**, 351].
Alferov, Zh. I., Korol'kov, V. I., Mikhailova-Mikheeva, I. P., Romanenko, V. N., and Tuchkevich, V. M. (1964). *Fiz. Tverd. Tela* **6**, 2353 [*Engl. Transl.:* (1965). *Sov. Phys. Solid State* **6**, 1865].
Allemand, R., Bouteiller, P., and Laval, M. (1977) *Rev. Phys. Appl.* **12**, 365.
Almasi, G. S., and Smith, A. C. (1968). *J. Appl. Phys.* **39**, 233.
Al'tshuler, A., Vekilov, Yu. Kh., Kadyshevich, A. E., and Rusakov, A. P. (1974). *Fiz. Tverd. Tela* **16**, 2860 [*Engl. Transl.:* (1975). *Sov. Phys. Solid State* **16**, 1852].
Arkad'eva, E. N., Matveev, O. A., Rud', Yu. V., Ryvkin, S. M., (1966a). *Zh. Tekh. Fiz.* **36**, 1146 [*Engl. Transl.: Sov. Phys. Tech. Phys.* **11**, 846].
Arkad'eva, E. N., Maslova, L. V., Matveev, O. A., Rud', Yu. V., and Ryvkin, S. M. (1967a). *Fiz. Tekh. Poluprovodn.* **1**, 805 [*Engl. Transl.: Sov. Phys. Semicond.* **1**, 669].
Arkad'eva, E. N., Matveev, O. A., and Rud', Yu. V. (1967b). *Fiz. Tverd. Tela* **8**, 2821 [*Engl. Transl.: Sov. Phys. Solid State* **8**, 2260].
Arkad'eva, E. N., Matveev, O. A., and Sladkova, V. A. (1969). *Fiz. Tekh. Poluprovodn.* **2**, 1514 [*Engl. Transl.: Sov. Phys. Semicond.* **2**, 1264].
Arkad'eva, E. N., Guseva, M. I., Matveev, O. A., and Sladkova, V. A. (1975). *Fiz. Tekh. Poluprovodn.* **9**, 853 [*Engl. Transl.: Sov. Phys. Semicond.* **9**, 563].
Arlt, G., and Quadflieg, P. (1968). *Phys. Status Solidi* **25**, 323.
Armstrong, J. A., Bloembergen, N., Ducuing, J., and Pershan, P. S. (1962). *Phys. Rev.* **127**, 1918.
Artobolevskaya, E. S., Afanas'eva, E. A., Vodop'yanov, L. K., and Sushkov, V. P. (1967). *Fiz. Tekh. Poluprovodn.* **1**, 1854. [*Engl. Transl.:* (1968). *Sov. Phys. Semicond.* **1**, 1531].
Auguston, R. H., and Reilly, T. D. (1974). "Fundamentals of Passive Nondestructive Assay of Fissionable Material." Los Alamos 5651-M USAEC Contract W-7405. Eng. 36.
Babonas, G. A., Kavalyauskas, Yu. F., and Shileika, Yu. (1965). *Litov. Fiz. Sb.* **5**, 395.
Babonas, G. A., Krivaite, E. Z., Raudonis, A. V., and Shileika, A. J. (1968). *Proc. Int. Conf. Phys. Semicond.*, 9th **1**, 400.
Babonas, G. A. Bendoryus, R. A., and Shileika, A. Yu. (1971). *Fiz. Tekh. Poluprovodn.* **5**, 449 [*Engl. Transl.: Sov. Phys. Semicond.* **5**, 392].
Bagaev, V. S., Belousova, T. Ya., Berozashvili, Yu. N., and Lordkipanidze, D. Sh. (1970). *Fiz. Tekh. Poluprovodn.* **3**, 1687 [*Engl. Transl.: Sov. Phys. Semicond.* **3**, 1418].
Bagnall, K. W. (1966). "The Chemistry of Selenium, Tellurium and Polonium." Elsevier, Amsterdam.
Baikova, D., Chizhikov, D. M., and Pliginskaya, L. (1966). *In* "Cadmium" (D. M. Chizhikov, ed.), p. 185. Pergamon, Oxford.

Bailly, F. (1968). *Phys. Status Solidi* **25**, 317.
Baĭmakov, A. Yu., and Petrova, Z. N. (1960). *Tsvetn. Met.* **33**, 43 [*Engl. Transl.: Sov. J. Non-Ferrous Met.*, 46].
Baj, M., Dmowski, L., Kończykowski, M., and Porowski, S. (1976). *Phys. Status Solidi A* **33**, 421.
Bajaj, K. K., Birch, J. R., Eaves, L., Hoult, R. A., Kirkman, R. F., Simmonds, P. E., and Stradling, R. A. (1975). *J. Phys. C* **8**, 530.
Baldereschi, A., and Lipari, N. O. (1970). *Phys. Rev. Lett.* **25**, 373.
Balkanski, M., and Beserman, R. (1968). *Proc. Int. Conf. Phys. Semicond.*, 9th **2**, 1042.
Barnes, C. E., and Kikuchi, C. (1968). *Nucl. Sci. Eng.* **31**, 513.
Barnes, C. E., and Kikuchi, C. (1970). *Radiat. Eff.* **2**, 243.
Barnes, C. E., and Kikuchi, C. (1975). *Radiat. Eff.* **26**, 105.
Barnes, C. E., and Zanio, K. (1975). *J. Appl. Phys.* **46**, 3959.
Barnes, C. E., and Zanio, K. (1976). *IEEE Trans. Nucl. Sci.* **23**, 177.
Baxter, R. D. (1976). *IEEE Trans. Nucl. Sci.* **23**, 493.
Bean, J. C. (1976). Ion Implantation in Cadmium Telluride., Tech. Rep. No. 4733-3. Nat. Sci. Foundation Grant No. GH31999. Stanford Electron. Lab., Stanford Univ., Menlo Park, California.
Beck, H. L., McLaughlin, J. E., and Miller, K. M. (1976). *IEEE Trans. Nucl. Sci.* **23**, 677.
Beer, A. C. (1963). *Solid State Phys. Suppl. 4.*
Bell, R. O. (1971). *Nucl. Instrum. Methods* **93**, 341.
Bell, R. O. (1974). *J. Electrochem. Soc.* **121**, 1366.
Bell, R. O. (1975). *Solid State Commun.* **16**, 913.
Bell, R. O., and Wald, F. V. (1972). *IEEE Trans. Nucl. Sci.* **19**, 334.
Bell, R. O., Hemmat, N., and Wald, F. (1970a). *Phys. Status Solidi A* **1**, 375.
Bell, R. O., Hemmat, N., and Wald, F. (1970b). *IEEE Trans. Nucl. Sci.* **17**, 241.
Bell, R. O., Wald, F. V., Canali, C., Nava, F., and Ottaviani, G. (1974a). *IEEE Trans. Nucl. Sci.* **21**, 331.
Bell, R. O., Entine, G., and Serreze, H. B. (1974b). *Nucl. Instrum. Methods* **117**, 267.
Bell, R. O., Wald, F. V., and Goldner, R. B. (1975a). *IEEE Trans. Nucl. Sci.* **22**, 241.
Bell, R. O., Serreze, H. B., and Wald, F. V. (1975b). *IEEE Photovoltaic Specialists Conf. 11th, Scotsdale, Arizona*, p. 497.
Bendow, B., Gianino, P. D., Mitra, S. S., and Tsay, Y-F. (1974). *Conf. High Power Infrared Laser Window Mater. 3rd* (C. A. Pitha and B. Bendow, eds.) **1**, 367.
Berlincourt, D., Jaffe, H., and Shiozawa, L. R. (1963). *Phys. Rev.* **129**, 1009.
Bernard, J., Lancon, R., Paparoditis, C., and Rodot, M. (1966). *Rev. Phys. Appl.* **1**, 211.
Berozashvili, Yu. N., and Dundua, A. V. (1975). *Fiz. Tekh. Poluprovodn.* **8**, 2008 [*Engl. Transl.: Sov. Phys. Semicond.* **8**, 1303].
Birch, J. A. (1975). *J. Phys. C* **8**, 2043.
Blank, Z., and Brenner, W. (1971). *J. Cryst. Growth* **11**, 255.
Bloom, S., and Bergstresser, T. K. (1968). *Solid State Commun.* **6**, 465.
Bodakov, Yu. A., Lomakina, G. A., Naumov, G. P., and Maslakovets, Yu. P. (1960). *Fiz. Tverd. Tela* **2**, 55 [*Engl. Transl.: Sov. Phys. Solid State* **2**, 49].
Bojsen, J., Rossing, N., Soberg, O., and Vaastrup, S. (1977). *Rev. Phys. Appl.* **12**, 361.
Bonek, E., Schiffner, G., Pisecker, R., and Kohl, F. (1974). *IEEE J. Quantum Electron.* **10**, 128.
Bonnet, D. (1970). *Phys. Status Solidi A* **3**, 913.
Bonnet, D., and Rabenhorst, H. (1971). *Proc. Int. Conf. Phys. Chem. Semicond. Heterojunctions Layer Struct.* **1**, 119.
Bonnet, D., and Rabenhorst, H. (1972). *IEEE Photovoltaic Specialists Conf. 9th.* Silver Springs, Maryland p. 129.
Borsari, V., and Jacoboni, C. (1972). *Phys. Status Solidi B* **54**, 649.

Borsenberger, P. M., and Stevenson, D. A. (1968). *J. Phys. Chem. Solids* **29**, 1277.
Bottger, G. L., and Geddes, A. L. (1967). *J. Chem. Phys.* **47**, 4858.
Boyd, G. D., and Patel, C. K. N. (1966). *Appl. Phys. Lett.* **8**, 313.
Brebrick, R. F. (1971). *J. Electrochem. Soc.* **118**, 2014.
Brebrick, R. F., and Strauss, A. J. (1964). *J. Phys. Chem. Solids* **25**, 1441.
Brelant, S., Entine, G., Elliott, M., and Chu, S. (1977) *Rev. Phys. Appl.* **12**, 141.
Broder, J. D., and Wolff, G. A. (1963). *J. Electrochem. Soc.* **110**, 1150.
Brodin, M. S., Kurik, M. V., Matlak, V. M., and Oktyabr'skii, B. S. (1968). *Fiz. Tekh. Poluprovodn.* **2**, 727 [*Engl. Transl.: Sov. Phys. Semicond.* **2**, 603].
Brodin, M. S., Gnatenko, Yu. P., Kurik, M. V., and Matlak, V. M. (1970). *Fiz. Tekh. Poluprovodn.* **3**, 991 [*Engl. Transl.: Sov. Phys. Semicond.* **3**, 835].
Brouwer, G. (1954). *Philips Res. Rep.* **9**, 366.
Browder, J. S., and Ballard, S. S. (1969). *Appl. Opt.* **8**, 793.
Browder, J. S., and Ballard, S. S. (1972). *Appl. Opt.* **11**, 841.
Bryant, F. J., and Totterdell, D. H. J. (1971). *Radiat. Eff.* **9**, 115.
Bryant, F. J., and Totterdell, D. H. J. (1972). *Phys. Status Solidi A* **10**, K75.
Bryant, F. J., and Webster, E. (1967). *Phys. Status Solidi* **21**, 315.
Bryant, F. J., Cox, A. F. J., and Webster, E. (1968). *J. Phys. C* **1**, 1737.
Bryant, F. J., Totterdell, D. H. J., and Hagston, W. E. (1972). *Phys. Status Solidi A* **14**, 579.
Bube, R. H. (1955). *Phys. Rev.* **98**, 431.
Bube, R. H., Fahrenbruch, A., Aranovich, J., Buch, F., Chu, M., and Mitchell, K. (1975). Semiann. Progr. Rep. NSF/RANN/SE/AER-75-1679/75/4, Dept. of Mater. Sci. and Eng., Stanford Univ., Stanford, California.
Bube, R. H., Fahrenbruch, A., Taheri, E. H. Z., Aranovich, J., Buch, F., Chu, M., Mitchell, K., and Ma, Y. (1976). Quart. Progr. Rep. NSF/RANN/SE/AER-75-1679/76/1. Dept. of Mater. Sci. and Eng., Stanford Univ., Stanford, California.
Bublik, V. T., Gorelik, S. S., and Smirnov, I. S. (1972). *Kristallografiya* **17**, 557 [*Engl. Transl.: Sov. Phys. Crystallogr.* **17**, 485].
Buch, F., and Ahlquist, C. N. (1974a). *J. Appl. Phys.* **45**, 1756.
Buch, F., and Ahlquist, C. N. (1974b). *Mater. Sci. Eng.* **13**, 194.
Bupp, F. E. (1976). Zero-G Flight Test of a Gauging System, Contract No. NAS 9-14349, January 1976. TRW Defense and Space Syst. Group, Redondo Beach, California.
Bupp, F., Nagel, M., Akutagawa, W., and Zanio, K. (1973). *IEEE Trans. Nucl. Sci.* **20**, 514.
Caillot, M. (1972). *Phys. Lett.* **38A**, 2.
Caillot, M. (1975). Inst. Phys. Conf. Ser. No. 23, Chap. 3, p. 280.
Camassel, J., Auvergne, D., Mathieu, H., Triboulet, R., and Marfaing, Y. (1973). *Solid State Commun.* **13**, 63.
Camphausen, D. L., Connell, G. A. N., and Paul, W. (1971). *Phys. Rev. Lett.* **26**, 184.
Canali, C., and Malm, H. L. (1976). *Nucl. Instrum. Methods* **134**, 199.
Canali, C., Martini, M., Ottaviani, G., and Zanio, K. (1970). *Phys. Lett.* **33A**, 241.
Canali, C., Martini, M., Ottaviani, C., Quaranta, A. A., and Zanio, K. R. (1971a). *Nucl. Instrum. Methods* **96**, 561.
Canali, C., Martini, M., Ottaviani, G., and Zanio, K. (1971b). *Phys. Rev. B* **4**, 422.
Canali, C., Martini, M., Ottaviani, G., and Zanio, K. (1971c). *Appl. Phys. Lett.* **19**, 51.
Canali, C., Nava, F., Ottaviani, G., and Zanio, K. (1973). *Solid State Commun.* **13**, 1255.
Canali, C., Ottaviani, G., Bell, R. O., and Wald, F. V. (1974). *J. Phys. Chem. Solids* **35**, 1405.
Canali, C., Nicolet, M. A., and Mayer, J. W. (1975a). *Solid State Electron.* **18**, 871.
Canali, C., Nava, F., Ottaviani, G., and Zanio, K. (1975b). *Phys. Status Solidi A* **28**, 581.
Cápek, V., Zimmerman, K., Koňák, C., Popova, M., and Polivka, P. (1973). *Phys. Status Solidi B* **56**, 739.
Cardona, M. (1961). *J. Appl. Phys.* **32**, 2151.

Cardona, M. (1963). *J. Phys. Chem. Solids* **24**, 1543.
Cardona, M. (1965). *J. Appl. Phys.* **36**, 2181.
Cardona, M., and Greenaway, D. L. (1963). *Phys. Rev.* **131**, 98.
Cardona, M., and Harbeke, G. (1962). *Phys. Rev. Lett.* **8**, 90.
Cardona, M., and Harbeke, G. (1963). *J. Appl. Phys.* **34**, 813.
Cardona, M., Shaklee, K. L., and Pollak, F. H. (1967). *Phys. Rev.* **154**, 696.
Carlsson, L., and Ahlquist, C. N. (1972). *J. Appl. Phys.* **43**, 2529.
Chadi, D. J., Walter, J. P., Cohen, M. L., Petroff, Y., and Balkanski, M. (1972). *Phys. Rev. B* **5**, 3058.
Chapnin, V. A. (1969). *Fiz. Tekh. Poluprovodn.* **3**, 566 [*Engl. Transl.: Sov. Phys. Semicond.* **3**, 481].
Chasmar, R. P., Durham, E. W., and Stuckes, A. D. (1960). *Proc. Int. Conf. Semicond. Phys.* U8, 1018.
Chelikowsky, J., and Cohen, M. L. (1976). *Phys. Rev. B* **14**, 556.
Chelikowsky, J., Chadi, D., and Cohen, M. L. (1973). *Phys. Rev. B* **8**, 2786.
Chern, S. S., and Kröger, F. A. (1974). *Phys. Status Solidi A* **25**, 215.
Chern, S. S., and Kröger, F. A. (1975). *J. Solid State Chem.* **14**, 44.
Chern, S. S., Vydyanath, H. R., and Kröger, F. A. (1975). *J. Solid State Chem.* **14**, 33.
Chester, R. O. (1967). *J. Appl. Phys.* **38**, 1745.
Chizhikov, D. M. (1966). "Cadmium" (translated by D. E. Hayler). Pergamon, Oxford.
Chizhikov, D. M., and Shchastliviyi, V. P. (1970). *In* "Tellurium and the Tellurides," (Chizhikov, D. M., and Shchastliviyi, V. P., eds.) Collet's Ltd., London.
Cho, K., Dreybrodt, W., Hiesinger, P., Suga, S., and Willmann, F. (1974a). *Proc. Int. Conf. Phys. Semicond., 12th*, p. 945.
Cho, Z. H., Ahn, I. S., and Tsai, C. M. (1974b). *IEEE Trans. Nucl. Sci.* **21**, 218.
Cline, C. F., and Stephens, D. R. (1965). *J. Appl. Phys.* **36**, 2869.
Cochran, W. (1959). *Proc. R. Soc. London* **253**, 260.
Cohen, M. L. (1967). *In* "II–VI Semiconducting Compounds" (D. G. Thomas, ed.). Benjamin, New York.
Cohen, M. L., and Bergstresser, T. K. (1966). *Phys. Rev.* **141**, 789.
Cohen, M. L., and Heine, V. (1970). *Solid State Phys.* **24**, 38.
Cohn, D. R., Larsen, D. M., and Lax, B. (1970). *Solid State Commun.* **8**, 1707.
Cohn, D. R., Larsen, D. M., and Lax, B. (1972). *Phys. Rev. B* **6**, 1367.
Corcoran, V. J., Martin, J. M., and Smith, W. T. (1973). *Appl. Phys. Lett.* **22**, 517.
Cornet, A., Siffert, P., Coche, A., and Triboulet, R. (1970a). *Appl. Phys. Lett.* **17**, 432.
Cornet, A., Siffert, P., and Coche, A. (1970b). *J. Cryst. Growth* **7**, 329.
Corsini-Mena, A., Elli, M., Paorici, C., and Pelosini, L. (1971). *J. Cryst. Growth* **8**, 297.
Costato, M., Jacoboni, C., and Reggiani, L. (1972). *Phys. Status Solidi B* **52**, 461.
Coulson, C. A., Rèdei, L. B., and Stocker, D. (1962). *Proc. R. Soc. London* **270**, 357.
Crowder, B. L., and Hammer, W. N. (1966). *Phys. Rev.* **150**, 541.
Čumpeik, R., Kargerova, J., and Klier, E. (1969). *Cesk. Cas. Fys.* [*Engl. Transl.: Czech. J. Phys. B* **19**, 1003].
Cusano, D. A. (1963). *Solid State Electron.* **6**, 217.
Cusano, D. A. (1966). *Rev. Phys. Appl.* **1**, 195.
Cusano, D. A. (1967). *In* "Physics and Chemistry of II–VI Compounds" (M. Aven and J. S. Prener, eds.), pp. 709–762. Wiley, New York.
Cusano, D. A., and Lorenz, M. R. (1964). *Solid State Commun.* **2**, 125.
Dąbrowski, A. J., Iwańczyk, J., and Triboulet, R. (1974). *Nucl. Instrum. Methods* **118**, 531.
Dąbrowski, A. J., Iwańczyk, J., and Triboulet, R. (1975). *Nucl. Instrum. Methods* **126**, 417.
Dąbrowski, A. J., Chwaszczewska, J., Iwańczyk, J., Triboulet, R., and Marfaing, Y. (1976). *IEEE Trans. Nucl. Sci.* **23**, 171; (1977) *Rev. Phys. Appl.* **12**, 297.

Davis, P. W., and Shilliday, T. S. (1960). *Phys. Rev.* **118**, 1020.
de Carolis, M., Dragnev, T., and Waligura, A. (1976). *IEEE Trans. Nucl. Sci.* **23**, 70.
Demidenko, A. F. (1969). *Izv. Akad. Nauk SSSR Neorg. Mater.* **5**, 252 [*Engl. Transl.: Inorg. Mater.* **5**, 210].
de Nobel, D. (1959). *Philips Res. Rep.* **14**, 361.
Desnica, U. V., and Urli, N. B. (1972). *Phys. Rev. B* **15**, 3044.
Deutsch, T. F. (1973). *J. Phys. Chem. Solids* **34**, 2091.
Deutsch, T. F. (1975). *J. Electron. Mater.* **4**, 663.
Devlin, S. S. (1967). *In* "Physics and Chemistry of II–VI Compounds" (M. Aven and J. Prener, eds.), pp. 551–606. North-Holland Publ., Amsterdam.
Devyatkova, E. D., and Smirnov, I. A. (1962). *Fiz. Tverd. Tela* **4**, 2507 [*Engl. Transl.:* (1963). *Sov. Phys. Solid State* **4**, 1836].
Dewey, C. F. (1977). *Rev. Phys. Appl.* **12**, 405.
Dewey, C. F., Jr., and Hocker, L. O. (1975). *Appl. Phys. Lett.* **26**, 442.
Dieleman, J., De Bruin, S. H., Van Doorn, C. Z., and Haanstra, J. H. (1964). *Philips. Res. Rep.* **19**, 311.
Dinger, R. J., and Fowler, I. L. (1977). *Rev. Phys. Appl.* **12**, 135.
Dixon, R. W. (1967). *J. Appl. Phys.* **38**, 5149.
Donnelly, J. P., Foyt, A. G., Hinkley, E. D., Lindley, W. T., and Dimmock, J. O. (1968). *Appl. Phys. Lett.* **12**, 303.
Droms, C. R., Langdon, W. R., Robison, A. G., and Entine, G. (1976). *IEEE Trans. Nucl. Sci.* **23**, 498.
Drowart, J., and Goldfinger, P. (1958). *J. Chim. Phys. Biol.* **55**, 721.
Dufresne, R., and Champness, C. H. (1973). *J. Cryst. Growth* **18**, 34.
Duke, C. B., and Segall, B. (1966). *Phys. Rev. Lett.* **17**, 19.
Dutt, B. V., Al-Delaimi, M., and Spitzer, W. G. (1976). *J. Appl. Phys.* **47**, 565.
Dutton, R. W., and Muller, R. S. (1968). *Solid State Electron.* **11**, 749.
Du Varney, R. C., and Garrison, A. K. (1975). *Phys. Rev. B* **12**, 10.
Eastman, D. E., Grobman, W. D., Freeouf, J. L., and Erbudak, M. (1974). *Phys. Rev. B* **9**, 3473.
Ebina, A., Koda, T., and Shionoya, S. (1965). *J. Phys. Chem. Solids* **26**, 1497.
Eckelt, P. (1967). *Phys. Status Solidi* **23**, 307.
Economou, T. E., and Turkovich, A. L. (1976) *Nucl. Instrum. Methods* **134**, 391.
Edwards, A. L., and Drickamer, H. G. (1961). *Phys. Rev.* **122**, 1149.
Eichinger, P., Halder, N., and Kemmer, J. (1974). *Nucl. Instrum. Methods* **117**, 305.
Fabre, E., Ngo-Tich-Phuoc, Martin, G. M., and Ortega, F. (1976). *IEEE Trans. Nucl. Sci.* **23**, 182.
Fahrenbruch, A. L., Vasilchenko, V., Buch, F., Mitchell, K., and Bube, R. H. (1974). *Appl. Phys. Lett.* **25**, 605.
Fahrenbruch, A. L., Buch, F., Mitchell, K., and Bube, R. (1975). *IEEE Photovoltaic Specialists Conf. 11th, Scotsdale, Arizona*, p. 490.
Fan, H. Y., Spitzer, W., and Collins, R. J. (1956). *Phys. Rev.* **101**, 566.
Farrell, R., Entine, G., Wilson, F., and Wald, F. V. (1974). *J. Electron. Mater.* **3**, 155.
Fenner, G. E., Slack, G. A., and Vallin, J. T. (1968) *Proc. Int. Conf. Phys. Semicond., 9th* **2**, 1180.
Figueira, J. F. (1974). *IEEE J. Quantum Electron.* **10**, 572.
Fisher, P., and Fan, H. Y. (1959). *Bull. Am. Phys. Soc.* **4**, 409.
Forster, D. C., Goodwin, F. E., and Bridges, W. B. (1972). *IEEE J. Quantum Electron.* **8**, 263.
Foss, N. A. (1968). *J. Appl. Phys.* **39**, 6029.
Foyt, A. G., Halsted, R. E., and Paul, W. (1966). *Phys. Rev. Lett.* **16**, 55.

Fraas, L. M. (1976). Private communication.
Fraas, L. M., Grinberg, J., Bleha, W. P., and Jacobson, A. D. (1976a). *J. Appl. Phys.* **47**, 576.
Fraas, L. M., Bleha, W. P., Grinberg, J., and Jacobson, A. D. (1976b). *J. Appl. Phys.* **47**, 584.
Frerichs, R. (1947). *Phys. Rev.* **72**, 594.
Furgolle, B., Hoclet, M., and Vandevyver, M. (1974). *Solid State Commun.* **14**, 1237.
Garcia, D. A., Entine, G., and Tow, D. E. (1974). *J. Nucl. Med.* **15**, 892.
Garcia, D. A., Frisbie, J. H., Tow, D. E., Sasahara, A. A., and Entine, G. (1976). *IEEE Trans. Nucl. Sci.* **23**, 594.
Garlick, G. F. J. (1959). *J. Phys. Chem. Solids* **8**, 449.
Garlick, G. F. J., Hough, J. M., and Fatehally, R. A. (1958). *Proc. Phys. Soc. London* **72**, 925.
Gaur, S. P., Vetelino, J. F., and Mitra, S. S. (1971). *J. Phys. Chem. Solids* **32**, 2737.
Gavini, A., and Cardona, M. (1970). *Phys. Rev. B* **1**, 672.
Gentile, A. L., Kiefer, J. E., Kyle, N. R., and Winston, H. V. (1973a). *Conf. High Power Infrared Laser Window Mater.* (C. A. Pitha, ed.), **2**, p. 625.
Gentile, A. L., Kliefer, J. E., Kyle, N. R., and Winston, H. V. (1973b). *Mater. Res. Bull.* **8**, 523.
Gettings, M., and Stephens, K. G. (1973). *Radiat. Eff.* **18**, 275.
Gettings, M., and Stephens, K. G. (1974). *J. Cryst. Growth* **22**, 50.
Ghezzi, C., and Paorici, C. (1974). *J. Cryst. Growth* **20**, 58.
Gippius, A. A., Panossian, J. R., and Chapnin, V. A. (1974). *Phys. Status Solidi A* **21**, 753.
Glang, R., Kren, J. G., and Patrick, W. J. (1963). *J. Electrochem. Soc.* **110**, 407.
Goldfinger, P., and Jeunehomme, M. (1963). *Trans. Faraday Soc.* **59**, 2851.
Goldstein, B. (1958). *Phys. Rev.* **109**, 601.
Goldstein, B., and Pensak, L. (1959). *J. Appl. Phys.* **30**, 155.
Górska, M., and Nazarewicz, W. (1974). *Phys. Status Solidi B* **65**, 193.
Greenough, R. D., and Palmer, S. B. (1973). *J. Phys. D* **6**, 587.
Grinberg, J. et al. (1975). *IEEE Trans. Electron Devices* **22**, 775.
Grois, A. Sh. (1969). *Fiz. Tekh. Poluprovodn.* **2**, 1083 [*Engl. Transl.: Sov. Phys. Semicond.* **2**, 910].
Gu, J., Kitahara, T., and Sakaguchi, T. (1973) *Jpn. J. Appl. Phys.* **12**, 1460.
Gu, J., Kitahara, T., Kawakami, K., and Sakaguchi, T. (1975a). *J. Appl. Phys.* **46**, 1184.
Gu, J., Kitahara, T., Fujita, S., and Sakaguchi, T. (1975b) *Jpn. J. Appl. Phys.* **14**, 449.
Guillien, R., Leitz, P., and Palz, W. (1971). *Proc. Int. Conf. Phys. Chem. Semicond. Heterojunctions Layer Struct.* **2**, 283.
Gul'tyaev, P. V., and Petrov, A. V. (1959). *Fiz. Tverd. Tela* **1**, 368 [*Engl. Transl.: Sov. Phys. Solid State* **1**, 330].
Gupta, R. P., Sinha, S. K., Walter, J. P., and Cohen, M. L. (1974). *Solid State Commun.* **14**, 1313.
Hall, R. B., and Woodbury, H. H. (1968). *J. Appl. Phys.* **39**, 5361.
Hall, E. L., Vander Sande, J. B., Lemaire, P. J., and Bowen, H. K. (1975). *Proc. Ann. Infrared Laser Window Mater., 4th* (C. R. Andrews and C. L. Strecker, eds.), p. 531.
Halsted, R. E., and Aven, M. (1965). *Phys. Rev. Lett.* **14**, 64.
Halsted, R. E., and Segall, B. (1963). *Phys. Rev. Lett.* **10**, 392.
Halsted, R. E., Lorenz, M. R., and Segall, B. (1961). *J. Phys. Chem. Solids* **22**, 109.
Halsted, R. E., Aven, M., and Coghill, I. D. (1965). *J. Electrochem. Soc.* **112**, 177.
Ham, F. S., Ludwig, G. W., Watkins, G. D., and Woodbury, H. H. (1960). *Phys. Rev. Lett.* **5**, 468.
Hartmann, H. (1975). *J. Cryst. Growth* **31**, 323.
Harvey, J. E., and Wolfe, W. L. (1975) *J. Opt. Soc. Am.* **65**, 1267.
Hayes, W., and Spray, A. K. L. (1969). *J. Phys. C* **2**, 1129.

Herman, F., Kortum, R. L., Kuglin, C. D., and Shay, J. L. (1967). *In* "II-VI Semiconducting Compounds" (D. G. Thomas, ed.), pp. 503–551. Benjamin, New York.
Herman, F., Kortum, R. L., Kuglin, Ch. D., Van Dyke, J. P., and Skillman, Sh. (1968). *In* "Methods in Computational Physics" (B. Alder, S. Fernbach, and M. Rotenberg, eds.), Vol. 8. Academic Press, New York.
Hiesinger, P., Suga, S., Willmann, F., and Dreybrodt, W. (1975). *Phys. Status Solidi B* **67**, 641.
Higinbotham, W., Zanio, K., and Akutagawa, W. (1973). *IEEE Trans. Nucl. Sci.* **20**, 510.
Hocker, L. O., and Dewey, C. F. (1976). *Appl. Phys. Lett.* **28**, 267.
Hoffer, P. B., and Beck, R. N. (1971). "The Role of Semiconductor Detectors in the Future of Nuclear Medicine" (P. Hoffer, R. Beck, and A. Gottschalk, eds.), p. 131. Soc. of Nucl. Medicine, New York.
Holt, D. B. (1962). *J. Phys. Chem. Solids* **23**, 1353.
Holt, D. B. (1964). *J. Phys. Chem. Solids* **25**, 1385.
Holt, D. B. (1966). *J. Mater. Sci.* **1**, 280.
Holt, D. B. (1969). *J. Phys. Chem. Solids* **30**, 1297.
Holt, D. B. (1974). *Thin Solid Films* **24**, 1.
Holt, D. B., and Abdalla, M. I. (1974). *Phys. Status Solidi A* **26**, 507.
Höschl, P. (1966). *Phys. Status Solidi* **13**, K101.
Höschl, P., and Koňák, Č. (1963a). *Cesk. Cas. Fys*, [*Engl. Transl.: Czech. J. Phys. B* **13**, 785].
Höschl, P., and Koňák, Č. (1963b). *Cesk. Cas. Fys*. [*Engl. Transl.: Czech. J. Phys. B* **13**, 850].
Höschl, P., and Koňak, Č. (1965). *Phys. Status Solidi* **9**, 167.
Hovel, H. J. (1975). "Semiconductors and Semimetals" (R. K. Willardson and A. C. Beer, eds.), Vol. 11. Academic Press, New York.
Huang, C. C., Pao, Y., Claspy, P. C., and Phelps, F. W., Jr. (1974). *IEEE J. Quantum Electron.* **10**, 186.
Ichimiya, T., Niimi, T., Mizuma, K., Mikami, O., Kamiya, Y., and Ono, K. (1960). *Solid State Phys. Electron. Telecommun. Proc. Int. Conf. 1958* **2**, 845.
Igarashi, O. (1971). *J. Appl. Phys.* **42**, 4035.
Ino, S., Watanabe, D., and Ogawa, S. (1964). *J. Phys. Soc. Jpn.* **19**, 881.
Inoue, M. (1969). *J. Phys. Soc. Jpn.* **26**, 1186.
Inoue, M., Teramoto, I., and Takayanagi, S. (1962). *J. Appl. Phys.* **33**, 2578.
Inoue, M., Teramoto, I., and Takayanagi, S. (1963). *J. Appl. Phys.* **34**, 404.
Ioffe, A. V., and Ioffe, A. F. (1960). *Fiz. Tverd. Tela* **5**, 781 [*Engl. Transl.: Sov. Phys. Solid State* **2**, 719].
Iseler, G. W., Kafalas, J. A., Strauss, A. J., MacMillan, H. F., and Bube, R. H. (1972). *Solid State Commun.* **10**, 619.
Iseler, G. W., Kafalas, J. A., and Strauss, A. J. (1974). Lincoln Lab. Rep., M.I.T. Lexington, Massachusetts 02173.
Iwańczyk, J., and Dąbrowski, A. J. (1976). *Nucl. Instrum. Methods* **134**, 1.
Jacoboni, C., and Reggiani, L. (1970). *Phys. Lett.* **33A**, 333.
Jäger, H., and Thiel, R. (1977). *Rev. Phys. Appl.* **12**, 293.
Jayaraman, A., Klement, W., Jr., and Kennedy, G. C. (1963). *Phys. Rev.* **130**, 2277.
Jenny, D. A., and Bube, R. H. (1954). *Phys. Rev.* **96**, 1190.
Jensen, B. (1973). *J. Phys. Chem. Solids*, **34**, 2235.
Johnson, C. J. (1968). *Proc. Inst. Electr. Eng.* **56**, 1719.
Johnson, C. J., Sherman, G. H., and Weil, R. (1969). *Appl. Opt.* **8**, 1667.
Joiner, R., Steier, W. H., and Christensen, C. P. (1976). *Ann. Conf. High Power Infrared Laser Window Mater.*, 5th.
Jones, L. T. (1977). *Rev. Phys. Appl.* **12**, 379.

REFERENCES

Jones, L. T., and Woollam, P. B. (1975). *Nucl. Instrum. Methods* **124**, 591.
Jordan, A. S. (1970). *Metall. Trans.* **1**, 239.
Jordan, A. S., and Zupp, R. R. (1969). *J. Electrochem. Soc.* **116**, 1285.
Justi, E. W., Schneider, G., and Seredynski, J. (1973). *Energy Convers*, **13**, 53.
Kachurin, G. A., Gorodetskii, A. E., Loburets, Yu. V., and Smirnov, L. S. (1967). *Fiz. Tverd. Tela* **9**, 494 [*Engl. Transl.: Sov. Phys. Solid State* **9**, 375].
Kachurin, G. A., Zelevinskaya, V. M., and Smirnov, L. S. (1968). *Fiz. Tekh. Poluprovodn.* **2**, 1836 [*Engl. Transl.*: (1969). *Sov. Phys. Semicond.* **2**, 1527].
Kalinkin, I. P., Muravyeva, K. K., Sergeyewa, L. A., Aleskowsky, V. B., and Bogomolov, N. S. (1970). *Krist. Tech.* **5**, 51.
Kamiyama, M., Haradome, M., and Kukimoto, H. (1962). *Jpn. J. Appl. Phys.* **1**, 202.
Kanazawa, K. K., and Brown, F. C. (1964). *Phys. Rev.* **135**, A1757.
Kane, E. O. (1969). *Phys. Rev.* **180**, 852.
Karmazin, V. V., and Miloslavskii, V. K. (1971). *Fiz. Tekh. Poluprovodn.* **5**, 1048 [*Engl. Transl.: Sov. Phys. Semicond.* **5**, 928].
Karmazin, V. V., Miloslavskii, V. K., and Mussil, V. V. (1973). *Fiz. Tekh. Poluprovodn.* **7**, 941 [*Engl. Transl.*: (1973). *Sov. Phys. Semicond.* **7**, 639].
Kato, H., and Takayanagi, S. (1963). *Jpn. J. Appl. Phys.* **2**, 250.
Kaufman, L., Gamsu, G., Savoca, C., Swann, S., Murphey, L., Hruska, B., Palmer, D., and Ullman, J. (1976). *IEEE Trans. Nucl. Sci.* **23**, 599.
Kiefer, J. E., and Yariv, A. (1969). *Appl. Phys. Lett.* **15**, 26.
Kiefer, J. E., Nussmeier, T. A., and Goodwin, F. E. (1972). *IEEE J. Quantum Electron.* **8**, 173.
Kireev, P. S., Martynov, V. N., and Vanyukov, A. V. (1969). *Fiz. Tekh. Poluprovodn.* **3**, 173 [*Engl. Transl.: Sov. Phys. Semicond.* **3**, 146].
Kireev, P. S., Nikonova, T. V., Artamonov, N. P., and Krolov, I. I. (1971). *Fiz. Tekh. Poluprovodn.* **5**, 2222 [*Engl. Transl.*: (1972). *Sov. Phys. Semicond.* **5**, 1939].
Kireev, P. S., Volkova, L. V., and Volkov, V. V. (1972a). *Fiz. Tekh. Poluprovodn.* **5**, 2080 [*Engl. Transl.: Sov. Phys. Semicond.* **5**, 1812].
Kireev, P. S., Volkova, L. V., and Volkov, V. V. (1972b). *Fiz. Tekh. Poluprovodn.* **5**, 2085 [*Engl. Transl.: Sov. Phys. Semicond.* **5**, 1816].
Kireev, P. S., Volkova, L. V., and Volkov, V. V. (1972c). *Fiz. Tekh. Poluprovodn.* **5**, 2090 [*Engl. Transl.: Sov. Phys. Semicond.* **5**, 1820].
Kirk, R. E., and Othmer, D. F., eds. (1964). "Encyclopedia of Chemical Technology," Vol. 3, 2nd ed., pp. 884–899. Wiley, New York.
Kirk, R. E., and Othmer, D. F., eds. (1969). "Encyclopedia of Chemical Technology," Vol. 19, 2nd ed., pp. 756–774. Wiley, New York.
Kittel, C. (1967). "Introduction to Solid State Physics." Wiley, New York.
Klausutis, N., Adamski, J. A., Collins, C. V., Hunt, M., Lipson, H., and Weiner, J. R. (1975). *J. Electron. Mater.* **4**, 625.
Klein, C. A., and Rudko, R. I. (1968). *Appl. Phys. Lett.* **13**, 129.
Knox, B. E., Geneczko, J., Gilbert, L., Howard, R., Mariner, G., and Vedam, K. (1975). *Ann. Conf. High Power Infrared Laser Window Mater.* 4th, p. 67.
Kobayashi, M. Z. (1911). *Anorg. Chem.* **69**, 1.
Kolb, E. D., Caporaso, A. J., and Laudise, R. A. (1968). *J. Cryst. Growth* **3**, 4, 422.
Koňák, Č. (1963). *Phys. Status Solidi* **3**, 1274.
Konorov, P. P., and Shevchenko, I. B. (1960). *Fiz. Tverd. Tela* **2**, 1134 [*Engl. Transl.: Sov. Phys. Solid State* **2**, 1027].
Kotov, E. (1966). *In* "Cadmium" (D. M. Chizhikov. ed.), p. 188. Pergamon, Oxford.
Kranzer, D. (1973a). *J. Phys. C* **6**, 2967.

Kranzer, D. (1973b). *J. Phys. C* **6**, 2977.
Kranzer, D. (1974). *Phys. Status Solidi A* **26**, 11.
Krapukhin, V. V., Tsokov, I. S., and Mamaev, Yu. O. (1967). *Fiz. Khim. Osn. Krist. Protsessov Glubokoi Ochistki Met. Mater. Soveshch.* **70**.
Kröger, F. A. (1965a). *J. Phys. Chem. Solids* **26**, 1707.
Kröger, F. A. (1965b). *J. Phys. Chem. Solid* **26**, 1717.
Kröger, F. A. (1974). "The Chemistry of Imperfect Crystals," 2nd ed. North-Holland Publ., Amsterdam.
Kröger, F. A. (1977). *Rev. Phys. Appl.* **12**, 205.
Kröger, F. A., and de Nobel, D. (1955). *J. Electron.* **1**, 190.
Kuhl, G. E. (1973). *Proc. Conf. High Power Infrared Laser Window Mater.* (C. A. Pitha, ed.) AFCRL-TR-73-0372 (11), p. 607.
Kujawa, R. v. (1963). *Phys. Status Solidi* **3**, 1089.
Kujawa, R. v. (1964). *Z. Physik. Chem. Leipsig* **227**, 416.
Kujawa, R. v. (1965). *Phys. Status Solidi* **12**, 169.
Kulwicki, B. M. (1963). Ph.D. Dissertation, Univ. of Michigan, Ann Arbor, Michigan.
Kurik, M. V. (1967). *Phys. Lett.* **24A**, 742.
Kurtin, S., McGill, T. C., and Mead, C. A. (1969). *Phys. Rev. Lett.* **22**, 1433.
Kyle, N. R. (1971). *J. Electrochem. Soc.* **118**, 1791.
Lambe, J., and Kikuchi, C. (1960). *Phys. Rev.* **119**, 1256.
Langer, D. (1964). *Proc. Int. Conf. Phys. Semicond., 7th* 241.
Lawaetz, P. (1971). *Phys. Rev. B* **4**, 3460.
Lawrenson, B., and Ray, B. (1971). *Phys. Status Solidi A* **5**, K101.
Lawson, W. D., Nielsen, S., Putley, E. H., and Young, A. S. (1959). *J. Phys. Chem. Solids* **9**, 325.
Lawson, W. D., Nielsen, S., and Young, A. S. (1960). *Solid State Phys. Electron. Telecommun. Proc. Int. Conf.* **2**, 830.
Lebrum, J. (1971). *Proc. Int. Conf. Phys. Chem. Semicond. Heterojunctions Layer Struct.* **4**, 163.
Legros, R., Marfaing, Y., and Triboulet, R. (1977). *Rev. Phys. Appl.* **12**, 245.
Ley, L., Pollack, R. A., McFeely, F. R., Kowalczyk, S. P., and Shirley, D. A. (1974). *Phys. Rev. B* **9**, 600.
Lipsett, J. J., and Stewart, W. B. (1976). *IEEE Trans. Nucl. Sci.* **23**, 321.
Lisitsa, M. P., Malinko, V. N., Nikonyuk, E. S., Novoseletskii, N. E., and Tsebulya, G. G. (1970). *Fiz. Tekh. Poluprovodn.* **3**, 1576 [*Engl. Transl.: Sov. Phys. Semicond.* **3**, 1321]].
Lisitsa, M. P., Zakharchuk, A. P., Malinko, V. N., Novoseletskii, S. F., Terekhova, S. F., and Tsebulya, G. G. (1974). *Fiz. Tekh. Poluprovodn.* **8**, 1361 [*Engl. Transl.:* (1975). *Sov. Phys. Semicond.* **8**, 883].
Llacer, J., and Cho, J. H. (1973). *IEEE Trans. Nucl. Sci.* **20**, 282.
Loferski, J. J. (1963). *Proc. IEEE* **51**, 667.
Lorenz, M. R. (1962a). *J. Phys. Chem. Solids* **23**, 939.
Lorenz, M. R. (1962b). *J. Phys. Chem. Solids* **23**, 1449.
Lorenz, M. R. (1962c). *J. Appl. Phys.* **33**, 3304.
Lorenz, M. R., and Blum, S. E. (1966). *J. Electrochem. Soc.* **113**, 559.
Lorenz, M. R., and Halsted, R. E. (1963). *J. Electrochem. Soc.* **110**, 343.
Lorenz, M. R., and Segall, B. (1963). *Phys. Lett.* **7**, 18.
Lorenz, M. R., and Woodbury, H. H. (1963). *Phys. Rev. Lett.* **10**, 215.
Lorenz, M. R., Aven, M., and Woodbury, H. H. (1963). *Phys. Rev.* **132**, 143.
Lorenz, M. R., Segall, B., and Woodbury, H. H. (1964). *Phys. Rev.* **134**, A751.
Lorimor, O. G., and Spitzer, W. G. (1965). *J. Appl. Phys.* **36**, 1841.
Losee, D. L., Khosla, R. P., Ranadive, D. K., and Smith, F. T. J. (1973). *Solid State Commun.* **13**, 819.
Lotspeich, J. F. (1974). *Appl. Opt.* **13**, 2529.

Low, F., and Pines, D. (1955). *Phys. Rev.* **98**, 414.
Ludeke, R., and Paul, W. (1967). In "II–VI Semiconducting Compounds" (D. G. Thomas, ed.), pp. 123–135. Benjamin, New York.
Ludwig, G. W. (1967). *IEEE Trans. Electron Devices* **14**, 547.
Ludwig, G. W., and Lorenz, M. R. (1963). *Phys. Rev.* **131**, 601.
Lynch, R. T. (1962). *J. Appl. Phys.* **33**, 1009.
Lyubin, V. M., and Fedorova, G. A. (1960). *Dokl. Akad. Nauk SSSR* **135**, 833 [*Engl. Transl.: Sov. Phys. Dokl.* **135**, 1343].
McArthur, D. A., and McFarlane, R. A. (1970). *Appl. Phys. Lett.* **16**, 452.
McCready, V. R., Parker, R. P., Gunnersen, E. M., Ellis, R., Moss, E., Gore, W. G., and Bell, J. (1971). In "The Role of Semiconductor Detectors in the Future of Nuclear Medicine" (P. Hoffer, R. Beck, and A. Gottschalk, eds.), p. 157. Soc. of Nucl. Med., New York.
McGill, T. C. (1974). *J. Vac. Sci. Technol.* **11**, 935.
McGill, T. C., Hellwarth, R. W., Mangir, M., and Winston, H. V. (1973). *J. Phys. Chem. Solids* **34**, 2105.
McSkimin, H. J., and Thomas, D. G. (1962). *J. Appl. Phys.* **33**, 56.
Madelung, O., and Treusch, J. (1968). *Proc. Int. Conf. Phys. Semicond., 9th* **1**, 38.
Magee, T. J., Peng, J., and Bean, J. (1974). *Proc. Ann. Electron Microsc. Soc. of Am., 32nd, St. Louis, Missouri* (C. J. Arceneaux, ed.).
Magee, T. J., Peng, J., and Bean, J. (1975). *Phys. Status Solidi A* **27**, 557.
Status Solidi A **27**, 557.
Maissel, L. I., and Glang, R. (1970). "Handbook of Thin Film Technology." McGraw-Hill, New York.
Malm, H. L., and Martini, M. (1973). *Can. J. Phys.* **51**, 2336.
Malm, H. L., and Martini, M. (1974). *IEEE Trans. Nucl. Sci.* **21**, 322.
Malm, H. L., Raudorf, T. W., Martini, M., and Zanio, K. (1973). *IEEE Trans. Nucl. Sci.* **20**, 500.
Malm, H. L., Canali, C., Mayer, J. W., Nicolet, M-A., Zanio, K. R., and Akutagawa, W. (1975). *Appl. Phys. Lett.* **26**, 344.
Malm, H. L., Litchinsky, D., and Canali, C. (1977). *Rev. Phys. Appl.* **12**, 303.
Manabe, A., Mitsuishi, A., and Yoshinaga, H. (1967). *Jpn. J. Appl. Phys.* **6**, 593.
Manasevit, H. M., and Simpson, W. I. (1971). *J. Electrochem. Soc.* **118**, 644.
Mandel, G. (1964). *Phys. Rev.* **134**, A1073.
Mandel, G., and Morehead, F. F. (1964). *Appl. Phys. Lett.* **4**, 143.
Marfaing, Y. (1977). *Rev. Phys. Appl.* **12**, 211.
Marfaing, Y., and Triboulet, R. (1971). *Proc. Int. Symp. Cadmium Telluride Mater. Gamma Ray Detect., Strasbourg, France*, No. 10. (P. Siffert and A. Cornet, eds.).
Marfaing, Y., Lascaray, J., and Triboulet, R. (1974). *Inst. Phys. Conf. Ser.* **22**, 201.
Mariano, A. N., and Warekois, E. P. (1963). *Science* **142**, 672.
Marple, D. T. F. (1963). *Phys. Rev.* **129**, 2466.
Marple, D. T. F. (1964). *J. Appl. Phys.* **35**, 539.
Marple, D. T. F. (1966). *Phys. Rev.* **150**, 728.
Marple, D. T. F. (1967). In "Physics and Chemistry of II–VI Compounds" (M. Aven and J. S. Prener, eds.), p. 333. Wiley, New York.
Marple, D. T. F., and Ehrenreich, H. (1962). *Phys. Rev. Lett.* **8**, 87.
Martin, G., Bach, P., Tranchart, J. C., and Fabre, E. (1975). *IEEE Trans. Nucl. Sci.* **22**, 226.
Martin, R. M. (1970). *Phys. Rev. B* **1**, 4005.
Martini, M. (1973). *IEEE Trans. Nucl. Sci.* **20**, 294.
Martini, M., Mayer, J. W., and Zanio, K. R. (1972a). In "Applied Solid State Science" (R. Wolfe, ed.), Vol. 3, pp. 182–257. Academic Press. New York.
Martini, M., Montano, H., and Zanio, K. (1972b). *Nucl. Instrum. Methods* **98**, 611.

Mason, W. D. (1950). "Piezoelectric Crystals and Their Application to Ultrasonics." Van Nostrand Reinhold, Princeton, New Jersey.
Matatagui, E., Thompson, A. G., and Cardona, M. (1968). *Phys. Rev.* **176**, 950.
Matlak, V. V., Nikonyuk, E. S., Savitskii, A. V., and Tovstyuk, K. D. (1972). *Fiz. Tekh. Poluprovodn.* **6**, 2065 [*Engl. Transl.*: (1973). *Sov. Phys. Semicond.* **6**, 1760].
Matsuura, K., Itoh, N., and Suita, T. (1967). *J. Phys. Soc. Jpn.* **22**, 1118.
Matsuura, K., Kawamoto, H., and Suita, T. (1970). *J. Phys. Soc. Jpn.* **29**, 946.
Mattox, D. M. (1976). *J. Vac. Sci. Technol.* **13**, 127.
Matveev, O. A., Prokof'ev, S. V., and Rud', Yu. V. (1969a). *Izv. Akad. Nauk SSSR Neorg. Meter.* **5**, 1175 [*Engl. Transl.*: *Inorg. Mater.* **5**, 1000].
Matveev, O. A., Rud', Yu. V., and Sanin, K. V. (1969b). *Fiz. Tekh. Poluprovodn.* **3**, 924 [*Engl. Transl.*: *Sov. Phys. Semicond.* **3**, 779].
Mayer, J. W. (1966). *Nucl. Instrum. Methods* **43**, 55.
Mayer, J. W. (1968). *In* "Semiconductor Detectors" (G. Bertolini and A. Coche, eds.), p. 445. North-Holland Publ., Amsterdam.
Meakin, J. D., Baron, B., Böer, K. W., Burton, L., Devaney, W., Hadley, H., Jr., Phillips, J., Rothwarf, A., Storti, G. and Tseng, W. (1976). *Sharing the Sun Conf.*, *Winnipeg, Manitoba, August* 15–20.
Mears, A. L., and Stradling, R. A. (1969). *Solid State Commun.* **7**, 1267.
Medvedev, S. A., Maksimovskii, S. N., Klevkov, Yu. V., and Shapkin, P. V. (1968a). *Izv. Akad. Nauk SSSR Neorg. Mater.* **4**, 2025 [*Engl. Transl.*: *Inorg. Mater.* **4**, 1759].
Medvedev, S. A., Shapkin, P. V., Maksimovskii, S. N., Klevkov, Yu. V., and Kuznetsov, A. V. (1968b). *Izv. Akad. Nauk SSSR Neorg. Mater.* **4**, 1782 [*Engl. Transl.*: *Inorg. Mater.* **4**, 1552].
Medvedev, S. A., Klevkov, Yu. V., Kiseleva, K. V., and Sentyurina, N. N. (1972). *Izv. Akad. Nauk SSSR Neorg. Mater.* **8**, 1210 [*Engl. Transl.*: *Inorg. Mater.* **8**, 1064].
Meiling, G. S., and Leombruno, R. (1968). *J. Cryst. Growth* **3**, **4**, 300.
Mertz, P. J. (1969). *Bull. Amer. Phys. Soc. (Ser. II)* **14**, 417.
Meyer, E., Martini, M., and Sternberg, J. (1972). *IEEE Trans. Nucl. Sci.* **19**, 237.
Meyer, O., and Lang, E. (1971). *Proc. Int. Symp. Cadmium Telluride Mater. Gamma Ray Detect.*, Strasbourg, France, No. 17 (P. Siffert and A. Cornet, eds.).
Milnes, A. G., and Feucht, D. L. (1972). "Heterojunctions and Metal–Semiconductor Junctions." Academic Press, New York.
Mitchell, K., Fahrenbruch, A. L., and Bube, R. H. (1975). *J. Vac. Sci. Technol.* **12**, 909.
Mitsuishi, A. (1961). *J. Phys. Soc. Jpn.* **16**, 533.
Mitsuishi, A., Yoshinaga, H., and Fujita, S. (1958). *J. Phys. Soc. Jpn.* **13**, 1235.
Miyasawa, H., and Sugaike, S. (1954). *J. Phys. Soc. Jpn.* **9**, 648.
Mizuma, K., Mikami, O., Ono, K., and Kamiya, Y. (1961). *Elec. Commun. Lab. Tech. J.* **10**, 895.
Mochalov, A., Urubkova, E., and Lepis, V. (1966). *In* "Cadmium" (D. M. Chizhikov, ed.), p. 187. Pergamon, Oxford.
Mooradian, A., and Wright, G. B. (1968). *Proc. Int. Conf. Phys. Semicond., 9th* **2**, 1020.
Mooser, E., and Pearson, W. B. (1959). *Acta Crystallogr.* **12**, 1015.
Morehead, F. F. (1967). *In* "Physics and Chemistry of II–VI Compounds" (M. Aven and J. S. Prener, eds.), pp. 613–654. Wiley, New York.
Morehead, F. F., and Mandel, G. (1964). *Phys. Lett.* **10**, 5.
Moss, T. S. (1959). *Proc. Phys. Soc. London* **74**, 490.
Movlanov, Sh., and Kuliev, A. A. (1963). *Tr. Inst. Fiz. Akad. Nauk Az. SSR* **11**, 36.
Muller, K. A., and Schneider, J. (1963). *Phys. Lett.* **4**, 288.
Muravjeva, K. K., Kalinkin, I. P., Aleskovsky, V. B., and Bogomolov, N. S. (1970). *Thin Solid Films* **5**, 7.
Musgrave, M. J. P., and Pople, J. A. (1962). *Proc. R. Soc. London A* **268**, 474.
Nakano, G. H., Imhof, W. L., and Kilner, J. R. (1976). *IEEE Trans. Nucl. Sci.* **23** **1**, 468.

Nakayama, N., Matsumoto, H., Yamaguchi, K., Ikegami, S., and Hioki, Y. (1976) *Jpn. J. Appl. Phys.* **15**, 2281.
Napolitano, P. P., and Mantell, C. L. (1964). *Trans. Metall. Soc. AIME* **230**, 133.
Naumov, G. P. (1967). *Fiz. Tverd. Tela* **9**, 64 [*Engl. Transl.: Sov. Phys. Solid State* **9**, 46].
Naumov, G. P., and Nikolaeva, O. V. (1961). *Fiz. Tverd. Tela* **3**, 3748 [*Engl. Transl.*: (1962). *Sov. Phys. Solid State* **3**, 2718].
Nikolaev, I. V., and Koblova, M. M. (1971). *Kvantovaya Élektron.* **1**, 57 [*Engl. Transl.: Sov. J. Quantum Electron.* **1**, 158].
Nikonyuk, E. S. Parfenyuk, O. A., Matlak, V. V., Tovstyuk, K. D., and Savitskii, A. V. (1975a). *Fiz. Tekh. Poluprovodn.* **9**, 1271 [*Engl. Transl.*: (1976). *Sov. Phys. Semicond.* **9**, 840].
Nikonyuk, E. S., Savitskii, A. V., Parfenyuk, O. A., Chiokan, I. P., and Zayachkivskii, V. P. (1975b). *Fiz. Tekh. Poluprovodn.* **9**, 1398 [*Engl. Transl.*: (1976). *Sov. Phys. Semicond.* **9**, 921].
Noblanc, J. P., Loudette, J., and Duraffourg, G. (1969). *Phys. Status Solidi* **32**, 281.
Noblanc, J. P., Loudette, J., and Duraffourg, G. (1970). *J. Lumin.* **1, 2**, 528.
Nolen, L. L., and Riter, Jr., J. R. (1973). *J. Phys. Chem. Solids* **34**, 1751.
Norris, C. B., and Barnes, C. E. (1977). *Rev. Phys. Appl.* **12**, 219.
Norris, C. B., Barnes, C. E., and Zanio, K. (1977). *J. Appl. Phys.* **48**, 1659.
Novik, F. T. (1962). *Fiz. Tverd. Tela* **4**, 3003 [*Engl. Transl.*: (1963). *Sov. Phys. Solid State* **4**, 2440].
Novik, F. T., Rumsh, M. A., and Zimkina, T. M. (1963). *Kristallografiya* **8**, 378 [*Engl. Transl.: Sov. Phys. Crystallogr.* **8**, 295].
Novikova, S. I. (1960). *Fiz. Tverd. Tela* **2**, 2341 [*Engl. Transl.*: (1961). *Sov. Phys. Solid State* **2**, 2087].
Nurmikko, A. V., Epstein, D. J., and Linz, A. (1975). In "Optical Properties of Highly Transparent Solids" (S. S. Mitra and B. Bendow, eds.), pp. 443-449. Plenum Press, New York.
Nussmeier, T. A., Goodwin, F. E., and Zavin, J. E. (1974). *J. Quantum Electron.* **10**, 230.
Nye, J. F. (1957). "Physical Properties of Crystals." Oxford Univ. Press (Clarendon), London and New York.
Oliver, M. R., and Foyt, A. G. (1967). *IEEE Trans. Electron Devices* **9**, 617.
Oliver, M. R., McWhorter, A. L., and Foyt, A. G. (1967). *Appl. Phys. Lett.* **11**, 111.
Ottaviani, G., Canali, G., Jacoboni, C., Quaranta, A., and Zanio, K. (1972). *Solid State Commun* **10**, 745.
Ottaviani, G., Canali, C., Jacoboni, C., Quaranta, A. A., and Zanio, K. (1973). *J. Appl. Phys.* **44**, 360.
Ottaviani, G., Canali, C., and Quaranta, A. A. (1975). *IEEE Trans. Nucl. Sci.* **22**, 192.
Owen, N. B., Smith, P. L., Martin, J. E., and Wright, A. J. (1963). *J. Phys. Chem. Solids* **24**, 1519.
Pakhomova, G. (1966). In "Cadmium" (D. M. Chizhikov, ed.), p. 184. Pergamon, Oxford.
Palatnik, L. S., Sorokin, V. K., and Marincheva, V. E. (1974). *Izv. Akad. Nauk SSSR Neorg. Meter.* **10**, 413 [*Engl. Transl.: Inorg. Mater.* **10**, 357].
Pankove, J. E. (1971). "Optical Processes in Semiconductors." Prentice-Hall, Englewood Cliffs, New Jersey.
Panossian, J. R., Gippius, A. A., and Vavilov, V. S. (1969). *Phys. Status Solidi* **35**, 1069.
Paorici, C., Pelosi, C., and Zuccalli, G. (1972). *Phys. Status Solidi A* **13**, 95.
Paorici, C., Attolini, G., Pelosi, C., and Zuccalli, G. (1973). *J. Cryst. Growth* **18**, 289.
Paorici, C., Attolini, G., Pelosi, C., and Zuccalli, G. (1974). *J. Cryst. Growth* **21**, 227.
Parker, G. H., and Mead, C. A. (1969). *Phys. Rev.* **184**, 780.
Pashinkin, A. S., Tishchenko, G. N., Korneeva, I. V., and Ryzhenko, B. N. (1960). *Kristallografiya* **4**, 261 [*Engl. Transl.: Sov. Phys. Crystallogr.* **5**, 243].
Patel, C. K. N. (1966). *Phys. Rev. Lett.* **16**, 613.
Paul, W. (1968). *Proc. Int. Conf. Phys. Semicond., 9th* **1**, 16.

Pauling, L. (1960). "The Nature of the Chemical Bond." 3rd ed., p. 246. Cornell Univ. Press, Ithaca, New York.
Perkowitz, S., and Thorland, R. H. (1974). *Phys. Rev. B* **9**, 545.
Petrescu, N., Protopopescu, M., and Drimer, D. (1960). *Acad. Repub. Pop. Rom. Stud. Cercet. Metal.* **5**, 51.
Pfann, W. G. (1955). *Trans. Met. Soc. AIME* **203**, 961.
Pfann, W. G. (1966). "Zone Melting." Wiley, New York.
Phelps, M. E., Hoffman, E. J., Mullani, N. A., Higgins, C. S., and Ter Pogossian, M. M. (1976). *IEEE Trans. Nucl. Sci.* **23**, 516.
Phillips, J. C. (1966). *Solid State Phys.* **18**, 55.
Phillips, J. C. (1968). *Phys. Rev. Lett.* **20**, 550.
Phillips, J. C. (1969). *Phys. Rev. Lett.* **22**, 645.
Phillips, J. C. (1970). *Rev. Mod. Phys.* **42**, 317.
Phillips, J. C. (1973a). *Solid State Commun.* **12**, 861.
Phillips, J. C. (1973b). "Bonds and Bands in Semiconductors." Academic Press, New York.
Phillips, J. C. (1974). *J. Vac. Sci. Technol.* **11**, 947.
Phillips, J. C., and Van Vechten, J. A. (1969a). *Phys. Rev. Lett.* **22**, 705.
Phillips, J. C., and Van Vechten, J. A. (1969b). *Phys. Rev. Lett.* **23**, 1115.
Phillips, J. C., and Van Vechten, J. A. (1970). *Phys. Rev. B* **2**, 2147.
Picus, G. S., DuBois, D. F., and Van Atta, L. B. (1968a). *Appl. Phys. Lett.* **12**, 81.
Picus, G. S., Van Atta, L., DuBois, D. F., and Yariv, A. (1968b). *Proc. Int. Conf. Phys. Semicond., 9th* **2**, 1171.
Piltch, M. S., Cantrell, C. D., and Sze, R. C. (1976). *J. Appl. Phys.* **47**, 3514.
Pinnow, D. A. (1970). *IEEE J. Quantum Electron.* **6**, 223.
Pitha, C. A., and Friedman, J. D. (1975). *Ann. Conf. High Power Infrared Laser Window Mater., 4th*, p. 150.
Planker, K. J., and Kauer, E. (1970). *Z. Angew. Phys.* **12**, 425.
Plumelle, P., and Vandevyver, M. (1976). *Phys. Status Solidi B* **73**, 271.
Pollak, F. H. (1967). *In* "II–VI Semiconducting Compounds" (D. G. Thomas, ed.), p. 552. Benjamin, New York.
Popa, A. E. (1973). *IEEE G-MTT Int. Microwave Symp. Dig. Tech. Papers*, p. 295.
Popa, A. E. (1974). *Gov. Microcircuit Appl. Conf. Dig. Papers*, p. 124. Department of the Navy, Arlington, Virginia.
Potykevich, I. V., Lyubchenko, A. V., and Boreiko, L. A. (1971). *Fiz. Tekh. Poluprovodn.* **5**, 1992 [*Engl. Transl.*: (1972). *Sov. Phys. Semicond.* **5**, 1729].
Price, D. L., Rowe, J. M., and Nicklow, R. M. (1971). *Phys. Rev. B* **3**, 1268.
Price, D. L., Sinha, S. K., and Gupta, R. P. (1974). *Phys. Rev. B* **9**, 2573.
Prokof'ev, S. V. (1970). *Izv. Akad. Nauk SSSR Neorg. Meter.* **6**, 1077 [*Engl. Transl.*: *Inorg. Mater.* **6**, 942].
Prokof'ev, S. V., and Rud' Yu. V. (1970). *J. Cryst. Growth* **6**, 187.
Przybyslawski, A., Sadowski, M., and Jakubowski, Z. (1967). *Rudy Met. Niezelav.* **12**, 162.
Quaranta, A. A., Canali, C., Ottaviani, G., and Zanio, K. (1970). *Lett. Nuovo Cimento Soc. Ital. Fis.* **4**, 908.
Ray, B., and Spencer, P. M. (1967). *Phys. Status Solidi* **22**, 371.
Reisman, A., Berkeublit, M., and Witzen, M. (1962). *J. Phys. Chem.* **66**, 2210.
Rode, D. L. (1970). *Phys. Rev. B* **2**, 4036.
Rooymans, C. J. M. (1963). *J. Inorg. Nucl. Chem.* **25**, 253.
Roth, W. L. (1967). *In* "Physics and Chemistry of II-VI Compounds" (M. Aven and J. S. Prener, eds.), pp. 119–160. Wiley, New York.
Rowe, J. M., Nicklow, R. M., Price, D. L., and Zanio, K. (1974). *Phys. Rev. B* **10**, 671.
Rubenstein, M. (1966). *J. Electrochem. Soc.* **113**, 623.

Rubenstein, M. (1968). *J. Cryst. Growth* **3, 4**, 309.
Ruch, J. G. (1972). *Appl. Phys. Lett.* **20**, 253.
Rud', Yu. V., and Sanin, K. V. (1969a). *Mater. Sci. Eng.* **4**, 186.
Rud', Yu. V., and Sanin, K. V. (1969b). *Fiz. Tekh. Poluprovodn.* **3**, 1089 [*Engl. Transl.*: (1970). *Sov. Phys. Semicond.* **3**, 919].
Rud', Yu. V., and Sanin, K. V. (1971a). *Fiz. Tekh. Poluprovodn.* **5**, 284 [*Engl. Transl.*: *Sov. Phys. Semicond.* **5**, 244].
Rud', Yu. V., and Sanin, K. V. (1971b). *Fiz. Tekh. Poluprovodn.* **5**, 1587 [*Engl. Transl.*: (1972). *Sov. Phys. Semicond.* **5**, 1385].
Rud', Yu. V., and Sanin, K. V. (1972). *Fiz. Tekh. Poluprovodn.* **6**, 886 [*Engl. Transl.*: *Sov. Phys. Semicond.* **6**, 764).
Rud', Yu. V., Sanin, K. V., and Shreter, Yu. G. (1971). *Fiz. Tekh. Poluprovodn.* **5**, 652 [*Engl. Transl.*: *Sov. Phys. Semicond.* **5**, 573].
Rumsh, M. A., Novik, F. T., and Zimkina, T. M. (1962). *Kristallografiya* **7**, 873 [*Engl. Transl.*: (1963). *Sov. Phys. Crystallogr.* **7**, 711].
Rusakov, A. P., and Anfimov, M. V. (1976) *Phys. Status Solidi B* **75**, 73.
Rusakov, A. P., Vekilov, Yu. Kh., Kadyshevich, A. E. (1970). *Fiz. Tverd Tela.* **12**, 3238 [*Engl. Transl.*: (1971). *Sov. Phys. Solis State* **12**, 2618].
Sagar, A., and Rubenstein, M. (1966). *Phys. Rev.* **143**, 552.
Samara, G. A., and Drickamer, H. G. (1962). *J. Phys. Chem. Solids.* **23**, 457.
Saraie, J., Akiyama, M., and Tanaka, T. (1972). *Jpn. J. Appl. Phys.* **11**, 1758.
Saravia, L. R., and Casamayou, L. (1972). *J. Phys. Chem. Solids* **33**, 145.
Scharager, C., Muller, J. C., Stuck, R., and Siffert, P. (1975). *Phys. Status Solidi A* **31**, 247.
Schaub, B., Gallet, J., Brunet–Jailly, A. Pelliciari, B. (1977). *Rev. Phys. Appl.* **12**, 147.
Schaub, B., and Potard, C. (1971). *Proc. Int. Symp. Cadmium Telluride Mater. Gamma Ray Detect. Strasbourg, France, No.* **2.**, (P. Siffert and A. Cornet, eds.).
Segall, B. (1966). *Phys. Rev.* **150**, 734.
Segall, B. (1968). *Proc. Int. Conf. Phys. Semicond., 9th* **1**, 425.
Segall, B., and Marple, D. T. F. (1967). *In* "Physics and Chemistry of II–VI Compounds" (M. Aven and J. S. Prener, eds.), pp. 317–381. Wiley, New York.
Segall, B., Lorenz, M. R., and Halsted, R. E. (1963). *Phys. Rev.* **129**, 2471.
Selim, F. A., and Kröger, F. A. (1977). *J. Electrochem. Soc.* **124**, 401.
Selim, F. A., Swaminathan, V., and Kröger, F. A. (1975). *Phys. Status Solidi A* **29**, 465.
Sella, C., Suryanaryanan, R., and Paparoditis, C. (1968). *J. Cryst. Growth* **3, 4**, 206.
Semiletov, S. A. (1956). *Kristallografiya* **1**, 306 [*Engl. Transl.*: *Sov. Phys. Crystallogr.* **1**, 236].
Semiletov, S. A. (1962). *Fiz. Tverd. Tela* **4**, 1241 [*Engl. Transl.*: *Sov. Phys. Solid State* **4**, 909].
Sennett, C. T., Bosomworth, D. R., Hayes, W., and Spray, A. R. L. (1969). *J. Phys. C* **2**, 1137.
Serreze, H. B., Entine, G., Bell, R. O., and Wald, F. V. (1974). *IEEE Trans. Nucl. Sci.* **21**, 404.
Shalimova, K. V., Bulatov, O. S., Voronkov, E. N., and Dmitriev, V. A. (1965). *Kristallografiya* **11**, 480 [*Engl. Transl.*: (1966) *Sov. Phys. Crystallogr.* **11**, 431].
Shay, J. L., and Spicer, W. E. (1967). *Phys. Rev.* **161**, 799.
Shay, J. L., Spicer, W. E., and Herman, F. (1967). *Phys. Rev. Lett.* **18**, 649.
Shiojiri, M., and Suito, E. (1964). *Jpn. J. Appl. Phys.* **3**, 314.
Shiozawa, L. R., Roberts, D. A., and Jost, J. M. (1973). *Conf. High Power Infrared Laser Window Mater.* (C. A. Pitha, ed.), p. 639.
Shuleshko, G. I., and Vigdorovich, V. N. (1968). *Tsvetn. Met.* **41**, 12 [*Engl. Transl.*: *Sov. J. Non-Ferrous Met.* 87].
Shvartsenau, N. F. (1960). *Fiz. Tverd. Tela* **2**, 870 [*Engl. Transl.*: *Sov. Phys. Solid State* **2**, 797].
Siffert, P., Gonidec, J. P., and Cornet, A. (1974). *Nucl. Instrum. Methods* **115**, 13.
Siffert, P., Cornet, A., Stuck, R., Triboulet, R., and Marfaing, Y. (1975). *IEEE Trans. Nucl. Sci.* **22**, 211.

Siffert, P., Berger, J., Sharager, C., Cornet, A., Stuck, R., Bell, R. O., Serreze, H. B., and Wald, F. V. (1976). *IEEE Trans. Nucl. Sci.* **23**, 159.
Silvey, G. A., Lyons, V. J., and Silvestri, V. J. (1961). *J. Electrochem. Soc.* **108**, 653.
Simmonds, P. E., Stradling, R. A., Birch, J. R., and Bradley, C. C. (1974). *Phys. Status Solidi B* **64**, 195.
Simov, S., Gantcheva, V., Kamadjiev, P., and Gospodinov, M. (1976). *J. Cryst. Growth* **32**, 133.
Singh, A. J., Mathur, B. S., Suryanarayana, P., and Tripathi, S. N. (1968). *Recent Develop. Non-Ferrous Met. Technol. Pap. Discuss. Symp.* **3**, 82.
Slack, G. A., and Galginaitis, S. (1964). *Phys. Rev.* **133**, A253.
Slack, G. A., Ham, F. S., and Chrenko, R. M. (1966). *Phys. Rev.* **152**, 376.
Slack, G. A., Roberts, S., and Vallin, J. T. (1969). *Phys. Rev.* **187**, 511.
Slapa, M., Huth, G. C., Seibt, W., Schieber, M. M., and Randtke, P. T. (1976). *IEEE Trans. Nucl. Sci.* **23**, 102.
Smith, D. L., and Davis, D. T. (1974). *IEEE J. Quantum Electron.* **10**, 138.
Smith, D. L., and Pickhardt, V. Y. (1975). *J. Appl. Phys.* **46**, 2366.
Smith, F. T. J. (1970). *Metall. Trans.* **1**, 617.
Smith, P. L., and Martin, J. E. (1963). *Phys. Lett.* **6**, 42.
Smith, T. F., and White, G. K. (1975). *J. Phys.* **8**, 2031.
Sokolova, A. A., Vavilov, V. S., Plotnikov, A. F., and Chapnin, V. A. (1969). *Fiz. Tekh. Poluprovodn.* **3**, 720 [*Engl. Transl.: Sov. Phys. Semicond.* **3**, 612].
Sorokin, G. P., Papshev, Yu. M., and Oush, P. T. (1965). *Fiz. Tverd. Tela* **7**, 2244 [*Engl. Transl.:* (1966). *Sov. Phys. Solid State* **7**, 1810].
Sparks, M., and Duthler, C. J. (1973). *J. Appl. Phys.* **44**, 3038.
Spears, D. L., and Strauss, A. J. (1977). *Rev. Phys. Appl.* **12**, 401.
Spears, D. L., Strauss, A. J., Chinn, S. R., Melngailis, I., and Vohl, P. (1976). Integrated Optics, Tech. Dig. Opt. Soc. of Am., T625.
Spînulescu–Carnaru, J. (1966). *Phys. Status Solidi* **15**, 761.
Spitzer, W. G., and Kleinman, D. A. (1961). *Phys. Rev.* **121**, 1324.
Spitzer, W. G., and Mead, C. A., (1964). *Phys. Chem. Solids* **25**, 443.
Stafsudd, O. M., and Alexander, D. H. (1971). *Appl. Opt.* **10**, 2566.
Stafsudd, O. M., Hack, F. A., and Radisavljević, K. (1967). *Appl. Opt.* **6**, 1276.
Stanley, A. G. (1975). *In* "Applied Solid State Science" (R. Wolfe, ed.) Vol. 5, p. 251. Academic Press, New York.
Steininger, J., Strauss, A. J., and Brebrick, R. F. (1970). *J. Electrochem. Soc.* **117**, 1305.
Strauss, A. J. (1971). *Proc. Int. Symp. Cadmium Telluride Mater Gamma Ray Detect.*, Strasbourg, France, No. 1, (P. Siffert and A. Cornet, eds.).
Strauss, A. J., and Iseler, G. W. (1974). Unpublished data, Lincoln Lab., M.I.T., Lexington, Massachusetts 02173.
Strauss, A. J., and Steininger, J. (1970). *J. Electrochem. Soc.* **117**, 1420.
Strehlow, W. H. (1969). *J. Appl. Phys.* **40**, 2928.
Stringfellow, G. B., and Greene, P. E. (1969). *J. Phys. Chem. Solids* **30**, 1779.
Strzalkowski, I., Joshi, S., and Crowell, C. R. (1976). *Appl. Phys. Lett.* **28**, 350.
Stuck, R., Cornet, A., Scharager, C., and Siffert, P. (1976). *J. Phys. Chem. Solids* **37**, 989.
Stuck, R., Cornet, A., and Siffert, P. (1977). *Rev. Phys. Appl.* **12**, 218.
Stuckes, A. D. (1961). *Br. J. Appl. Phys.* **12**, 675.
Stuckes, A. D., and Farrell, G. (1964). *J. Phys. Chem. Solids* **25**, 477.
Suga, S., Dreybrodt, W., Willman, F., Hiesinger, P., and Cho, K. (1974). *Solid State Commun.* **15**, 871.
Suito, E., and Shiojiri, M. (1963). *J. Electron Microsc.* **12**, 134.
Svechnikov, S. V., and Aleksandrov, V. T. (1957). *Zh. Tekh. Fiz.* **27**, 919 [*Engl. Transl.:* (1958). *Sov. Phys. Tech. Phys.* **2**, 842].

Swaminathan, V., Selim, F. A., and Kröger, F. A. (1975). *Phys. Status Solidi A* **30**, 721.
Swank, R. K. (1967). *Phys. Rev.* **153**, 844.
Swierkowski, S. P. (1976). *IEEE Trans. Nucl. Sci.* **23**, 131.
Taguchi, T., Shirafuji, J., and Inuishi, Y. (1973). *Jpn. J. Appl. Phys.* **12**, 1558.
Taguchi, T., Shirafuji, J., and Inuishi, Y. (1974a). *J. Appl. Phys.* **13**, 1003.
Taguchi, T., Shírafuji, J., and Inuishi, Y. (1974b). *Jpn. J. Appl. Phys.* **13**, 1169.
Taguchi, T., Shirafugi, J., and Inuishi, Y. (1977). *Rev Phys. Appl.* **12**, 117.
Talwar, D. N., and Agrawal, B. K. (1973). *Phys. Rev. B* **8**, 693.
Talwar, D. N., and Agrawal, B. K. (1974). *Phys. Rev. B* **9**, 2539.
Talwar, D. N., and Agrawal, B. K. (1975). *Phys. Rev. B* **12**, 1432.
Tang, C. L., and Flytzanis, C. (1971). *Phys. Rev. B* **4**, 2520.
Taziev, Zh. Sh., Esyutin, V. S., and Tseft, A. L. (1965). *Tr. Inst. Metall. Obogashch. Akad. Nauk Kaz. SSR* **13**, 11.
Teramoto, I. (1963). *Philos. Mag.* **8**, 357.
Teramoto, I., and Inoue, M. (1963). *Philos. Mag.* **8**, 1593.
Thomas, D. G. (1960). *J. Phys. Chem. Solids* **15**, 86.
Thomas, D. G. (1961). *J. Appl. Phys. Suppl.* **32**, 2298.
Thomas, D. G., and Hopfield, J. J. (1959). *Phys. Rev.* **116**, 573.
Thomassen, L., Mason, D. R., Rose, G. D., Sarace, J. C., and Schmitt, G. A. (1963). *J. Electrochem. Soc.* **110**, 1127.
Title, R. S. (1967). In "Physics and Chemistry of II–VI Compounds" (M. Aven and J. S. Prener, eds.), pp. 267–312. Wiley, New York.
Toušek, J., Kargerová, J., and Klier, E. (1967). *Cesk. Cas. Fys.* [*Engl. Transl.: Czech. J. Phys. B* **17**, 462].
Toušková, J., and Kužel, R. (1972). *Phys. Status Solidi A* **10**, 91.
Toušková, J., and Kužel, R. (1973). *Phys. Status Solidi A* **15**, 257.
Toušková, J., and Toušek, J. (1971). *Proc. Int. Conf. Phys. Chem. Semicond. Heterojunctions Layer Struct.* **5**, 363.
Tranchart, J. C., and Bach, P. (1976). *J. Cryst. Growth* **32**, 8.
Treusch, J., Eckelt, P., and Madelung, O. (1967). In "II–VI Semiconducting Compounds" (D. G. Thomas, ed.), pp. 588–597. Benjamin, New York.
Triboulet, R. (1968). *C. R. Acad. Sci. Paris B* **266**, 796.
Triboulet, R., and Rodot, H. (1968). *C. R. Acad. Sci. Paris Ser. B* **266**, 498.
Triboulet, R. (1971). *Proc. Int. Symp. Cadmium Telluride, Mater. Gamma Ray Detect.*, Strasbourg, France, No. 5, (P. Siffert and A. Cornet, eds.).
Triboulet, R., and Marfaing, Y. (1973). *J. Electrochem. Soc.* **120**, 1260.
Triboulet, R., Marfaing, Y., and Burgos, M. E. (1970). *Phys. Status Solidi A* **2**, 851.
Triboulet, R., Marfaing, Y., Cornet, A., and Siffert, P. (1973). *Nature (London) Phys. Sci.* **245**, 12.
Triboulet, R., Marfaing, Y., Cornet, A., and Siffert, P. (1974). *J. Appl. Phys.* **45**, 2759.
Tsay, Y. F., Mitra, S. S., and Vetelino, J. F. (1973). *J. Phys. Chem. Solids* **34**, 2167.
Tyagai, V. A., Snitko, O. V., Bondarenko, V. N., Vitrikhovskii, N. I., Popov, V. B., and Krasiko, A. N. (1973). *Fiz. Tverd. Tela* **16**, 1373 [*Engl. Transl.:* (1974). *Sov. Phys. Solid State* **16**, 885].
Ueda, R. (1975). *J. Cryst. Growth* **31**, 333.
Urli, N. B. (1966). *J. Phys. Soc. Jpn.* **21**, 259.
Urli, N. B. (1967). In "II–VI Semiconducting Compounds" (D. G. Thomas, ed.), pp. 1335–1347. Benjamin, New York.
Van Atta, L. B., Picus, G. S., and Yariv, A. (1968). *Appl. Phys. Lett.* **12**, 84.
Vandekerkhof, J. (1971). *Proc. Int. Symp. Cadmium Telluride Mater. Gamma Ray Detect.*, Strasbourg, France, No. 8, (P. Siffert and A. Cornet, eds.).

Van Doorn, C. Z., and de Nobel, D. (1956). *Physica* **22**, 338.
Van Vechten, J. A. (1969a). *Phys. Rev.* **182**, 891.
Van Vechten, J. A. (1969b). *Phys. Rev.* **187**, 1007.
Van Vechten, J. A. (1972). *Phys. Rev. Lett.* **29**, 769.
Van Vechten, J. A. (1974). *Phys. Rev. B* 10, 4222.
Van Vechten, J. A. (1975). *J. Electrochem. Soc.* **122**, 419.
Van Vechten, J. A., and Phillips, J. C. (1970). *Phys. Rev. B* **2**, 2160.
Vanyukov, A. V., and Krotov, I. I. (1971). *Kristallografiya* **16**, 460 [*Engl. Transl.*: *Sov. Phys. Crystallogr.* **16**, 392.
Vanyukov, A. V., Kozhitov, and Bulanov, A. I. (1969). *Zh. Prikl. Khim.* (*Lenningrad*) **42**, 2044 [*Engl. Transl.*: *J. Appl. Chem.* **42**, 1920].
Vanyukov, A. V., Baronenkova, R. P., Krotov, I. I., and Indenbaum, G. V. (1974). *Kristallografiya* **19**, 1025 [*Engl. Transl.*: (1975). *Sov. Phys. Crystallogr.* **19**, 635].
Vavilov, V. S., and Nolle, E. L. (1965). *Dokl. Akad. Nauk SSSR* **164**, 73 [*Engl. Transl.*: (1966). *Sov. Phys. Dokl.* **10**, 827].
Vavilov, V. S., and Nolle, E. L. (1966). *Fiz. Tverd. Tela* **8**, 532 [*Engl Transl.*: *Sov. Phys. Solid State* **8**, 421].
Valilov, V. S., and Nolle, E. L. (1968). *Proc. Int. Conf. Phys. Semicond., 9th* **1**, 600.
Vavilov, V. S., Nolle, E. L., Egorov, V. D., and Vintovkin, S. I. (1964). *Fiz. Tverd. Tela* **6**, 1406 [*Engl. Transl.*: *Sov. Phys. Solid State* **6**, 1099].
Vavilov, V. S., Nolle, E. L., and Maksimovskii, S. N. (1965). *Fiz. Tverd. Tela* **7**, 1558 [*Engl. Transl.*: *Sov. Phys. Solid State* **7**, 1253].
Vavilov, V. S., Dzhioeva, S. G., and Stopachinskii, V. B. (1969). *Fiz. Tekh. Poluprovodn.* **3**, 727 [*Engl. Transl.*: *Sov. Phys. Semicond.* **3**, 617].
Vekilov, Yu. Kh., and Rusakov, A. P. (1971). *Fiz. Tverd. Tela* **13**, 1157 [*Engl. Transl.*: *Sov. Phys. Solid State* **13**, 956].
Vetelino, J. F., Gaur, S. P., and Mitra, S. S. (1972). *Phys. Rev. B* **5**, 2360.
Vigdorovich, V. N., and Selin, A. A. (1971). *Izv. Akad. Nauk SSSR Met.* **5**, 106 [*Engl. Transl.*: *Bull. Acad. Sci. USSR Met.*].
Vigdorovich, V. N., Vol'pyan, A. E., and Shuleshko, G. I. (1970). *Izv. Akad. Nauk SSSR Neorg. Meter.* **6**, 1041 [*Engl. Transl.*: *Inorg. Mater.*]
Vinogradov, E. A., Vodop'yanov, L. K., and Oleinik, G. S. (1972). *Fiz. Tverd. Tela* **14**, 452 [*Engl. Transl.*: (1973). *Sov. Phys. Solid State* **15**, 322].
Visvanathan, S. (1960). *Phys. Rev.* **120**, 376.
Vitrikhovskii, N. I., and Mizetskaya, I. B. (1959). *Fiz. Tverd. Tela* **1**, 996 [*Engl. Transl.*: *Sov. Phys. Solid State* **1**, 912].
Vodakov, Yu. A., Lomakina, G. A., Naumov, G. P., and Maslakovets, Yu. P. (1960). *Fiz. Tverd. Tela* **2**, 3 [*Engl. Transl.*: *Sov. Phys. Solid State* **2**, 1].
Vodop'yanov, L. K., and Abramov, A. A. (1968). *In* "Cadmium Telluride" (V. S. Vavilov and B. M. Vul, eds.), p. 122. Nauka, Moscow.
Vodop'yanov, L. K., Vinogradov, E. A., Blinov, A. M., and Rukavishnikov, V. A. (1972). *Fiz. Tverd. Tela* **14**, 268 [*Engl. Transl.*: *Sov. Phys. Solid State* **14**, 219].
Vodop'yanov, L. K., Vinogradov, E. A., and Vinogradov, V. S. (1974). *Fiz. Tverd. Tela* **16**, 849 [*Engl. Transl.*: *Sov. Phys. Solid State* **16**, 545].
Vogel, J. M., Ullmann, J., and Entine, G. (1977). *Rev. Phys. Appl.* **12**, 375.
Vul, B. M., and Chapnin, V. A. (1966). *Fiz. Tverd. Tela* **8**, 256 [*Engl. Transl.*: *Sov. Phys. Solid State* **8**, 206].
Vul, B. M., Plotnikov, A. F., Sal'man, V. M., Sokolova, A. A., and Chapnin, V. A. (1968). *Fiz. Tekh. Poluprovodn.* **2**, 1243 [*Engl. Transl.*: (1969). *Sov. Phys. Semicond.* **2**, 1045].
Vul, B. M., Sal'man, V. M., and Chapnin, V. A., (1970). *Fiz. Tekh. Poluprovodn.* **4**, 67 [*Engl. Transl.*: *Sov. Phys. Semicond.* **4**, 52].

REFERENCES

Wagman, D. D., Evans, W. H., Parker, V. B., Halow, I., Bailey, S. M., and Schumm, R. H. (1968). Nat. Bur. Std. Tech. Note No. 270-3 U.S. Govt. Printing Office, Washington, D.C.
Wagner, R. J., and McCombe, B. D. (1974). *Phys. Status Solidi B* **64**, 205.
Wald, F. V., and Bell, R. O. (1972). *Nature (London) Phys. Sci.* **237**, 13.
Wald, F. V., and Bell, R. O. (1975). *J. Cryst. Growth* **30**, 29.
Waldman, J., Larsen, D. M., Tannenwald, P. E., Bradley, C. C., Cohn, D. R., and Lax, B. (1969). *Phys. Rev. Lett.* **23**, 1033.
Walford, G. V., and Parker, R. P. (1973). *IEEE Trans. Nucl. Sci.* **20**, 318.
Walter, J. P., and Cohen, M. L. (1971). *Phys. Rev. B* **4**, 1877.
Wannier, G. H. (1937). *Phys. Rev.* **52**, 191.
Wardzyński, W. (1970). *J. Phys. C* **3**, 1251.
Warekois, E. P., Lavine, M. C., Mariano, A. N., and Gatos, H. C. (1962). *J. Appl. Phys.* **33**, 690.
Watkins, G. D. (1971). *In* "Radiation Effects in Semiconductors" (J. W. Corbett and G. D. Watkins, eds.), p. 301. Gordon & Breach, New York.
Watts, R. K., and Holton, W. C. (1967). *Phys. Lett.* **24A**, 365.
Weidel, J. Z. (1954). *Naturforsch* **9A**, 697.
Weil, R. B., and Sun, M. J. (1971). *Proc. Int. Symp. Cadmium Telluride Mater. Gamma Ray Detect.*, Strasbourg, France, No. 19, (P. Siffert and A. Cornet, eds.).
Weinstein, M., Wolff, G. A., and Das, B. N. (1965). *Appl. Phys. Lett.* **6**, 73.
Weiss J. A. (1975). *In* "Optical Properties of Highly Transparent Solids" (S. S. Mitra and B. Bendow, eds.), pp. 339–350. Plenum Press, New York.
Wernick, J. H. and Thomas, E. E. (1960). *Trans. Metall. Soc. AIME* **218**, 763.
Whelan, R. C., and Shaw, D. (1967). *In* "II-VI Semiconducting Compounds" (D. G. Thomas, ed.), p. 451. Benjamin, New York.
Whelan, R. C., and Shaw, D. (1968). *Phys. Status Solidi* **29**, 145.
White, A. M., Dean, P. J., Day, B., and Mansfield F. (1976). *Phys. Status Solidi B* **74** K9.
Williams, M. G., Tomlinson, R. D., and Hampshire, M. J. (1969). *Solid State Commun.* **7**, 1831.
Williams, R. T., and Schnatterly, S. E. (1975). *In* "Optical Properties of Highly Transparent Solids" (S. S. Mitra and B. Bendow, eds.), pp. 145–160. Plenum Press, New York.
Wolff, G. A., and Broder, J. D. (1959). *Acta Crystallogr.* **12**, 313.
Woodbury, H. H. (1967). *In* "Physics and Chemistry of II-VI Compounds" (M. Aven and J. S. Prener, eds.), pp. 225–259. Wiley, New York.
Woodbury, H. H. (1974). *Phys Rev. B* **9**, 5188.
Woodbury, H. H., and Aven, M. (1964). *Proc. Int. Conf. Phys. Semicond.*, 7th, p. 179.
Woodbury, H. H., and Aven, M. (1968). *J. Appl. Phys.* **39**, 5485.
Woodbury, H. H., and Aven, M. (1974). *Phys. Rev. B* **9**, 5195.
Woodbury, H. H., and Hall, R. B. (1967). *Phys. Rev.* **157**, 641.
Woodbury, H. H., and Lewandowski, R. S. (1971). *J. Cryst. Growth* **10**, 6.
Woolley, J. C., and Ray, B. (1960). *J. Phys. Chem. Solids* **15**, 27.
Yamada, S. (1960). *J. Phys. Soc. Jpn.* **15**, 1940.
Yamada, S. (1962). *J. Phys. Soc. Jpn.* **17**, 645.
Yamada, S., Kawasaki, Y., and Nishida, O. (1968). *Phys. Status Solidi* **26**, 77.
Yamaguchi, K., Nakayama, N., Matsumoto, H., Hioki, Y., and Ikegami, S. (1975). *Jpn. J. Appl. Phys.* **14**, 1397.
Yamaguchi, K., Masumoto, H., Nakayama, N., and Ikegami, S. (1976). *Jpn. J. Appl. Phys.* **15**, 1575.
Yariv, A., Nussmeier, T. A., and Kiefer, J. E. (1973). *IEEE J. Quantum Electron.* **9**, 594.
Yezhovsky, Yu. K., and Kalinkin, I. P. (1973). *Thin Solid Films* **18**, 12.
Yokozawa, M., Otsuka, S., and Takayanagi, S. (1965). *Jpn. J. Appl. Phys.* **4**, 1018.

Yu, P. Y., and Cardona, M. (1973). *J. Phys. Chem. Solids* **34**, 29.
Zaitsev, E. Ya. (1971). *Fiz. Tekh. Poluprovodn.* **5**, 1619 [*Engl. Transl.*: (1972). *Sov. Phys. Semicond.* **5**, 1413].
Zanio, K. (1969a). *Appl. Phys. Lett.* **14**, 56.
Zanio, K. (1969b). *Appl. Phys. Lett.* **15**, 260.
Zanio, K. (1970a). *J. Appl. Phys.* **41**, 1935.
Zanio, K. (1970b). *Nucl. Instrum. Methods* **83**, 288.
Zanio, K. (1971). *Proc. Int. Symp. Cadmium Telluride Mater. Gamma Ray Detect.*, Strasbourg, France, No. 21, (P. Siffert and A. Cornet, eds.).
Zanio, K. (1972). *Isot. Radiat. Technol.* **9**, 456.
Zanio, K. (1974). *J. Electron. Mater.* **3**, 327.
Zanio, K. (1977). *Rev. Phys. Appl.* **12**, 343.
Zanio, K., Akutagawa, W., and Kikuchi, R. (1968). *J. Appl. Phys.* **39**, 2818.
Zanio, K., Neeland, J., and Montano, H. (1970). *IEEE Trans. Nucl. Sci.* **17**, 287.
Zanio, K., Akutagawa, W., and Montano, H. (1972). *IEEE Trans. Nucl. Sci.* **19**, 257.
Zanio, K., Krajenbrink, F., and Montano, H. (1974). *IEEE Trans. Nucl. Sci.* **21**, 315.
Zanio, K., Montano, H., Krajenbrink, F., and Peterson, G. (1975a). *IEEE Trans. Nucl. Sci.* **22**, 422.
Zanio, K., Montano, H., and Krajenbrink, F. (1975b). *Appl. Phys. Lett.* **27**, 159.
Zayachkivskii, V. P., Savitskii, A. V., Nikonyuk, E. S., Kitsa, M. S., and Matlak, V. V. (1974). *Fiz. Tekh. Poluprovodn.* **9**, 1035 [*Engl. Transl.*: *Sov. Phys. Semicond.* **8**, 675].
Ziman, J. M. (1960). "Electrons and Phonons," p. 208, Oxford Univ. Press, London and New York.
Zitter, R. N. (1971). *Surf. Sci.* **28**, 335.
Zitter, R. N., and Chavda, D. L. (1975). *J. Appl. Phys.* **46**, 1405.
Zosimovich, D. P., Kladnitskaya, K. B., and Grisevich, A. N. (1961). *Zh. Prikl. Khim. (Lenningrad)* **34**, 1764 [*England. Transl.*: *J. Appl. Chem.* **34**, 1680].
Zvára, M., Žaloudek, F., and Prosser, V. (1966). *Phys. Status Solidi* **16**, K21.

Index

A

Absorption coefficient
  gamma ray, 164
  optical, 74, 93, 192
    free carrier, 89, 192
    intraband, 89
    temperature dependence of, 96
    at 10.6 μm, 191
  x-ray, 164
Absorption edge, (optical) effect of
  doping, 94
  excitons, 90
  pressure, 88
  temperature, 98
Acceptors, 148, 162, *see also* Impurities in CdTe
  cadmium vacancies, 118, 147, 155, 161
  double acceptor, 152
  tellurium interstitials, 120
  0.15 eV complex, 139, 156, 161
  0.06 eV level, 147
Activation, energy for
  diffusion, 125
  non-Γ donors, 134
Alloys
  CdS_xTe_{1-x}, 202
  $Hg_{1-x}Cd_xTe$, 209
Annealing, 115, 122, 126, 142, 145, 149, 153, 154, 158
Antireflective coatings, 209
Association of defects, 121, 145, 155, 158

B

Band structure, 77
Barrier heights, 184
Birefringence, 186
Bonding, 59
  covalent, 59
  ionic, 59, 195
  metallic, 64
Brillouin zone, 78
Brout's sum rule, 75
Brouwer plots, 117, 128, 154

C

Capture cross-section, 139, 142, 180
Characteristic temperature, 154
Coherence length, 206
Compensation, 130, 146, 155
Complex, *see* Acceptors
Conductivity
  electrical, 116, 123, 128
  thermal, 111
Constitutional supercooling, 12, 22, 126
Contacts
  blocking, 168, 183, 205
  ohmic, 203
Coupling constant, 103, 109
Crystal growth, 11
  Bridgman method, 11
  chemical transport, 31
  congruent melt, 11
  Czochralski technique, 12
  foreign hosts, 38
  molecular beam epitaxy, 36
  seeding, 11
  solution, 18
  vacuum deposition, 33
  vapor phase, 24
Crystal structure, 53

D

d electrons, 59, 61, 64, 81
Damping constant, 72
Debye temperature, 77, 105
Debye–Waller factors, 100
Defects, 115
Deformation potential, 105, 108
Degeneracy, 117, 130
Delocalization, 132
Detectors
  gamma ray, 164
    gauging, 170
    nuclear medicine, 171
    safeguards, 168
  infrared, 206, 209
  x-ray, 164, 167
Dielectric constants, 61, 72
Dielectric theory, 61
Diffusion, 119
  ambipolar, 125
  chemical, 123
  impurity, 203
  ring mechanism, 122
  surface, 30
Donors, 129
  cadmium interstitials, 116
  deep, 135
  hydrogen-like, 129
  non-Γ, 133
  tellurium vacancies, 116

E

Edge emission, 149
Effective mass
  electrons, 91
  holes, 91
Elastic constants, 65
Electron affinity, 184
Electron irradiation, 143
Electron spin resonance, 132, 142, 146
Electrooptic coefficients, 188
Epitaxy, 30, 209
Equilibrium constant
  atomic defects, 117, 156
  solid–vapor, 6

Crystallographic polarity, 55
Cyclotron resonance, 104

Etch pits, 55
Excitons
  in absorption, 94
  binding energy, 91
  effect on absorption edge, 95
  free, 90
  in luminescence, 131
  oscillator strengths of, 93
  in reflectance, 91
  "saddle point", 93

F

Faraday effect, 96, 107
Franz–Keldyshi effect, 191
Free energy
  of formation, 4, 9
  of sublimation, 63
Frenkel defects, 161

G

Grain boundaries, 30, 57
Grüneisen factor, 76
Gunn effect, 208

H

Hall coefficient, 110
Heat of formation, 65
Heat of fusion, 4
Heat of mixing, 4

I

Impact ionization, 208
Impurities in CdTe
  Ag, 49, 141, 144, 148, 193
  Al, 48, 90, 120, 121, 133, 139, 158, 193
  Ar, 155, 203
  As, 146, 148, 203
  Au, 49, 122, 144, 148, 155, 193
  B, 48
  Be, 192
  Bi, 49
  Br, 133, 141
  C, 148
  Ca, 50

INDEX

Cl, 31, 48, 107, 127, 129, 133, 138, 141, 150, 194
Co, 49, 146
Cr, 49, 146, 148, 203
Cu, 48, 122, 143, 144, 148, 155, 159, 193
Er, 146
F, 50
Fe, 49, 146, 159
Ga, 107, 120, 126, 133
Ge, 147, 159
Hg, 48
I, 31, 48, 90, 133, 140
In, 49, 90, 107, 120, 127, 130, 133, 138, 141, 150, 155, 194
K, 48
Kr, 203
Li, 48, 141, 145, 148
Mg, 49, 159, 193
Mn, 113, 146
N, 48
Na, 48, 148
Nd, 146
O, 48, 147
P, 51, 122, 146, 148, 203
Pb, 48, 148, 159
Pd, 51
Pt, 49
S, 48, 147, 193
Sb, 49, 148, 194
Se, 48, 147, 193
Si, 50, 147, 159
Sn, 48
Tm, 146
Tl, 48, 132, 146, 155
Y, 51
Yb, 146
Zn, 48, 113
Zr, 51
Impurity band, 130, 146
Index of refraction, 72, 83
Interchange energy, 2
Interstitials, *see* Acceptors, Donors, Native defects
Ion channeling, 122
Ion implantation, 155, 203
Ionic charge, effective, 68, 72
Ionicity, degree of, 62, 68
Ionization energies
  acceptors, 148
  donors, 129

J

Junctions,
  hetero-, 199, 204
  homo-, 198

K

KKR method, 81
$k \cdot p$ method, 81
Kramers–Kronig relationship, 83

L

Lattice constant, 54
Lattice dynamics, 69
  dispersion relationship, 69
  local mode vibrations, 192
Lifetime
  minority, 198
  radiative, 198
Liquid crystal, 204
Liquidus, 2
Luminescence
  cathodo-, 139, 142, 160
  electro-, 208
  excitation spectra of, 101
  photo-, 100, 131, 138
Lyddane–Sachs–Teller relationship, 74

M

Magnetic circular dichroism, 97
Mechanical properties, 195
  dislocations, 27, 128, 195
  fracture, 191, 195
  yield stress, 195
Melting point, 1, 7
Mobility
  anomalous, 106, 131
  drift, 107, 111
  Hall, 103, 109
  negative differential, 108
  relaxation time, 104
Modulators
  acoustooptic, 189
  electrooptic, 186

N

Native defects, 115
Neutrons
  defect generation by, 143, 155
  inelastic scattering of, 69
Nonlinear optics, 206

O

One-electron approximation, 78
  nearly free electron model, 79
  tight binding theory, 79
Optical elements, 191
Orthogonal plane wave, 79, 81
Oxide films, 185

P

Phase diagram, 1, 10
Phonons
  acoustical, 70, 95
  optical, 70, 95, 100, 103, 130, 139, 192
Photoconductivity, 99, 107, 130, 206
Photoemission, 86
Photovoltage, 57, 198
Piezobirefringence, 102
Piezoelectric effect, 67, 190
Piezoresistance, 107
Polarization, 182
Polaron, 103
Polymorphism, 55
Poole–Frenkel effect, 134, 136
Precipitation, 123, 125, 154, 194
Pseudopotential method
  determination of band structure, 78
  determination of charge densities, 60
Purification
  Cd, 39
  CdTe, 46
  Te, 43

Q

Quenching, 123

R

Recrystallization, 30, 33
Reflection coefficient, 82

Regular association solution theory, 3
Relaxation time, 125
Reststrahl frequency, 72, 192
Retrograde solid solubility, 6
Rock salt structure, 10, 58

S

Scattering, Compton, 173
Scattering, free carrier
  deformation potential, 105, 108
  intervalley, 108
  ionized impurity, 106, 145
  optical phonon, 104
  piezoelectric, 106
Schottky disorder, 115
Segregation coefficient of impurities
  in Cd, 42
  in CdTe, 48, 49
  in Te, 45
Solar cells, 197
Solid solutions, 209
Solidus, 5
Space-charge-limited currents, 139
Space charge region, 182
Space group, 53
Specific heat, 77
Spin–orbit interaction, 84
Stacking faults, 56
Stark effect, 130
Stoichiometry, deviation from, 5, 12, 115
Surface recombination velocity, 198
Surface states, 184

T

Thermal conductivity, 111
Thermal expansion, 76, 99
Thermal quenching of luminescence, 140
Thermally stimulated currents, 139, 141, 147, 155
Thermodynamic factor, 125
Thin films, 30, 55
Three-phase boundary, 7
Threshold energy, 149, 153
Time of flight measurements, 135, 141
Transmission electron microscopy, 126
Trapping
  electrons and holes, 136, 139, 141, 176
  excitons, 131

INDEX

Triple point, '0
Tunneling, 136, 185, 202
Twinning, 13, 16, 57, 206

U

Urbach's rule, 96

V

Vacancies, *see* Acceptors, Donors, Native defects
Vapor pressure of CdTe, Cd, and Te, 6

W

White tin structure, 10, 58
Wurtzite structure, 33, 56

Z

Zeeman splitting, 130
Zinc blende structure, 33, 56
Zone refining
 Cd, 41
 CdTe, 14, 17, 46
 Te, 43